PRAISE FOR

Bioshelter Market Garden

Darrell Frey's inspirational book gives you all you need to know to create an energy-saving, food-producing bioshelter. It not only offers plenty of detail on the nuts and bolts of construction and maintenance, but it also provides the big picture: the concepts and principles behind these innovative structures. *Bioshelter Market Garden* covers everything you need to understand, build, or simply admire these important tools for sustainability.

—TOBY HEMENWAY, author of *Gaia's Garden: A Guide to Home-Scale Permaculture*

Darrell Frey has mapped out the possibility of regeneration of individuality and of nature, of liberty, of community, of food security and of ethics such as the modern agricultural literature has never known — a harmony with nature, with the earth, with one another in a sharing society such as the world has often dreamed.

—PROF. DECLAN KENNEDY - Chairman, Advisory Board, www.gaiauniversity.org

Darrell Frey's *Bioshelter Market Garden* is a welcome addition to the North American permaculture lexicon. Darrell has pondered and observed his systems — large and small — for a long time, and I'm glad he has shared his successes and lessons with us. The book is sensible, grounded, and practical, while offering a wide view of how to put the pieces together in a multifunctional way, on the ground, as a business, and as a way of life. The world needs more bioshelters, and this book contributes substantially to reinvigorating the development and deployment of this technology.

—DAVE JACKE, Dynamics Ecological Design

The increasingly complex ecological problems we face today will require increasingly complex solutions, leaving the average person to wonder where to even start. Darrell Frey's example, as exhibited in this book, is as simple as it is essential . . . one must start here and now, with what nature has provided. Frey is no dreamer, but a practitioner, who shows us how the world can and must be changed, one farmer, and one small plot of land at a time.

—BR␣␣␣␣␣␣␣␣␣␣␣ctor,
Pennsylvania As␣␣␣␣␣␣␣␣␣riculture (PASA)

It is no simple matter to create livelihood on a few acres and to preserve the biodiversity of that place. In *Bioshelter Market Garden*, author Darrell Frey shares the wisdom of his 20-plus years of doing so, and invites readers to work with the sun, the dragonflies, the muskrats, and the complexities of human enterprise to engage their own blessed swatch of land.

—Terril L. Shorb, Ph.D.,
Founder, Prescott College Sustainable Community Development Program

Some of us learn by reading, some by asking questions, and some of us just forge ahead and take risks. Darrell Frey is in that last category, and the lessons he has gleaned the hard way over the past 30 years, are now available to anyone who wants to plan for an uncertain future. Frey shows, with well-illustrated case studies, how to grow a four-season garden, turn sunlight and rainfall into sources of income, and live in the comfort of an architectural ecosystem that nourishes the planet as it shelters the family.

—Albert Bates, author of *The Biochar Solution: Carbon Farming and Climate Change*.

Darrell Frey is at once keen observer, adept teacher and consummate communicator, demonstrating to the reader in a style reminiscent of Aldo Leopold how we are all connected to the web of life and that our daily choices matter. *Bioshelter Market Garden* is equal parts inspired storytelling and how-to manual for living creatively and responsibly upon the earth. Through his own experience as a permaculture instructor and market gardener, Frey sets forth a detailed roadmap by which we might all reach a sustainable future and celebrate our interdependence.

—Dan Sullivan, managing editor *BioCycle* magazine and
former senior editor with NewFarm.org and *Organic Gardening* magazine.

Bioshelter Market Garden is a must read for home gardeners and farmers alike. Darrell Frey shares the secrets of creating a great year round growing system developed through a lifetime of farming. Best of all, he teaches readers simply how to grow things the right way, organically. The lavishly illustrated book is an essential tool for sustainable growing.

—Doug Oster, *Pittsburgh Post-Gazette* Garden Columnist and
Co-host The Organic Gardeners Radio Show

Bioshelter
Market Garden

A PERMACULTURE FARM

Darrell Frey

NEW SOCIETY PUBLISHERS

Cataloging in Publication Data:
A catalog record for this publication is available from
the National Library of Canada.

Copyright © 2011 by Darrell Frey. All rights reserved.

Cover design by Diane McIntosh. Main photograph: David Travis
Lineart illustration: Bob Kobet, Architect. ©iStock, Carole Gomez (seeds)
Liliya Zakharchenko (vegetables)/Alexander Bolbot (frog).

Printed in Canada. First printing December 2010

Paperback ISBN: 978-0-86571-678-0
eISBN: 978-1-55092-457-2

Inquiries regarding requests to reprint all or part of *Bioshelter Market Garden* should
be addressed to New Society Publishers at the address below.

To order directly from the publishers, please call toll-free (North America)
1-800-567-6772, or order online at www.newsociety.com

Any other inquiries can be directed by mail to:

New Society Publishers
P.O. Box 189, Gabriola Island, BC V0R 1X0, Canada
(250) 247-9737

New Society Publishers' mission is to publish books that contribute in fundamental
ways to building an ecologically sustainable and just society, and to do so with the least
possible impact on the environment, in a manner that models this vision. We are committed to doing this not just through education, but through action. Our printed, bound
books are printed on Forest Stewardship Council-certified acid-free paper that is **100%
post-consumer recycled** (100% old growth forest-free), processed chlorine free, and
printed with vegetable-based, low-VOC inks, with covers produced using FSC-certified
stock. New Society also works to reduce its carbon footprint, and purchases carbon
offsets based on an annual audit to ensure a carbon neutral footprint. For further information, or to browse our full list of books and purchase securely, visit our website at:
www.newsociety.com

Library and Archives Canada Cataloguing in Publication

Frey, Darrell
    Bioshelter market garden : a permaculture farm / Darrell Frey.

Includes index.
ISBN 978-0-86571-678-0

    1. Permaculture. 2. Farms, Small--Management. 3. Solar greenhouses.
4. Gardening. 5. Sustainable horticulture. 6. Human ecology. I. Title.

S494.5.P47F74 2011   631.5'8   C2010-906969-2

NEW SOCIETY PUBLISHERS

FSC

Mixed Sources
Product group from well-managed forests,
controlled sources and recycled wood or fiber
www.fsc.org Cert no. SW-COC-000952
© 1996 Forest Stewardship Council

Books for Wiser Living
recommended by *Mother Earth News*

TODAY, *MORE THAN EVER BEFORE,* our society is seeking ways to live more conscientiously. To help bring you the very best inspiration and information about greener, more sustainable lifestyles, *Mother Earth News* is recommending select New Society Publishers' books to its readers. For more than 30 years, *Mother Earth* has been North America's "Original Guide to Living Wisely," creating books and magazines for people with a passion for self-reliance and a desire to live in harmony with nature. Across the countryside and in our cities, New Society Publishers and *Mother Earth* are leading the way to a wiser, more sustainable world.

Join the Conversation

Visit our online book club at www.newsociety.com to share
your thoughts about *Bioshelter*. Exchange ideas with other readers,
post questions for the author, respond to one of the sample
questions or start your own discussion topics. See you there!

Contents

Acknowledgments

THE MAKING OF THIS BOOK goes far beyond the last six years of researching, writing and documenting. It is impossible to properly acknowledge everyone who has contributed. Co-workers, permaculture teachers, interns, permaculture students and all our old friends who spent time working in our gardens contributed insight and effort.

Thanks are due to Linda Susan Strawbridge Frey for her years of artful gardening and loving dedication to good stewardship. Linda has contributed key parts to this book and greatly forwarded the art of permaculture farm management with her intuition and keen observation.

Zackary Thor, Christopher Ra and Terra Kachina Frey grew with the farm. Their good work ethic and cheerful lives have made the farm a family home. Cody Noel kept the fires burning as the manuscript was revised twice.

Thanks to Wayne Frey for showing his children how to contribute to the betterment of one's community; Deonne Frey for endless support and encouragement; Marlin Hartman for a lifetime of inspiration in gardening; Dawn and Frank Hyeldahl Shiner and family for two years of their lives helping us germinate Three Sisters Farm; and Jack Schmidt for half a lifetime of vision, support and good neighborliness.

Earl Barnhart and Hilde Maingay provided essential design and management advice in 1988 and generously share their work today. Anna Edey gave us great inspiration and personally showed us how to improve our building's productivity.

Bruce Fulford provided highly detailed and valued consultation about composting greenhouses. Dan Desmond and Phil Schuler, representing the Commonwealth of Pennsylvania, proved their vision for a sustainable future by funding our initial project. Bob Kobet provided architectural consultation on our original design. Robert A. Macoskey offered encouragement and inspired confidence to dream big and pursue our dreams.

Bill Mollison gave us in-person advice and inspiration in 1982 and gracious encouragement when he visited our fledgling farm in 1990.

Thank you to permaculture teacher and publisher Dan Hemenway for allowing reuse of the temperate zone wetland species list originally published in *The 1986 Permaculture Seed Yearbook*.

I must also thank Bob Benek for his many hours of effort and consultation on planning and redesigns. Special thanks go to Ellen Benek and Jeanine Jenkins, who shared their wondrous gardens and generous spirits.

Prescott College's Terril Shorb and Jeanine Canty allowed me to stack functions and earn an undergraduate degree while pulling together the first draft of this manuscript. Jill Calderone provided tea, encouragement and her most excellent editorial advice. Nancy Martin Silber and Daninne Egizio-Hughes gave moral support, editorial assistance, insightful advice, and friendship. Thanks go to Ben Watson of Chelsea Green Publishing for his early review and input into the book's structure and content.

Thanks to Corinne Ogrodnic, Nancy Martin-Silber, Terri MacCartney, Vivas Macoskey and the ALTER Project, and Chuck McDougal of Mountain Meadow Farm for funding support and inspiration to complete this book.

Permaculture teachers Chris McHenry-Glenn, Christopher Leininger and Susana Lein contributed greatly to farm plans and schemes. And also, thanks to Chris McHenry-Glenn for her enduring patience and effort in creating the illustrations for this book.

Design Course students Scott Martens and Don McCarty provided thorough research into composting options.

Thanks are due to the many other mentors, interns and farm coworkers who have contributed to our successes, and the permaculture teachers and students who have aided our endless unfolding.

Finally, I am grateful to editor Linda Glass and the staff of New Society Publishing for their skillful yet modest revisions and clarifications.

Introduction

THIS BOOK IS ABOUT RELATIONSHIPS. All things, from the minerals of the earth to the energy of the sun and everything between, interact in a vast web of energy and elements, organisms and events. Each organism occupies a niche, a position within the ecosystem's network. The cumulative interactions of organisms, soils, landforms and climates evolve over time to a balanced, self-regulating state.

Over the course of human history, traditional cultures around the world have found a right relationship to the rest of creation. For many thousands of years, people have been sustainably harvesting and managing ecosystems by understanding and respecting the basic laws of nature. They developed stories and myths to help them remember and pass on these laws. They recognized sacred spaces to preserve the character and health of the ecosystem. They set limits to harvests and guided their lives by the cycles of the seasons.

Today the dominant world culture is barely beginning to recognize the relationships that make up the web of life. But our education systems teach ecology as a "personal value" rather than a science. The "environmentalist" is described as a special interest rather than a person with special understanding. Concern for the natural world is viewed as a philosophy rather than scientific reality. The world is seen as a commodity to be traded. But in reality, it is a commonwealth to be shared and cared for with informed stewardship.

As a people, we have lost our native sense of place. We have become tangled in our own web of science and technology. Our tools have become our master. The challenge ahead is to rediscover our land and ourselves as a part of the great web of life. Permaculture design is a tool to help us in the journey of rediscovery.

Permaculture design is a system of land-use planning that incorporates concepts of ecosystem dynamics, ecologically appropriate technologies, and an ethic of care of the earth into a comprehensive design system. It is a process of analysis and design that can unwind inappropriate technologies and unsustainable practices. Like nature, it has its own cycles — of study, design, planting, harvesting, more study, redesign and refining — that can lead us, as individuals and as a society, back to a right relationship in *all* our relations. When we choose technologies and materials appropriate to our locale and design our cities, homes and gardens to co-exist with and enhance the natural world around us, we re-enter — as co-participants — into the sacred dance of life.

Finding our own personal relationships to the planet begins at home. Choices we make as consumers, including what materials we use and what food we eat, are votes we cast and investments we make in the systems that surround us. In the 21st century, we, as consumers, can make many decisions with global impact.

I hope this book will inspire some of you to think big. Together, we can create the Garden. Dream your dreams. Then live them.

As a child I was captivated by wild fruit. It was my favorite of all foods, free for the harvest. Sweet, juicy raspberries and black mulberries grew along the stream in the town park. Delicious blue huckleberries and tiny, intensely flavored wild strawberries grew on my uncles' farms. Teaberries were a rare treat from the woods, and tart, wild grapes hung there in the trees. Early in my life I developed a personal vision of paradise. On a family trip to Niagara Falls, as my souvenir I bought a copy of Maurice Kains' *Five Acres and Independence*. My future homestead, I dreamed, would be filled with wild food.

Another early childhood memory is my father's garden in the summer following the flood of 1964. I was six years old. A fast thaw of an extra heavy snow brought high water to our town. Soon, our house (safe on the hill) was full of displaced relatives. For several weeks we had cots in the living room and huge pots of food on the kitchen stove. That summer, my father planted a large garden in a vacant lot beside an uncle's house. The bountiful harvest from the flood-fertilized field was mostly given away to folks still recovering from the trauma of the spring.

As a young adult, these early influences combined with a growing disillusionment with the sciences. It was the 1970s. Nuclear waste and the threat of nuclear winter, rising cancer rates, and the industrial pollution epitomized by Love Canal, were, to me, issues urgently needing to be addressed. The growing body of information on how humans were impacting the biosphere likewise required our attention.

Like many others of the time, my partner, Linda, and I sought a life closer to the earth. We learned homesteading from friends and neighbors. Old-time farmers became sources of "heirloom knowledge." We devoured books and periodicals for clues on how to find the "good life" as described by Helen and Scott Nearing. Our free time was spent gardening, tending several goats and a small flock of chickens, learning to forage for wild foods and medicinal plants, making cider from wild apples, and canning, drying and otherwise preserving food.

Our search for right living led us to study alternative and energy technologies. We were inspired by the work of the New Alchemy Institute and their solar greenhouses and bioshelter prototypes. We were drawn to low-cost, owner-builder options and green design.

Early in the 1980s, we learned of the work of Bill Mollison and David Holmgren in permaculture design. Permaculture design pulled it all together for us, integrating all we had been studying into a coherent system of planning and development. We understood from the start that permaculture offered not an answer to the dilemma of humans on the earth, but a strategy for finding the answer. Permaculture presents a vision of humans striving for harmony with the earth. That vision provided us with a goal to move toward.

After ten years of intensive studies, six of which were devoted to learning permaculture design, I found myself teaching an annual permaculture course at a state university. The growing urge to teach by doing, rather than lecturing about the work of others, led to Linda and me making preliminary plans to create a bioshelter market garden farm. A *bioshelter* is greenhouse operated as an ecosystem. A *market garden* is a commercial-scale farm that supplies fresh produce to a regional market. We wanted our farm and bioshelter to reflect the permaculture vision. Our hopes were to create a research and demonstration facility for ecological farming in the 21st century. The farm we dreamed of would be diverse, small-scale, intensive and organic, and it would enhance the quality of life in our corner of the earth. It would be, we knew, a lifetime commitment to principles of caring for the earth.

On New Year's Day 1988, many points converged in our lives: Five acres of excellent farmland were available for our use; five friends were committed to the permaculture farm project; funding assistance was available; and we had some money in the bank. We took a dive, so to speak, into the future. We have not looked back.

The story of our lives in the time since 1988 is a common one in the permaculture world. We committed all our resources and time to what was, for a while (and to some extent still is), a self-funded research project. Many people have helped us on the way. Some of them we still see daily; some early associates have moved on to other, equally interesting projects. We have weathered downpours and droughts, late frosts and heat waves, confidence and burnout. When the burnout came, something or someone always showed up to revive us: the customer who appreciated our chemical-free produce; the child fascinated with a shamrock spider in its web and a monarch chrysalis on the milkweed; the first fruit of the season; and (especially) the exhilaration of being in a sunny bioshelter on a cold winter day. Through it all, we continued to develop the farm — reviewing, planning and expanding our plantings and gardens. Crop mixes were developed, relationships with customers established and ecological relationships encouraged. As you read these words we're probably working on some new phase of farm development, identifying a new species, designing a new garden or greeting a new visitor. A permaculture farm is an evolving, changing thing. With thoughtful, responsible stewardship it will, each year, get closer to our vision of integration with the local ecology.

CHAPTER 1

Toward a Permanent Culture

Life — all life — is in the service of life. Necessary nutrients are made available to life by life in greater and greater richness as the diversity of life increases. The entire landscape comes alive, filled with relationships and relationships within relationships.

—FRANK HERBERT, *Dune*

Prologue

VAST CHALLENGES FACE OUR RELATIONSHIP to our planet. Although these challenges are now topics on the evening news, they are not new. Many people have been working for decades to alert the general public to the myriad human-caused environmental problems and social and economic inequities, that are, in a word, unsustainable.

To the student of permaculture, problems are signposts pointing to solutions. Permaculture attempts to find solutions in cultural and ecological systems rather than technology. The environmentalist adage that technological solutions breed new technological problems has proven true. The converse can be true of ecological solutions. Thoughtful application of ecological design for problem solving can set in motion regeneration of soil, watersheds and local ecosystems that in turn help heal regional and global environments.

It is easy to forget that everything *is* connected. The choices we make in how we grow and sell our food, the energy we use to produce our products, and the way we manage our diminishing resources all have effects for the entire planet.

All the tools and information needed to design and plan sustainable communities are available now. The question is whether we decide to use those tools or not. And the consequences won't affect only us. As developing nations seek to emulate Western culture, Western culture needs to demonstrate stewardship based on scientific understanding and environmental consciousness.

One cannot predict what a long-term sustainable future will look like. But we believe it will be rooted in the land. It will come as an organic outgrowth of a rekindled dynamic relationship between people and their landscape. The book in your hands is intended to be a tool for those who want to participate in the continuing evolution of a sustainable society.

Book Outline

This book is roughly divided into two parts. The first part, chapters 1 through 6, introduces permaculture concepts and methods for designing a permaculture market garden farm. The second part, chapters 7 through 15, provides a detailed study of permaculture design applied to Three Sisters Farm's activities, landscapes, and the bioshelter.

This first chapter introduces the concept of a sustainable food system and permaculture design with a first look at Three Sisters Farm. Chapter 2 examines issues relevant to safe, healthful and secure food systems. Chapter 3 looks at marketing options and strategies for the small farm. (Note that in this book the word "marketing" does not mean advertising. For us "marketing" and "selling" are almost interchangeable, but the word "marketing" evokes our role as farmers who produce crops to sell at a particular community's local *market*.) Chapter 4 details

the use of permaculture to design and plan a small-scale, year-round, intensive farm, with staged development. Chapter 5 looks closely at sustainable energy systems on the farm. The final introductory chapter, Chapter 6, presents ecological strategies and methods for controlling pests and disease.

Chapter 7 and 8 begin a closer study of Three Sisters Farm, looking at the details of farm design and management, crops and tools and seasonal work cycles. Chapters 9 and 10 finally come to the study of bioshelter design and management. A bioshelter is essentially a solar greenhouse managed as an indoor ecosystem. Our bioshelter is the heart of our farm, and the preceding chapters provide the necessary context for understanding its role on the farm. Chapters 11 and 12 examine details of the role of compost and chickens in the bioshelter and on the farm. Chapter 13, "Permaculture for Wetlands," is offered as a tool for preserving wetlands and learning to use the special capabilities of wetland plants in our landscapes. Chapter 14 presents the farm as an education center, both as a source of revenue and as a service to the community. And finally, chapter 15, "Knowing Home," and the epilogue look at the ongoing development of the farm and the application of permaculture to our own homestead. The appendix provides facts and figures related to solar design.

Five Acres and Interdependence

Three Sisters Farm began as a five-acre field of bare soil and corn stubble. The soil was good silt loam, but had been depleted of life and nutrients by decades of conventional agriculture. A scrubby tree line on the abandoned barbed wire fencerow defined the space but added little diversity. The year was 1983. We were starting with a clean slate. On this open field, we could put into practice the permaculture theory we wanted to explore. Our youthful idealism was influenced by philosopher Stephen Gaskin, of The Farm, an intentional community in rural Tennessee. To paraphrase Gaskin's Zen-inspired directive, "If you see a problem, you probably should fix it." Thus empowered, we began our preparations to become part of the solution — as permaculture market gardeners.

The chapters that follow provide a narrative of the changes that have occurred in this small field since 1983. Though our work is framed by human intent, our palette and canvas is provided by nature. As we've followed an unfolding vision of what we wanted the farm to be, we have been guided by the principle of caring for the earth so she will care for us. Our farm has been an experiment in permaculture design. And it has been an attempt to forge a right livelihood out of the universal struggle to survive and prosper.

This book functions on several levels. While it closely examines Three Sisters Farm and the bioshelter concept, it also presents a broader picture of sustainable food systems and the larger role farms can play in a sustainable society. Three Sisters is just one regional project among many. In this book, I also examine the work that others are doing developing gardens, landscapes and bioshelters, and integrating community and educational activities into the farm plan. There is a growing momentum toward the development of sustainable food systems. People of all ages, but especially young people, are drawn to a life closer to the earth and closer to the source of fresh food.

The Value of the Small-Scale Farm

The small-scale intensive farm can offer many benefits to a community and a region: food security is enhanced, organic matter and excess fertilizer are removed from the waste stream, healthy soil is built, streams are cleaner, and groundwater is recharged. The organic farm, with its diversity of crops and other plantings, enhances local biodiversity and helps create and preserve critical habitat for wildlife, especially birds and pollinating insects. Farms offer jobs and training. They can also engender many related enterprises, including craft production and value-added processing. The small-scale intensive farm can offer a pleasant space for social gatherings and community events, and there are many opportunities for educational activities and for reconnecting the community to the earth. Local culture as a whole is enhanced when local foods are available, events are seasonal, and people have direct access to nature and agriculture.

Food

Fresh, local, and organic food should be accessible to everyone. The fact that this is a political statement is unfortunate. "Be on good terms with all people," say the great teachers. But declaring oneself organic apparently implies that something is wrong with how the "conventional farmer" farms. Similarly, to promote the superiority of fresh, local food somehow impugns the entire food industry and the global economy. Although it's true we would prefer that no one use chemical fertilizer, herbicides or insecticides, we know that national and international trade is here to stay. And we recognize and applaud the moves toward more sustainable agriculture that many conventional farmers are making.

What we *do* mean to imply, or rather, to state clearly, is that much of our food can be produced locally, in a manner that helps preserve and heal the planet, build strong communities and make us healthier in the process.

Biodiversity

We live in a world of expanding populations and diminishing natural resources. A changing climate is putting further stresses on the natural world. As wildlands continue to be developed, remaining habitat becomes ever more critical to the health of the planet. We can address these issues through thoughtful and informed land use. The ecologically designed agricultural landscape is a contribution to the preservation of natural habitat and biodiversity.

Protecting biodiversity includes protecting the quality of water in our streams and rivers, preserving critical habitat and undeveloped wildlands, managing forests sustainably, using and preserving native plants in our landscapes, protecting native pollinators and promoting habitat for the entire web of the natural world. Nature is the matrix within which we must develop our production systems.

Spring Peepers.

Education

A farm is a valuable educational facility that offers many different types of instruction — organic gardening, diet and nutrition, nutrient cycling, diverse agricultural enterprises and the farm's interconnection with the natural world. Many young people seek internships on working farms to learn the trade. Home gardeners seek out advice and opportunities to learn. At a farm, children can learn to love the earth and begin to understand its processes. It all begins with personal contact with nature. All of these things are services the farm can provide to a community while at the same time earning income to keep the farm viable. As visitors learn the value of the farm on the landscape and of fresh local food in their diet, they become customers of the farm.

Creation of Sustainable Local Food Systems

Challenges to the revitalization and re-creation of regional food systems are many — and formidable — but they are not insurmountable. The economic viability of small farms is the foremost challenge. Competition

Benefits of Small-Scale Intensive Agriculture

• Food security
• Employment
• Nutrient cycling
• Biodiversity
• Pollinator preservation
• Water cycling
• Education
• Social space for gatherings
• Building a local food system
• Right livelihood in a community
• Sustainable food systems

with global food systems requires creative design, management and marketing. Start-up cost support and innovative partnership funding is often required. Other issues include access to land and resources, zoning and tax laws, insurance costs and the seasonal nature of most farm incomes. Because we are in the process of re-creating local and regional food systems, there are many missing links in support structures and market access.

Making a living from agriculture has always been difficult. One is dependent on the weather, the market and the quality of the labor force. All are variable. Because cash flow is seasonal and not always dependable, most agricultural enterprises rely on the farmer to wear a lot of hats. Besides being able to plan and manage crop production, soil fertility and other farm resources, the farmer needs to be a capable mechanic, carpenter, electrician, plumber, supervisor, planner, accountant, marketer, office manager — and gardener. An organic and permaculture farmer also needs to be an ecologist, an ecological designer and planner, a natural builder and an alternative energy specialist. Knowledge about regulations for organic certification, compost management, health and safety requirements and tax codes also is required.

Farming is rightfully said to be a lifestyle choice. You do not get rich farming, but you can profit in many ways. People are drawn to a rural lifestyle to be close to nature, to benefit from the slower pace of traffic and life in general and to engage in the care and nurturing of plants and animals. However, you need more than to eat well and have a roof over your head and shoes on your feet. Because health care and education are primary needs for most families, the majority of small-scale farmers are either part-time farmers or they have other income. Many people retire on a pension or savings account to become market gardeners as a second career. Part-time, seasonal work or a working spouse is often inevitable for the small farmer trying to compete with the global economy.

A local food system that truly supports the sustainable farm's economic viability requires not only creativity and innovation but also the support and assistance of community leaders, local government, national policy and consumers.

In a very real sense, a return to regional food systems is a return to life of a century ago. Home food systems, supported and supplemented by neighborhood farmers markets, grocers, butchers, and the milkmen, fed the nation well. As we attune to the role of being good earth stewards in our daily lives, we will see the value of lower energy-use models of the 18th and early 19th centuries.

All these issues are examined in the chapters ahead. The picture I portray, of a diverse agricultural enterprise, demonstrates permaculture farms' role in the regional food system and local community.

In the sustainable society we seek, each farm will be unique, with its own mix of crops, enterprises and relationships. To achieve this sustainable society we will need many more small-scale farmers, literally one in every neighborhood. Farmers will need stable, profitable markets and the organizations and infrastructure to support a diverse and economically viable local agriculture. The issues are complex. Many gaps in local food systems and many disconnects in our modern lifestyles discourage both local consumers and producers. However, as discussed in the next chapter, many groups and agencies are working to promote a resurgence of local agriculture.

Integrated Systems: Permaculture Design

Permaculture (a.k.a. permanent culture or permanent agriculture) is a system of land-use planning that incorporates concepts of ecosystem dynamics, ecologically appropriate technologies and an ethic of care of the earth into a comprehensive design system.

Permaculture is also a growing and evolving network of individuals and organizations. Practitioners of permaculture are dedicated to searching for, creating and exploring sustainable solutions to the dilemma of our confused human relationship to the natural world.

My study of permaculture began in 1981, when I read an interview with Bill Mollison in *Mother Earth News*. Mollison described in detail a system for the design of agricultural ecosystems. He succinctly explained the concept of permaculture and summarized its main goals and principles, presenting a way for modern societies to rediscover cultural links to a healthy landscape. This is done by designing our homes, towns and regions with an understanding of ecology, energy dynamics and a logical ethic of caring for the earth — so it can care for us.

This system was first articulated in the book *Permaculture One*, which Mollison co-authored with David Holmgren, and further developed in *Permaculture Two*. In the 1980 *Mother Earth News* interview, Mollison stated that the essence of permaculture was to "apply the principles of environmental science to our production systems." In explaining the permaculture concepts of designing integrated systems, that is, of placing design "elements" into functional relationships to maximize productivity, Mollison presented the idea of combining a chicken coop, a forage yard and a greenhouse. Such an integration sets up ecological relationships: the exchange of oxygen and carbon dioxide; a nutrient cycle between plants and poultry; and poultry regulating their own temperature by being able to move outside during the day and inside at night.

Reading this interview forged new synapses in my mind, connecting many lessons of homestead management that my wife, Linda, and I had been learning from rural neighbors. I immediately knew I wanted to be a permaculture design consultant. For the next six years we did a thorough study of permaculture, beginning with the references Mollison and Holmgren used to write *Permaculture One* (first published in 1978) and those Mollison used in *Permaculture Two* (published in 1979). We studied ecology, design, native and useful plants, landscaping, and many other facets of sustainable design. In 1984, we began to specifically apply permaculture to the development of the land that became Three Sisters Farm and the ten-acre property our house was on. In 1986 I enrolled in Permaculture Design at Slippery Rock University taught by teacher, writer and publisher Dan Hemenway. This intensive course was a three-week, 120-hour class that provided an in-depth study and the opportunity to apply the permaculture design process. After my completion of the course work, Linda and I were re-invigorated to continue work on the development of the Three Sisters Farm property.

In the years since Mollison and Holmgren first articulated the principles of permaculture, many voices have been added to the permaculture design field. In 2002, in *Permaculture: Principles and Pathways Beyond Sustainability*,

Illustration of chicken and bioshelter exchanges.

CO₂
Heat
Manure

Oxygen
Food

David Holmgren redefined the field with his 12 permaculture design principles. The principles built upon and expanded the literature on permaculture design.

Permaculture design has spread globally. The pages of *Permaculture Activist,* published in the US, and the UK's *Permaculture Magazine* feature the work of hundreds of permaculture designers around the world.

Ecological Design

In the course of our studies, we soon learned about the developing field of ecological design. As with permaculture, ecological design draws its inspiration from nature and ecology. But, whereas permaculture was a grassroots effort that hadn't yet reached academia, ecological design as a field was already developing in colleges and universities and through the work of progressive architects and research institutes. Among these was the New Alchemy Institute on Cape Cod. Happily, the New Alchemists thoroughly documented their work in aquaculture, bioshelter design and sustainable agriculture in their journals, quarterly newsletters and several books, so there was solid information available.

A number of authors have developed principles of ecological design as tools to guide planners. John Todd and Nancy Jack Todd, writing in *From Eco-Cities to Living Machines: Principles of Ecological Design,* defined ecological design as "design for human settlements that incorporate principles inherent in the natural world in order to sustain human populations over the long span of time." They present nine "Precepts for Emerging Biological Design."

William McDonough, with Michael Braungart, developed a set of ecological design principles known as the Hannover Principles. These ten principles offer concepts to consider as design criteria. Fundamental to the Hannover Principles is the need to evaluate materials, and designs to reduce and eliminate negative environmental impact.

The Ark Bioshelter at New Alchemy Institute. Original design.

EARLE BARNHART

McDonough has also spoken of the need to keep the products of technology away from the natural world. The products of technology tend to disrupt and degrade the natural world, but they are too valuable to waste by allowing them to simply become pollutants. In effect, he calls for two "nutrient" cycles, one that produces and recycles the products of technology and industry, the other the natural biological cycle that nurtures life.

Sym Van der Ryn, with Stuart Cowan in their book *Ecological Design* call for design that focuses on *local* solutions implemented by *local* people using *regional* materials. Their approach necessitates an understanding of the surrounding environment and nature's processes.

These various approaches to ecological design have many similarities: a call for human-scaled and socially equitable planning, a need to set the design in the local and bioregional landscape, and the need to reduce negative impact on the biosphere. Beyond the need to reduce negative impact, ecological design calls for our buildings and landscapes to become integrated into and actually enhance the surrounding environment.

Since the 1980s, ecological design and permaculture design have become intertwined. Permaculture is taught at colleges and universities. Permaculture designers and teachers have integrated various schools of sustainable and ecological design into their work. In turn, the field of ecological design has borrowed from permaculture concepts.

In this book, the terms *permaculture, sustainable design* and *ecological design* are used interchangeably. Both permaculture and ecological design are young fields. As new understanding of ecology and nature emerge and new insights into sustainable resource management develop, permaculture design continues to adapt and expand. Principles and concepts of sustainable design are tools, rather than dogma. Humility regarding the limits of our knowledge about nature's processes is perhaps the most critical principle in permaculture design.

Sustainability

I once heard a Pennsylvania state energy official tell a group: "Sustainability is a society living in harmony with nature forever." But what does *sustainable* really mean and how can we achieve it?

Webster's defines "sustainable" as relating to a way of using a resource that does not permanently deplete or damage the resource.

In 1987, The United Nations' World Commission on Environment and Development (known the Brundtland Commission) presented this definition: "Sustainable development is development that meets the needs of the present without compromising the ability of future generations to meet their own needs."

Sustainable design, therefore, is design that focuses on sustainable *resource management* for any development project. Planning for sustainable development requires an understanding of the ecology of the region and the human impact on the local and planetary environment. In our global society, daily choices we make in our consumption can affect the entire globe. In a search for sustainability we have two needs: first, to live within our region in a manner that honors and protects its diverse ecology; and second, to choose goods, technologies and fuel that will keep the global ecology healthy. As the developing world seeks a Western lifestyle, it falls on Western countries to lead the way to a more ecological, truly sustainable way of life. As environmental educator and author David Orr has shown, we would need four planet Earth's for all humans to live as we do in the United States. We have a choice: to fight a global battle over diminishing resources, or to join a global effort to seek sustainable and self-renewing forms of agriculture, technology and development.

The unfortunate truth is that national governments will not act to protect and restore the environment nor will they work hard to develop sustainable systems unless citizen groups continue to demand and promote these alternatives. Industry, business interests, politicians and facilities managers are only beginning to take into consideration the need to reduce their impact on the local and global environment. The best way to promote a broad-scale sustainability movement is to join with others into influential blocks of voters, to engage with groups doing good work in the area, and to use the power of your purchases to support the development of sustainable societies.

Concepts for a Sustainable Design System: Care of the Earth

Following the logic that human well-being is totally dependent on the health of our planet, the ethic of care of the earth is basic to permaculture design. This ethic of stewardship requires the permaculture designer to cultivate intensively and with ecological methods, allowing, as much as possible, for native wildness to return to the rest of the land. Wildness promotes the health of the farm system and enhances the health of the bioregion — and the planet we live on.

Gaia

The Gaia hypothesis, developed by James Lovelock (with contributions from microbiologist Lynn Margulis) in the late 1970s, proposes that life on earth interacts with the chemical, geological and energy cycles of the planet (and perhaps the solar system) to maintain conditions necessary for life. Lovelock and Margulis suggested that certain "functions" of ecosystems are vital for the stability of earth's climate. These functions include carbon sequestration in the oceans, tropical forests and temperate soils; recycling of nutrients; oxygen cycling; and climate modification. Each bioregion is thus one of the many faces of Gaia. For example, mid-latitude temperate forests and grasslands and tropical forests act as reservoirs of biodiversity and modify the climate and atmosphere. As stewards of the earth, we must consider the impact of our lives on all different levels: on our local environment as part of a bioregion; on the ecology of other bioregions through international trade; and on the Gaian system through our impact on earth.

The most critical regions of the biosphere, including forests, wetlands, coastal estuaries, and the atmosphere, are also the most impacted by human activity. The damage to the planet's forests is well documented and continues. Our impact on the atmosphere is also becoming widely recognized. Our impact on the seas and oceans is from several fronts, primarily overharvesting of seafood and continuing pollution of our waters. Climate change is altering the acidity, salinity and the temperature of the oceans. And there is ongoing damage to biologically rich polar seas from excess ultraviolet rays because of ozone depletion. In each area, human activities are measurably weakening these systems, and therefore their ability to respond to climate change and overexploitation.

The March of Civilization Continues

Modern agriculture is the result of a misguided science that cannot see the whole while peering through a microscope. The compartmentalization of disciplines and the war against nature raged by modern agriculture is a downward spiral, leading to further disruption of the natural world. We in the "developed" world are still on the path of environmental over-exploitation that our ancestors started down 10,000 years ago.

Good stewardship of resources and our environment is ultimately the only way to develop permanent agriculture and permanent cultures. Following the teachings of the Iroquois Nation, we must begin to think of the seventh generation to come. This is the generation we will not live to see. People living now are the receivers of choices made seven generations ago, when the petroleum era and Industrial Revolution began. So, now we face global climate change, overpopulation and mass extinctions. What will our great-grandchildren's great-grand children face? We hope it is a legacy of renewed respect for Mother Earth and a truer understanding of our place in the larger web of life.

Concepts of Ecology

Certain concepts of ecology are essential to the understanding of a permaculture farm. Below, I examine some of the key concepts in ecology applied to the farm: community, flows and cycles, diversity, succession, patterns, edge effect, and entropy.

Community

All life on earth exists in communities. To the ecologist, a community is a *group of organisms living in dynamic relationships in a shared environment*. Between the microenvironment of the garden soil and the community of customers the farm serves lay an intricate web of interactions. The goal of the farm manager is to establish, encourage and nurture these interactions so as to both reap an abundant harvest and enhance the health and stability of the total community. Each species has needs to be met by its community and yields that it contributes.

A permaculture farm is created with an awareness of its relation to the larger environment; it is itself an ecological system nested in the web of life.

Flows and Cycles

Sunlight strikes the leaf. A photon's energy is absorbed and stored by the plant as sugars and other carbohydrates. The plant, assisted by symbiotic fungi, gathers nutrients from the soil. Plants process this energy and the nutrients into an incredible variety of compounds and substances. Animals, including us, consume plants to build tissue and fuel our lives. Animal waste, dead plants and bodies are broken down by a succession of fungus, animals and bacteria that feed off the remaining energy store, returning the nutrients to the soil. And so the cycle continues as a new day's sunlight strikes a leaf and the roots reach into the living soil. The seasons come and pass in an endless cycle, generating other cycles of birth, growth, decline, death, decomposition and renewal.

Nature has evolved within a complex of flows and cycles. Limited nutrients and erratic cycles disrupt the health and stability of natural systems and agricultural alike. With the ongoing changes to global and regional climates — and the local instabilities they cause — our ecosystems are under increasing stress.

An understanding of the flow of energy and cycling of materials through the landscape and through the year is the essence of permaculture. As Bill Mollison has repeatedly stated, the goal of good design is to maximize useful stores of energy, water and nutrients in a system.

Diversity

Diversity ensures balance. Ecological communities are most stable where a diversity of native species co-evolve within an environment. A community of plants, each in its niche, mobilizes a full range of nutrients, keeping essential soil minerals available to the whole community. Diversity of animals gives resilience to a web of foragers, predators and parasites; diversity of fungi, bacteria and other decomposers ensures thorough reprocessing and return of nutrients to the soil. A diversity of niches and microclimates allow for the diversity of organisms to find their place in the system. On the farm, a diversity of crops provides security against disease and weather extremes and allows the farmer to exploit each season. Placing crops in purposeful relationships and arrangements can further increase yield or the health of the system.

Diversity is even relevant to farm economics. Diverse marketing strategies help stabilize farm income; diverse activities on the farm, including production, processing and education events can balance the farm economy.

Succession

After fire levels a forest, pioneer species take root and flourish because of decreased competition. Grasses succeed to brambles, brambles to shrubs and pioneer trees. Young trees create conditions for more shade-tolerant species. Through the year, a succession of flowers and fruits feed a succession of insects.

Permaculture designers use succession in crop rotation and soil building. They also use succession on a longer timescale by implementing designs in logical sequence. Gardens are protected from wind by berry bush hedges as perennials succeed annuals. Tree crops are established and the farm is steadily diversified into a network of gardens, hedges, wild areas and orchards. Succession is also managed to maintain a specific ecological state as desired. Some meadow species, such as goldenrods, suppress the germination and development of other plants and can keep a field relatively tree-free for decades.

Pattern illustration.

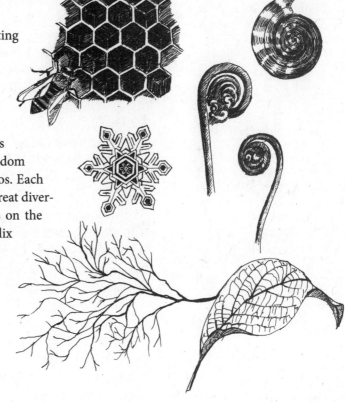

Patterns

Patterns are forms created by energy interacting with matter. Nature is manifest as patterns: Spirals of tropical storms and galaxies, branching of trees and rivers, hexagonal beehive cells and snowflakes, sine waves of rivers and radiant energy, elliptical orbits and fractal coastlines — these are all patterns of nature. Nature gathers and scatters, random chaos leads to new order, new order to chaos. Each ecosystem is nested in larger systems. The great diversity of life is, at its base, simply variations on the patterns set by the ladder-bonded double helix of DNA.

Time is also an element in patterns of nature and design. On the farm, interlocking patterns of daily routines, seasonal change, and ecological interactions define our world. Much of good ecological design is working with these patterns in order to increase both the productivity and stability of the farm ecosystem. The patterns of our days and seasons and the pace of our lives make or break the success of the permaculture farm.

Edge Effect

Edges are the interface of two or more systems. Whether it's the edge of the forest or the edge of the sea, the edge is a dynamic and productive place. Some species live only in the forest or the meadow, some move across systems, some inhabit only the edge. Habitat fragmentation by development can create *too much edge*. For example, some forest species require large areas — sometimes thousands of acres — of uninterrupted forest. When gaps appear and edges increase, there is an inevitable decline of native species within the forest. In permaculture, we try to use our understanding of edge to inform our design. An increasing edge can be a good thing in the permaculture garden when useful plantings of herbs, brambles, shrubs, trees and vines are used for windbreaks or at the edge of woodlots.

Humans evolved on the edges of forest and grassland. Our original habitat, the savannah, gave us access to the resources of the grassland and the fruits and shelter of the forest. The permaculture landscape can mimic the savannah, providing us with a mix of trees, shrubs, pastures and croplands to forage and tend.

Entropy

One definition of entropy is "a measure of unavailable energy in a closed system." In the world of physics, entropy tends to increase. Entropy is related to the tendency of the universe to move toward a state of equilibrium. For example, life on earth feeds off the dissipating energy of the sun. As this energy moves through the chain of life — from capture by plants, to herbivore to carnivore and finally to the soil through decomposition — a quantity of energy is lost with each transaction. Every resource we use is diminished and degraded to some extent in the process. A micro-example: with each rainstorm, a small amount of soil is lost to erosion.

We generally do not see the disorder that industry has inflicted on the environment; in recent years, North Americans have literally made it invisible by shifting industrial production and its accompanying environmental damage to developing countries. From mountaintop removal for coal mining, to accumulation of plastic in the oceans, to excess nutrients in our waters and the decline of native habitat, our attempts at creating order in our world is creating greater disorder in the environment.

In his comprehensive examination of the role of entropy in all endeavors, Jeremy Rifkin describes life as a form of *negative entropy*. Life creates and builds order and structure from the process of decay. The community of life works together, in the face of entropy, to maximize the cycling of material and the use of available energy. A permanent agriculture must consider the physics of entropy and negative entropy to sustain the health of the farm and its environment. Our struggle with entropy includes complex tasks, such as using biological resources to build soil, but it also includes simpler acts, such as properly storing tools to prevent rot and rust.

Integrating Concepts

Permaculture design applies these ecological concepts to farm design and planning. Farm components are examined for needs and yields. Crops and animals are chosen to fit local conditions. Plants and animals are linked to promote natural, beneficial interactions. Buildings are integrated into the landscape with earthen berms, windbreaks and other climate control plantings. Resources of climate and microclimate, soil, wind, water and sun are viewed in terms of their ecological associations. Native biodiversity is promoted by allowing as much area as possible to be left in, or restored to, a natural state. Farm components are laid out to maximize production, conserve time and energy and to promote good patterning and nutrient cycles. Building materials and supplies are chosen to minimize environmental impact. Materials with low embodied energy (the energy used to create the material) are preferred. When possible and practical, alternative forms of energy are developed and used. Planning allows for site evolution and maturation.

Permaculture Planning

Zone System

The zone system is a conceptual tool to help place elements in the system to save time, energy and labor. There are six zones ranging from Zone 0, the interior of the home, to Zone 5, uncultivated wildlands. The plants, animals and structures visited or managed daily are placed in

Zone 1 to allow the easiest access. Zone 2 has plantings and structures visited every few days or weeks. Zone 3 includes orchards and pastures, visited or managed seasonally. Zone 4 includes managed woodlots and low-maintenance perennial plantings. In a foraging sense, Zone 5 is the larger world; many resources can be gathered with permission, or purchased, from other people's land.

Sector Planning

Sector planning deals with energies from beyond the site that affect the site. Wind, sunlight, streams, rain, brush fires and wildlife all can enter, affect and then leave a site. Understanding the dynamics of these energies allows us to design our buildings, farms and landscapes to control and effectively use these energies. Sector analysis also includes assessing views, sound, privacy and pollution. Farm planning using sector analysis gives an understanding of place and leads to design that capitalizes on special features of a specific location.

Other Concepts

Several other concepts are common to permaculture and good design in general. These are concepts gleaned from many sources, including observation of nature, traditional wisdom and common sense.

Multiple Functions: Most elements in a system provide more than one function. A tree may provide fruit, shade, nectar and pollen for insects, a windbreak, nesting sites for birds, medicine and inspiration. Poultry provide body heat, CO_2, fertilizer and much more besides eggs and meat.

Stacking Functions: Every use of time and energy should be planned to integrate more than one purpose. A trip to town for the shopping one day and the bank the next and the post office the next is simply wasteful. Combining trips saves time and money.

Relative Location: Placing elements in the right relationship to each other, through zone and sector analysis is a concept related to function in that it similarly increases efficiency and productivity of the system. Our namesake, the three sisters — corn, beans and squash — as companion plants, exemplify this concept.

Redundant Systems: For any important farm product or resource, we should have multiple ways of providing for it. The bioshelter has multiple and backup heating systems and various ventilation modes. Diverse crops ensure harvest if some crops fail. Wells, ponds and rain barrels all provide irrigation.

Appropriate Scale: Whether we garden with a trowel, shovel, tiller, or tractor depends on the size of the garden and the scope of the work. Appropriate tools are important design considerations; every tool is appropriate for a garden of a different scale.

Biological Resources: Permaculture tries to make use of biological resources whenever possible. For example, a windbreak reduces heat loss and thus fuel costs; recovering heat from compost in a hot bed can get seedlings off to a great start in the spring; and poultry can add CO_2 to the greenhouse.

Energy efficiency and time efficiency are always considered in the planning stages and are reflected in choice of building design, materials and building management. Again, the aim of placing design elements with the zone system is a tool for enhancing efficiency.

Permaculture and Agriculture

When permaculture design is applied to the small-scale intensive farm, the result is an evolving, diversified, ecologically integrated agricultural system. Permaculture farms are as varied as the landforms and bioregions of the earth because ecological design is site specific. Farm design begins with an assessment of resources and a survey of the factors impacting the farm. The farm's relationship to the surrounding community and ecosystem is reviewed. Farm

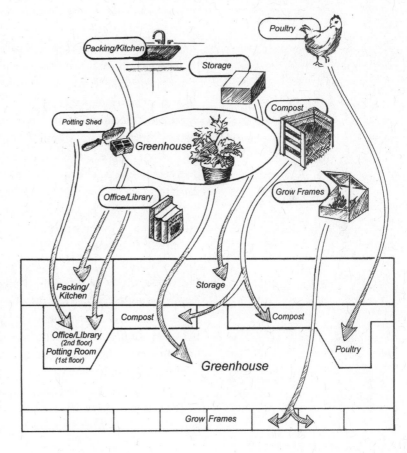

The integration of various farm components under one roof allows for efficient use of building materials and for efficient farm management.

products are chosen to make the best use of the specific soil types, aspect and exposure of the land, climate and microclimates, and other factors. Farm products are also determined by the tastes and demands of the farm's customers.

The word "sustainable" as applied to agriculture can be confusing. In the broadest sense the term includes everything from certified organic production to integrated pest control (which seeks to reduce rather than eliminate the use of synthetic chemicals in agriculture). The National Campaign for Sustainable Agriculture defines sustainable agriculture as being "economically viable, environmentally sound, socially just and humane."

To be permanent and sustainable, a culture must be able to interact with the Gaian system by maintaining a healthy ecosystem. This involves restoring, preserving and protecting natural areas of plants and other species unique to the bioregion. Farms can play a vital role in this preservation. When designed as part of the mosaic of the regional ecology, permaculture farms can provide habitat and restored ecosystems, protect groundwater and fix carbon by building soil — all while producing products for market.

The permaculture farm is an organism: the various systems (the inside propagation areas, animal housing, compost chambers and growing beds; and the outside gardens, nurseries, orchards, windbreaks, beekeeping and poultry forage) interact for the benefit of the whole by gathering, transforming and storing available energy. Life on earth is manifest as a vast network of systems within systems. These systems interact in a web of biochemical and energetic feedback cycles. A permaculture farm tries to develop and feed these cycles and work within them to promote the health and productivity of the farm. The compost piles, soil ecology, garden ecosystems, buildings and relationships to the larger community all interconnect.

Attainment of a permanent agricultural system requires understanding local conditions and which crops are suited to those conditions. Observation, design and labor are used to establish and guide the farm organism through a natural evolution of crops and enterprises to a steady state of mixed perennial and annual crops. Developing a seasonal cycle of cultivation and harvests brings steady income year-round while spreading the workload evenly throughout the year.

Farm Ecology

The permaculture farm is set in the matrix of nature. This concept is the first of the major ecological design principles presented by John and Nancy Todd in their book *From Eco-Cities to Living Machines*. Ecological design proceeds from a study of natural systems and Indigenous cultures that at are tied closely to their ecosystems. Crops are selected to suit the local climate. Microclimates are identified or developed to extend the crop selection and growing season. Beneficial insects, birds and other predators are encouraged with habitat plantings. Soil life is nurtured.

The Story of the Three Sisters

According to Iroquois legend, the three sisters — corn, beans, and squash — were gifts from the Great Spirit to the Native Americans. The three sisters are traditionally grown together as beneficial companions. The corn provides a trellis for the bean vine; the bean replaces nitrogen in the soil; and the squash's large leaves provide a living mulch, shading the ground and conserving moisture. Grown in this way, the three sisters are an excellent example of a sustainable agriculture system.

Corn, beans and squash can also be the basis of a healthy bioregional diet. The Iroquois knew them as "sustainers of life." When combined, the amino acids in corn and beans provide balanced protein, and both are good sources of vitamins and minerals. Squash is high in vitamin A and potassium and provides a complex carbohydrate that is low in calories. Squash seeds are high in protein and minerals.

Together, the three sisters were the basis of a nutritious, low-calorie, high-fiber diet for the original Americans.

The farm is viewed in its relationships to other systems: The neighborhood heron visits the farm pond on its regular rounds; the hummingbirds that live at our farm in summer spend winters in Central America; the source of our morning coffee can impact the hummingbird's winter home; the monarch butterflies born here in September winter in Mexico; rain comes from the west and drains via Mill Creek and French Creek to the Allegheny River, through the Ohio and Mississippi rivers and on to the Gulf of Mexico.

Three Sisters Farm

Three Sisters Farm was first conceived in the early 1980s. Our friend and neighbor, John Schmidt, purchased ten acres of farmland and we purchased ten acres of forestland several miles away in Mercer County, Pennsylvania. We all intended to establish our respective homesteads on our two properties; Linda and I had vague plans to develop five acres of the Schmidt property into an organic permaculture farm.

The first step was easy, but very important. We simply allowed the land to remain fallow for, as it turned out, six years. During this time, a

Three Sisters Farm

natural diversity of plants was allowed to grow. Red clover, yarrow, goldenrod, grasses, biennial taproots such as wild carrot, docks, dandelions, and other plants worked to heal the damage done from conventional agriculture. Inspired by the book, *Weeds: Guardians of the Soil*, by Joseph Cocannouer, and by Masanobu Fukuoka's writing on natural farming, we trusted in nature's ability to regenerate the soil. Both writers said that damaged land is best regenerated by a mix of what are usually considered weeds. From Cocannouer, we learned that annual and biennial plants such as purslane, ragweed, lambsquarters, amaranth, dandelion, docks and Queen Anne's lace will rejuvenate depleted soils. With their various root systems penetrating the soil and accumulating minerals, these and other weeds will, over time, increase fertility, loosen compacted soils and support a diverse soil ecology.

Allowing mixed species to grow started the process of healing the land. Decades of plowing, herbicide application and continuous cropping of corn or soybeans had depleted soil nutrients and damaged the soil structure. The clover provided nitrogen and organic matter. The biennial taproots delved deep into the soil to improve drainage by breaking up the plow pan and bringing minerals back to the surface. The successive increase in the field's diversity nurtured an increase in soil health. We also allowed the tree line to expand. Native dogwood shrubs and other plants seeded by birds or the wind became established and spread.

Following discussions with Bill Mollison at a weekend workshop in 1982, the concept for Three Sisters Farm became clear. Mollison suggested that a person wanting to bring permaculture to a region should collect useful plants and establish a permaculture nursery. Following his lead, we reasoned that market gardening would bring income while we waited for perennial plantings to grow into windbreaks, hedgerows, forage, habitat and crops. A permaculture farm could evolve to follow a seasonal cycle of cultivation and harvests that would bring steady income year-round while spreading the workload over time. Such a farm would serve as a demonstration of permaculture design, a resource for nursery stock, and a teaching center for the community.

Resource Survey

The study of available resources is another important first step in designing a permaculture farm. Identifying a site's unique features and resources reveals a site's potentials and limitations. In our case, assessment of our local resources revealed the availability of massive quantities of horse manure and waste hay and straw from surrounding stables and farms, and sawdust and bark mulch from local hardwood sawmills.

The farm site was a five-acre field that sloped gently to the south and was bordered by a country road to the west. There was a mature tree line with shrubby understory to the north and east; and a young woodlot and wet floodplain and stream to the south. The rich, sandy loam glacial soil was ideal for gardening. With a natural pH of 5.5, the soil would only need an occasional application of pulverized limestone. The field had a pond site fed by a spring on a neighboring property. Groundwater was plentiful, with well depths of 30 to 60 feet in the area. When the most recent ice sheets plowed across the Lake Erie basin from Canada, they literally plastered northwest Pennsylvania and northern Ohio with sand, gravel, clay, rocks and boulders. At the site of Three Sisters Farm, this layer of compacted glacial till, 60 feet thick over bedrock, holds a pure freshwater aquifer.

We also counted as a resource our growing knowledge of intensive cultivation, agronomy, permaculture design and our desire to develop a model system.

Once we had performed our resource survey, we took several years to think over our plans and do the necessary research. During this time we collected plants, established homes, had children and practiced our designs on smaller scales.

Interior view of planting in the Solviva Bioshelter, built by Anna Edey on Martha's Vineyard, MA.

DARRELL FREY

Early Stages and Planning

From 1981 through 1987, we collected, purchased and planted various useful plants for trial. These included pea shrubs, honey locust, bamboo, hazelnuts, rosehips, currants, apples, pears, grapes, plums, numerous berries, herbs and other plants.

All the while, we studied and discussed permaculture topics. We were particularly inspired by Anna Edey's Solviva Bioshelter on Martha's Vineyard and the New Alchemy's Institute's work with bioshelters and composting greenhouses. Our plans began to coalesce around the idea of building a composting bioshelter. With freely available horse manure (and our 4WD dump truck), we could use compost to enrich a passive solar bioshelter with supplemental heat and CO_2. Finished compost would then be used to fertilize crops in the bioshelter and outdoor gardens.

A Bioshelter for Three Sisters Farm

After six years of study and planning, we were ready when a funding opportunity came along. Federal courts had ruled that the major oil companies needed to refund profits from oil overcharges in the 1970s. The US Department of Energy was charged with distributing the funds through qualifying state agencies. The Pennsylvania Energy Office requested grant proposals for projects demonstrating ways to save energy in agriculture. We responded with a proposal to build our bioshelter.

When we received the grant to build the bioshelter, we were thrust into doing what we had so long studied and planned for. The bioshelter is the center of farm activities. It's where seedlings are started for the garden, tools and equipment are stored, and produce is processed for sale. A complete description of our bioshelter's design and management is given in subsequent chapters.

Gardens and Landscape

The bioshelter is built on the highest and most level part of the field, roughly centered on the property. Its central location allows for easy access to the main production gardens. Each year since 1988, we have

Crops

Our basic crop is a salad mix chosen for our particular bioregion. Self-seeding wild edible greens are part of our mix. These are combined with cultivated greens and lettuces to create a salad unique to our farm. The diversity of seasonal ingredients provides crop security and keeps customers interested. Our other crops include herbs, cut flowers, edible flowers, fennels, radicchio, lettuces, tomatoes, greens, root crops, peas, beans, squash, and potatoes. This crop mix has evolved as we attempt to cater to the needs of our customers: chefs, caterers, grocers and household subscribers. We use a system of sales of mostly pre-ordered produce; this reduces unnecessary plantings and crop waste. Excess produce is either consumed, sold or given away. Recently, we started selling berries, crab apples, pears and apples.

expanded our gardens and our production, so we have also used some of the adjoining land available to us. Today (2010) the bioshelter is surrounded on three sides by gardens. Each garden is composed of raised beds laid out on contour. Pathways between beds collect rainwater or can be flooded with water pumped from the pond. Wide central paths allow access to the contour paths. As we develop gardens, we allow uncultivated areas of goldenrod and wildflowers within or near them. These biological islands, together with herbs and flowers and other insectary plants, provide a healthy balance of predatory creatures. Birdhouses, rock piles and perennial plantings of fruits, nuts, berries, vines, shrubs and flowers are added each year. Plants create windbreaks and shade zones.

Garden beds laid out on contour at Three Sisters Farm.

As we continue to develop our plant nursery enterprise, we are enjoying our trial varieties of filberts, hazelnuts, currants, grapes, plums, apples, pears, Juneberries, rosehips and more. Other fruits are wild-crafted from neighboring properties. Our ten-acre woodland, with three diverse forest systems, provides forest products for farm use and contains some of our plant collections.

Energies

Sun: Much of the sunlight entering the bioshelter is absorbed — by plants, surfaces and materials (as thermal mass) — to provide our primary heat source. We also capture the sun's energy by using a photovoltaic panel to power an irrigation pump.

All our gardens receive full sun. Adding shade trees and shrubs to compost areas and some gardens is a continuing endeavor. Along the bioshelter's north wall, we have established a 2-foot-wide shade garden of native woodland wildflowers, ferns, vines and other useful plants. Fruit trees are mulched and planted with an understory of herbs and flowers. Other shade zones include a visitor's camp, picnic area and hammocks in the tree line. A kiwi arbor near the bioshelter kitchen provides a cool place to work or rest. A 600-square foot barn adjoining the bioshelter provides 4,800 cubic feet of shaded space. Cool air from this space is drawn into the bioshelter, naturally cooling the structure during summer months.

Wind : Wind is a major factor at Three Sisters Farm. Situated on the western slopes of the Allegheny Plateau, our site receives daily wind, often steady and strong. Reduction of garden level wind is a priority. The tree line to the north provides our primary windbreak, sheltering the bioshelter and our gardens from the northwest winds. We have added a secondary windbreak of roses, shrubs and small trees 100 feet west of the bioshelter. A third, mixed species windbreak provides extra shelter to the herb and flower gardens next to the bioshelter. The trellised kiwi arbor between the bioshelter and herb garden provides a fourth wind deflector for the bioshelter.

Water: The pond provides irrigation for the gardens via pumps and sprinklers. The pond also provides fish, cattails, duckweed, algae and other useful plants. We continue to add useful species. Snakes, toads, turtles, dragonflies and many wild birds make the pond their home or feeding ground, as do muskrats. Generally, we receive 40–45 inches of rain each year. Roof run-off is caught in swales and contour paths to recharge our well. The 60-foot-deep well provides water for the bioshelter and our home.

Strategies

Good implementation strategies are imperative for successful permaculture design. Primary among these are staging and timing. We knew when we started that it would take up to five years to see income. Rather than go deeply in debt (beyond a loan for matching grant funds), we chose a more natural pay-as-you-go expansion. Each year we have expanded our gardens and increased sales. For the first ten years we averaged 20 percent growth in sales annually. After that, we began to diversify our activities to include more educational outreach. The bioshelter developed as we earned cash and learned more. Generally, bigger construction projects are planned for fall, when more money and time is available.

Our landscaping strategies follow Bill Mollison's suggestion that it is better to plant only what you can care for each year rather than over-plant and lose trees. We've found that limiting our spring planting of new tree and shrubs to ten per year works well with the demands of our annual gardens.

Season Extension

Season extension is creating and using microclimates and structures to extend the growing season. Spring season extension allows the gardener to get a head start in crop production by starting seedlings for the spring and summer gardens. Fall season extension expands the production time for summer and fall crops. Winter season extension allows the year-round production and harvest of crops.

Season extension has been an important tool for market gardeners for centuries. In northern climates, the ability to extend the growing and harvest season is vital to the development of sustainable regional food systems.

Season extension can be as simple as mulching your carrot beds so you (and the voles) can harvest fresh vegetables from under the winter snow; it can be raising a bit of salad or cooking greens in a cold frame; or it can be winter gardening in a bioshelter.

Outreach

Fritjof Capra, writing in *The Web of Life*, declares the *network* to be the basic pattern of life. Networks of systems within systems form the basis of our biological existence, from the micro flora and fauna of our bodies and the soil to the biomes of the whole of the biosphere. We see our farm as part of several networks: sustainable agriculture in our bioregion; regional outlets for organic products; sites demonstrating sustainable practices; and the many farms and citizens in our region exploring sustainable organic gardening.

Since 1988, several hundred people have visited Three Sisters Farm. They come for tours, workshops, internships and celebrations. They come to buy produce and plants. We hope they learn a little about ecology and permaculture.

Three Sisters Farm is a working model of permaculture design. It is not a model in the sense of a completed project, but a model of the living laboratories needed to fully develop sustainable permaculture farms in temperate climates. We see bioshelter market gardens as a key to sustainable living in our climate. As we continue to observe, create and implement our designs, we move closer to the idea of permanent agriculture.

Back to the Earth

The best way to describe the changes permaculture development has brought to our farm is to compare it to what came before. A small field that was previously plowed each year and planted in corn or soybeans now contains many dozens of species of plants. It is home to a thriving natural diversity of animal life. Bats, moles, mice voles, and weasels live between our gardens and along the edges. Swallows, orioles, bluebirds, catbirds, wrens, redwing blackbirds and many others songbirds nest here. Herons, ducks, geese, owls, and hawks forage our pond and fields. Turtles, snakes, frogs, toads, spiders, bees, wasps, and other native creatures live here as well. Wildflowers and new insects constantly amaze us as they join our evolving system. Together the gardeners, gardens and nature manifest the essence that is greater than the sum of its parts.

PROFILE

Homescale: The Jenkins Homestead

While the garden at Joe and Jeanine Jenkins' Homestead evolved and grew as a result of their own ideas and planning, it is laid out in a classic permaculture landscape, from a Zone 1 kitchen garden to the Zone 3 orchard and Zone 4 and 5 woodlots beyond. Fifteen paces from the kitchen door, after passing through a tree-shaded dining deck, one enters the garden. The 4,200 square foot space has one side facing the house and three sides surrounded by a double-fenced poultry run.

The east side of the garden, bordering the house, is edged with flowers, ferny asparagus and small shrubs. A cooking pit and picnic table are placed under the shade of a friendly young maple tree. A slate-roofed chicken house, sheltering a small flock of hens, is in the northwest corner. The chickens have free access to a 6-foot-wide run, where weeds and unusable produce are tossed for the chickens to forage. Two fences provide double protection from deer, rabbits and woodchucks, and they give the chickens the chance to intercept slugs and other crawling garden invaders. Joe Jenkins explains that the poultry run design evolved from a need to keep deer out of the garden. He had read that a deer would not cross a double fence. Connecting the area between the fences to the poultry yard was a logical way to keep the weeds down in the space and to use the birds' natural foraging habit to keep the fence lines clean. The inner fence doubles as a trellis for beans, peas and flowers. Grapes hang from the outer fence. The northern third of the garden, adjoining the chicken house, is weeded and fertilized by foraging chickens during the fall and winter months. Each summer, sweet corn and companion crops grow well in this annually renewed plot.

Jenkins' zone one garden.

Apples, peaches, peonies, gladiolus and other perennial flowers are planted along the outer garden fence. Gates through the fences provide easy access to compost and mulch stockpiles and to the orchard area. The half-acre orchard includes apples, pears, hardy kiwi, plums, peaches, hazelnuts, raspberries, blueberries and chestnuts. Mulberries are located near the chicken house, where newly hatched chicks can peck at the fallen fruit each June. Native

Jenkins' garden with grape arbor.

patches of Joe Pye weed, ironweed and jewelweed are encouraged in the few wet spots in the orchard, creating excellent habitat for butterflies and beneficial insects. The orchard and yard area is surrounded by forest edge to the north and west and a perennial border to the south. Dogwood, and other shrubs and trees provide winter forage for birds. A dwarf hydrangea on the garden border gives summer shade for understory plantings of daffodils and other spring flowers, and provides seeds for cardinals, juncos and doves well into the winter. The birds are also provided with fully stocked feeders most of the year.

When Jeanine Jenkins describes her garden, she mentions the *I Ching,* in which stable lines move to change, and a balance is achieved between natural wildness and cultivated beauty. Jeanine explains that the garden, developed and worked by both her and her husband, Joe, is a result of the dynamic tension between garden labor and the organic evolution of the growing space. Her plants form a natural balance instead of being forced from a plan. Beauty and bounty meet in simple grace and elegance. Self-seeded scarlet poppies and biennial purple-throated foxgloves intermingle with cabbages and beans. Permanent, stone-edged beds line a gravel path leading to a garden bench. These stone beds protect alyssum, yarrow and thyme, marigold and sage, calendula and cosmos, gladiolus, and daisy, as well as kitchen and tea herbs. Branching off at right angles from this central path are smaller paths and beds of various sizes. A second bench and small grape arbor are placed in the southwest corner of the garden.

Most of the beds are fluid, changing size and shape each year as crops are rotated. Wild edible purslane, lambsquarters, and other plants self-seed. Large, crispy cabbage, sweet kale and broccoli, tangy tomatoes, peppers (both sweet and hot), eggplant, potatoes, summer and winter squash, cucumbers, peas, beans, carrots, onions, garlic and beets trade companions and positions annually and seasonally in a colorful fairy dance of succession, season and gardener's whim.

From numerous birdhouses and nests, wrens, phoebes, robins, bluebirds and swallows patrol the gardens and orchards for caterpillars and other pests. Bees and other beneficial insects forage on the succession of flowers in the stone-lined beds and garden borders. Ducks, fenced out of the main garden, forage and fertilize the orchard. The ducks also love to clean up fallen fruit, helping to break the life cycle of pests and keep the orchard clean.

This garden and landscape evoke a sense of balance, bounty and wholeness.

Jeanine Jenkins gives a tour of her poultry forage area along her zone one garden.

DARRELL FREY

Its encircling symmetry and colorful beauty combine to give the gardeners and their guests a feeling of satisfied well being. Sitting on the deck, sipping iced herb tea in the dappled tree shade and afternoon sun, while hummingbirds flash past to sip the gladiolus or foxglove, is as relaxing an experience as one may find, and it is just outside the kitchen door.

A Nutrient Cycle — Full Circle

Readers may be familiar with the Jenkins's garden from Joe's book, *The Humanure Handbook,* wherein he details the humanure toilet and safe composting of human manure: A toilet receptacle (usually a five-gallon bucket) containing feces, urine and sawdust is added to the wood-framed compost pile every few days. Each addition to the pile is covered with straw, garden weeds or other fibrous organic material. The compost generates sufficient heat and micro-biodiversity to destroy pathogens as it composts and ages. One bin is filled over the course of a year. The two-bin compost system ensures a bin will age for a year after the initial one-year thermophilic composting period.

The finished compost is used to fertilize the Jenkins's garden and landscape each spring, returning nutrients to the soil. Every summer solstice, a new pile is started in the emptied bin. The Jenkins' continually monitor the temperature and biological activity of their composting system and have developed an elegant and effective system to ensure the safety of their crops and the quality of their compost. The result is thriving, healthy gardens; thriving, healthy people and a wondrous and magical garden ecosystem. (Note: Most farms and facilities with composting toilets restrict humanure compost use to trees and other Zone 3 and 4 perennials. One reason for this is to prevent misunderstandings with customers who buy produce. At Three Sisters Farm we do not use humanure in our production gardens.)

Permaculture principles are very well demonstrated in the Jenkins's Homestead garden. Every element is in a good location relative to the house and each other. The poultry house and run are designed to serve multiple functions and make good use of the biological resource of chickens as foragers and recyclers. David Jacke, writing in his book *Edible Forest Gardens, Vol. II,* calls this type of double-fenced chicken run a "chicken moat." The poultry house, fencing and border plantings break the wind and create a sheltered, sun-trapping microclimate. As mentioned above, the entire landscape is laid out in a zone system. Everything is designed for efficient use and management. The fencing and chicken house are built to last (the slate roof on the chicken house mirrors the care taken on all the Jenkins's buildings). Nature is nurtured, even celebrated, with birdhouses and feeders and natural art. Living space and garden have been integrated into one system. The nutrient cycle between gardener and garden is complete. The Jenkins's home embodies the sense of permanence we seek in good permaculture design.

Chapter 2

Sustainable Food Systems: Safe, Healthful and Secure

Tending my garden, I saved the world.
— Zen wisdom, paraphrased

A FOOD SYSTEM IS ALL THE ACTIVITIES THAT BRING FOOD to your table. That includes food production, processing, transportation, storage and distribution. It may be further defined to include preparation, cooking, consumption and waste management. Our food system should provide food that is safe to eat, promotes good health (for consumers as well as soil), and provides food security for all consumers. With this chapter, I seek to place the market garden farm in the larger context of the regional food system by looking at three aspects of the food system: food safety, healthful diets and food security.

Sustainability for Food Systems

A permaculture perspective on what a sustainable food system really is takes into account the ecological and cultural effects of the food system, both local and global. A system should be able to provide sustenance for consumers, a profitable livelihood for producers, support for healthy biodiversity, and support for the cultural vitality of a region.

While some level of international food trade is desirable, the need to support and develop local food systems is vital for a sustainable society. The global economy provides an incredible array of foods to the far corners of the planet. But it does so at a high cost to the biosphere and local cultures. High energy inputs are required to produce, process, transport and store both basic commodities and specialty foods. Excess nutrients and agricultural chemicals pollute surface water and groundwater. Many people rightfully object to the mass production of milk, eggs, beef, pork and poultry in the mega-farm industry on the basis of personal philosophy and out of concern for environmental impacts. There is great potential for disruption, contamination and slow accountability in the system. Overfishing, soil erosion, deforestation, groundwater depletion, population increase and climate change already undermine the foundations of the global food supply. The need for biogenic energy sources (such as biodiesel and ethanol) to replace fossil fuels is removing crops and cropland from the food system.

Understanding and acknowledging these issues is crucial to the public debate. Also critical is the need for working examples of sustainable systems, and a broad understanding of the economic, social, health and ecological benefits of a sustainable food system. We cannot affect positive change without proposing positive solutions.

There are countless examples of work toward sustainable systems in towns, cities and the countryside around the world. Whether the efforts are large or small, each is a piece of the puzzle, and together they provide a glimpse of a sustainable future. Small farms incorporate alternative energy into their operations; environmental centers build worm bins to recycle organic matter; municipalities offer mulched leaves to residents; urban gardeners grow produce and teach age-old skills to new generations. Hope grows for a green and healthy planet.

Yet, on closer study, most sustainable projects are struggling. New market gardens start up and fade away. Customer turnover in a Community Supported Agriculture program (CSA) may be 20 percent each year. Farmers have to rely on off-farm income and innovators rely on grants and tax breaks. A disparity can exist in the cost of managing a small farm and the market value of the farm's products. Trying to compete with the highly subsidized global food system will keep local sustainable agriculture at a disadvantage as long as our personal shopping choices support the former and our political systems hide the true costs with subsidies and infrastructure.

Permaculture offers a design system to help bridge this divide. Clear assessment of a farm's assets, innovative design strategies, and planning to integrate sustainable systems all help to reduce a farm's energy and maintenance costs. Planning for diversity and the long-term development of ecological relationships and community service help create sustainability.

The permaculture ethics, care of the earth and care of people, imply, at the very least, an organic approach to agriculture and food production. Permaculture applied to agriculture is a process of design and management for the farm to be productive, culturally relevant and positively integrated into the larger environment. Permaculture design for a bioregional food system implies a sensitivity and response to cultural diversity, seeking to develop relationships with customers by satisfying their tastes and traditions. These approaches in turn can increase a farm's long-term viability.

For the near future, the development of sustainable agriculture will continue to need financial support, including subsidies and grants, as we put into place new local food systems. True national security depends on the strength of these systems. Education of consumers and political leaders must also continue. It is far better to see the future coming and prepare for it than to react to a crisis.

In the long term, these farms must sustain themselves. A garden is more than a place to grow food. It is a place for nature to reside and for us to learn about our intricate web of connections with nature.

Local Food

Fresh, local, organic food is obtained either by growing your own or buying it from local growers, grocers and restaurants. (Here, and throughout, local is considered to be within 100 miles; regional is within 300 miles to 500 miles.)

Home food production is a widely practiced cultural tradition. While the percentage of a household's food produced by gardening has generally declined in recent decades, gardening continues to be a popular activity. Fresh seasonal vegetables from your own organic garden cannot be improved on for flavor and nutrition.

For urban dwellers and others without their own land, community gardens are an important asset. As demand for fresh produce grows, demand for community garden space will increase.

Market Gardens in the Sustainable Food System

Next to growing your own produce, market gardens provide the most direct access to locally grown fresh produce. From the time the first

cities were founded, until sometime in the past century, market gardeners and regional farmers supplied people with most of the food they ate.
But the local farmer has become almost unknown because farmland
near urban centers is disappearing. Today, urban sprawl and commercial development continue to transform former farmland into bedroom
communities and shopping centers. Moreover, many issues complicate
the preservation and maintenance of farmland. Zoning, taxation, land
value, public awareness and government regulations all influence the
ability of farms to compete for land and markets. If land is to be preserved
and access to local food ensured, it will only be as a result of concerted
efforts by community groups and sustainable agriculture organizations.

As we build a sustainable society, the role of market gardens in local
food systems will continue to grow. Market garden farms tend to be
small and diverse, offering a wide range of seasonal products. Because
of their smaller size, diverse crops and local distribution, these farms
can be highly responsive to the needs of their customers. Seasonal
and cultural specialties can be grown to suit local tastes. As you will
see in following chapters, market garden farms established on small
parcels can provide a wide range of products and services to the local
community.

Small- to medium-scale commodity producers will play an expanding role in the sustainable food system. For example, shiitake and other
mushroom production, flour milling, maple syrup and honey production all can be done on a family farm. Dairy, livestock, eggs, and cheese
are also well suited to small-scale production.

Season extension for vegetable crops is an essential component of
any local and regional food system. Winter crop production techniques
that are profitable and do not rely on fossil fuel will increasingly supply
salads, greens, herbs and other products.

Value of Local Agriculture

The reasons for preserving local agriculture are numerous, chief among
them, nutrition and quality. Vegetables begin losing nutrients when
they are harvested; flavors are richer in fresh foods. A fruit or vegetable
shipped thousands of miles, from across the country or from a different
country, will not be the same quality — or even look as good — as food
purchased from local growers. A more direct link between consumers

and local producers creates a positive feedback loop: local customers ensure markets for farm products, which encourages local production, thus ensuring a continued supply of locally produced food. Because they are more closely linked to the consumer, local farms have more incentive to be responsible in ensuring the safety of the food they produce; thus, they are more likely to be concerned with the reduction and elimination of pesticide residues and food-borne pathogens through organic production and sanitary practices.

The potential disruption of food supplies by natural and political disasters is another reason to promote local food systems. And energy costs for processing and transporting food is greatly reduced with local production.

Food Safety and Public Awareness

"Are you organic?"

When we began selling produce from our farm in 1988, no customers asked us this question. When we did mention that our crops were grown with organic methods, customers showed vague disinterest. Price and quality were the primary concerns. But change came soon. After a 1989 Natural Resources Defense Council report on potential health effects on children of the chemical alar was publicized, many of our customers began to ask: "Are you organic?"

Alar, the trade name for a chemical called daminozide, was sprayed on apples to promote uniform ripening. Researchers concluded that alar degraded into a chemical that was a potential carcinogen. The ensuing outcry was perhaps the most significant shift in public awareness of chemicals in the food system since the publication of *Silent Spring* in 1963. Alar was banned by the EPA in 1989. The organic food market has been growing ever since.

Similar shifts in public awareness and concern about the safety of the food system continue to occur. In September 2006, between 50 and 60 million dollars worth of spinach was recalled due to fear of contamination with *E. coli* bacteria. According to the Centers for Disease Control, this particular bacteria, *E. coli* 0157:H7, originates in the intestines of livestock, including cattle, sheep, goats and deer; it can cause severe diarrhea and, in young children and the elderly, kidney failure. There were 200 reports of illness in at least 20 states. The nationwide recall of spinach greatly raised awareness of the concept of "local food." The fact that the FDA is often unable to trace the origins of such contaminations points to the need for traceability in the food distribution system. When the customer buys directly from the farmer, accountability is immediate. Also, should problems occur in locally produced products, the impact isn't so widespread.

It is up to the farmer to understand food safety issues and ensure that the farm's composting methods, irrigation, harvest and packaging are safe. By learning food safety issues and practices — and keeping the consumer informed — the small-scale farmer is better positioned to ensure product safety than the giant commercial producers are.

Healthy Food

Providing an interesting and healthful diet with locally grown food is a challenging prospect for most people in the United States. Consumer trends, shopping patterns and government policies of the last 50 years have led to centralization of our food systems. For example, most produce consumed in the northeast is imported from Mexico, California, and a few other states. This, in turn, has decreased the availability of locally grown foods.

Nutritional science and diet planning are continually evolving fields. Some controversy revolves around various theories of consumption of fats, carbohydrates and proteins in relation to obesity, heart disease, and other diet-related illness. However, consensus is emerging: unrefined whole grains and legumes, with plenty of fresh vegetables and fruits, are the ideal basis for a healthy diet. Dairy products, meats, eggs and plant oils are needed — in moderation — to provide protein and energy. Use of saturated fats, refined sugars and refined grains should be minimized. Most researchers recommend low-fat dairy products.

A healthful diet consists of eating the right amount of a variety of foods. Correctly chosen, these foods should provide the proper amounts of the necessary elements of nutrition. Elements of nutrition include water, protein, carbohydrates, fats, vitamins (and their precursors) and minerals. Recent research has begun to identify the role of micronutrients, including phytochemicals (phyto = plant) and other antioxidants in a healthy diet. Dietary fiber, found in plant-based foods, is also vital to a healthful diet.

When developing a diet based on local foods, one must select from among the region's products to provide adequate nutrients from local foods. A wide variety of locally raised meat and dairy products are readily available. Milk and cheese are important sources of calcium and are a complete protein. Eggs are low in fat and are a valuable source of protein, especially for growing children. From late spring until fall, a good variety of locally grown fruits and vegetables are available to those who seek them. Regionally produced grains and beans can also be found. However the quantities of produce, grains and beans available regionally can be small compared to the population to be served.

During the winter, when fresh food is limited, stored root crops, cabbages and squashes can be supplemented with canned and frozen foods and greenhouse-grown salads and greens.

Large-scale transition to local food systems will require a concerted effort among producers, consumers and government agencies. Developing a local food system that addresses regional health needs will require both increased consumer demand for local products and increased local food production. In the meantime, conscious consumers can enjoy the benefits of local foods through careful shopping and menu planning.

Food for Permanent Culture: Embracing Cultural Traditions

As permaculture design seeks to forge a path to sustainable living, we must propose solutions that people will love. Food choices result from a complex blend of cultural, social, religious, behavioral and personal influences. For example, many of the world's cultures derived sustenance and pleasure from adding insects to their diet, Europeans and North Americans outside of Mexico cringe at the thought. Milk and dairy products, considered an essential food group in the US, are indigestible to most of the world's adults.

When considering food systems, understanding cultural influences is essential. Food often conjures pleasant memories and provides a touchstone in our lives and a connection with our families. Every Easter, Grandma Reeder hosted the traditional baked ham dinner. On New Year Day, we had to have baked sauerkraut ("very brown" — Mother insists!) and pork. Christmas and Thanksgiving featured the traditional turkey dinner with all the fixings. In our family, all these meals included mashed potatoes and gravy and several vegetables, usually green beans, corn, brussels sprouts, cauliflower or broccoli. Being raised with Pennsylvania Dutch staple meals, like beef and noodles, chicken pot pie, and boiled ham and cabbage, it was a cultural surprise to experience Georgia hospitality and cuisine as a teenager. I had never even seen black-eyed peas, grits, or greens, much less in such generous helpings! When a brother married into an Italian family, we all learned that great Italian food extended far beyond spaghetti and lasagna.

Discovering new foods and food traditions is a pleasant way to learn about other cultures. Tofu, miso, tempeh and other Asian foods are now a regular part of our diet. Chinese-American, Thai, Mexican, fine Italian, European nouvelle cuisine and the new American Fusion cuisine all have inspired our own home cooking. The work of Alice Waters and her restaurant Chez Panisse in developing new American cuisine has influenced a generation of chefs and helped create restaurant demand for fresh local ingredients across the nation.

Even in small cities, restaurants have menus that read like lists of UN member countries. The US is truly the melting pot of culture; each addition to the ethnic diversity of the country brings a unique cuisine.

Gardening and Health

Exercise goes hand in hand with diet for maintaining good health. It would be wonderful if everyone got a free health club membership as part of a national health care system. (Those who participate in weekly exercise regimens would get a discount on their premiums.) In the meantime, we have to find our own health clubs. The ideal health club would provide a variety of activities that allow members to exercise their entire body: muscles, lungs and heart. It would include outdoor activities in the fresh air. It would be in a psychologically and spiritually stimulating environment that allowed for relaxation and reflection amid beauty — along with the exercise. Included would be a reward mechanism, providing positive feedback for invested effort. What if this place also provided its members with the freshest food possible, especially a plentiful assortment of seasonal fruits and vegetables? Does such a place exist?

Yes. It's as close as your home garden. Gardening at home can provide all these benefits: occasional cardiovascular workouts and overall exercise in a stimulating environment with the reward of fresh food. (Of course as with any workout, exercise through gardening requires a few warm-up stretches, proper equipment, and proper use of that equipment.) Additionally, gardening relieves stress, channels energies, and helps one maintain a youthful health and vigor.

Providing the ingredients for these diverse cuisines is an opportunity for the enterprising farmer to develop relationships with chefs and create marketing niches. Targeting production to supply special foods for specific holidays, like the traditional Thanksgiving turkey, also offers local farmers marketing opportunities.

Seeking a Native Diet

As we seek a permanent agriculture for any bioregion, it is instructive to study the traditional food systems of the indigenous people of that region. We cannot soon restore the native ecology and biodiversity as it existed in 1492, but study of the traditional diet of earlier inhabitants can provide insight into planning a healthy and ecologically sound diet based on local foods.

Our farm is located in the former territory of the Iroquois. The Iroquois Nation, also known as the Hau-de-no-sau-nee (People of the Long House), are native to western New York, northwestern Pennsylvania and southern Ontario. The Six Nations of the Iroquois Confederacy are the Seneca, the Onodaga, the Mohawk, the Oneida, the Cayuga and the Tuscarora peoples. At the time of first contact with Europeans, in the 17th century, the Iroquois were primarily an agricultural people. They had lived in a sustainable manner in the north central Appalachians for

many hundreds of years, practicing natural resource management without depleting their resource base. In many ways, their diet was typical of the Eastern Woodland Indians of that era.

The Iroquois diet at the time of first European contact was a complex mix of both cultivated and foraged plants, fish, shellfish and hunted game. Writing in the book *Native Harvest*, E. Barrie Kavasch reports that the Iroquois generally had one large mid-day meal and one or two smaller meals throughout the day. They used little or no salt, though various ashes, which provide ☞

The Three Sisters are corn, beans and squash. They are considered gifts of the Great Spirit by the Iroquois Indians, and have been traditionally grown together as beneficial companions. The corn provides a trellis for the bean vine, the bean replaces nitrogen to the soil, and the squash's large leaves provides a living mulch, shading the ground and conserving moisture. Corn, beans and squash can also be the basis of a healthy bioregional diet.

The three sisters represent to us the vision of native agriculture we strive to rediscover here at Three Sisters Farm.

minerals, were used in cooking (these included ashes of plants, such as mineral-rich coltsfoot leaves, and various wood ashes, including cedar and hickory).

The primary crops of the Iroquois were the Three Sisters: corn, beans and squash. To this day, the Iroquois promote the use of these foods and maintain traditional varieties of some of their ancient strains.

Corn alone is inadequate as the basis of a healthy diet. Corn is deficient in the amino acid lysine, one of the nine amino acids people must obtain from food. Corn eaten as a primary protein source will lead to protein deficiency. Additionally, 95–97% of the niacin in corn is in a form which is unavailable to humans. Niacin deficiency, which can be fatal, is common in "corn-dependent people." Similarly, beans are low in the amino acid methionine. Native Americans solved this dilemma in two ways. One was to combine foods with complementary proteins to provide a balance of amino acids. Corn was often eaten together with beans or combined with meat, shellfish or nuts. The other method was to prepare corn with ashes or lime. The lime or ash treatment made niacin available and balanced the overall quality of the protein. Hominy grits, a southern staple, are traditionally prepared this way.

Corn, beans and squash were eaten year round, either fresh or dried. Many traditional recipes include combinations of the three, often with other foods. For instance, there is a traditional corn bread that is baked with beans included.

Cultivated sunflowers were another high-protein food and source of oil for the early Native Americans. Sunflower seeds are a very good source of the minerals phosphorus, potassium and iron and are known as a rich source of vitamin B-6. Sunflower oil is high in polyunsaturated fat and, therefore, one of the more healthful oils.

The Iroquois ate a great variety of seasonal foods. In addition to cultivated crops, they foraged wild foods. The succession of roots, shoots and fruits utilized was extensive. Nut trees provided a source of protein and healthful oils. American chestnuts were abundant before the chestnut blight all but eliminated their presence in the early 20th century. Many other nuts were used as well. Acorns, hazelnuts, beechnuts, walnuts and hickory nuts all provided high-quality protein and oils. Easily stored, nuts were especially important in the winter, when other foods were scarce. Nut trees do not generally bear every year, however. They typically give a heavy yield every second or third year. Nuts were and still are an important mast crop, providing food for wild game populations.

The list of plant shoots, leaves, pods and roots used as vegetables is long and includes cattails, milkweed, Jerusalem artichoke, ground nuts, ramps, jewelweed, ☞

The sunflower is another traditional crop of Native Americans.

ferns, violets, nettles, bulrush and coltsfoot. Many semi-wild plants introduced by Europeans after the 16th century were adopted, including watercress, dandelion, daylily, plantain, sheep sorrel, chicory and lambsquarters. These and other wild foods contributed essential vitamins, minerals and other micronutrients to the diet. An abundance of spring greens quickly rejuvenated the body after a long winter diet of stored foods. Pine needle tea supplied vitamin C through the winter. A variety of mushrooms also were gathered in their particular seasons.

A succession of fruits were eaten fresh and dried for storage. Juneberry, strawberry, blueberry, cranberry, raspberry and blackberry are all native to the central Appalachians and were used by the Iroquois. Wild plum, grape, elderberry, currants, gooseberry, mulberry, nannyberry, strawberry and wild apple were also harvested for fresh use and stored for the lean winters. Like the vegetables, these various fruits provided important vitamins and other micronutrients for the Iroquois diet. As with vegetables, the Iroquois adopted some European fruits. Apples, peaches, pears and European plums were grown after contact with Europeans. The Iroquois had extensive orchards and cornfields around their settlements in the Susquehanna Valley of southern New York at the time of the American Revolution.

There is some evidence that the pre-historic Native Americans maintained agroforestry systems. An account by a pioneer in western Pennsylvania, Captain A. McGill, in *In French Creek Valley* describes an abundant forest garden at the convergence of Woodcock Creek and French Creek:

The banks of French Creek were fringed to the water's edge with evergreen bushes and trees,

Hazel nuts and other native nuts, grown in the farm's tree line and windbreaks, are important bioregional food crops.

while ranged along on the higher bank was a row of stately pines beautiful in their majesty as the cedars of Lebanon. In rear of the pines half a mile in extent was a very gently undulating plain on which grew great old oak trees with spreading tops, the rare old oak that tells of Centuries, a variety that now seems to be extinct. They grew with ample space between without underbrush or obstruction to the view, to the limits of this wonderful park. Around the outer semi-circle of the park there arose a little plateau, not 10 feet in elevation, and from its base flowed springs of pure cool soft water, which fed a circlet of mighty elms, unrivaled in size and beauty ... there were hundreds of these great trees with wide spreading branches supplementing in grandeur the great oaks they encircled. Beneath these grew ☞

hazel bushes, blackberry and raspberry bushes, hawthorn and crabapple trees and many varieties of beautiful shrubs and plants while near the northern extremity there was a veritable orchard of wild plums bearing a great variety of large red and yellow fruit. The ground rose from the river margin in regular successive plateaus of easy grade covered with the finest timber of the most valuable and useful kind. The view was enchanting and they moored the canoe to the bank to make further explorations. Here they were met by John Fredebaugh, who had located a claim that took in Woodcock Creek and joined on the north the land that had attracted their attention. His land ... was naturally alluvial and very rich ... a forest of white walnut (butternut) with here and there a great sycamore towering above and extending its weird white arms over the umbrageous growth beneath. The wild grape vine interlaced the trees and hung in festoons from the branches, forming arboreal recesses of rare and inviting beauty.

This scene would have been a paradise of food to the Iroquois and their game animals and was most probably managed, if not created by them prior to the arrival of Europeans. It serves us today as a model for the development of productive forest gardens, yielding building material, nuts, fruits, wildlife forage and medicinal plants.

Fish and game were used year round by the Iroquois. The abundance of game described by early explorers of the 18th century was astounding. Bear, deer, elk, woodland buffalo, ground hog, beaver, turkey, passenger pigeon, quail, and rabbits were all hunted. Because it provided valuable skins for clothing and other crafts, venison was the primary meat in their diet. Bear was hunted in the fall for its high quality fat, used both for its food energy and to make bear oil. Bear fat and oil was used for lamps, softening leather and other crafts. Rivers provided abundant species of mussels, turtles, eels, and a great variety of fish. The protein and fats from fish and game supplemented crop nutrients.

The Iroquois practiced a seasonal cycle of sustainably using the resources around them. In the spring they would plant their gardens, fish and forage for wild foods. After the crops were planted, many people would leave the villages for nearby hunting grounds. In late summer, the villagers would gather again to harvest the crops and gather fruits and nuts for winter. In the colder fall and winter months, hunters would supplement the village's stored crops and fruits with game. In the late winter, maple sap as well as the sap of various nut trees was gathered and boiled for sugar. Villages were moved as frequently as every ten years as croplands became depleted and game scarce.

Botanically based condiments such as wild leeks, wild allspice and bergamot, supplemented the Iroquois diet. These plants provided minerals, vitamins and other micronutrients. Many plants were used medicinally as well.

The diet of the Iroquois was a healthy one. Wild game tends to be leaner and richer in omega-3 fatty acids, and lower in saturated fat than modern grain-fed meat products. The abundant fish and shellfish also provided high-quality protein and omega-3 fatty acids. The cultivated crops, including corn, beans, squash and sunflowers also provided a balance of protein, vitamins and minerals. The sunflower's seeds provided valuable polyunsaturated oils as well. Variety was a key component ☞

of the diet. Fresh plants in their season and stored plants during winter provided many nutrients.

Most modern Americans eat many of these native foods, including corn, beans, squash, sunflower seeds and some of the same wild fish and game. Although life is different for us (we certainly have lower calorie requirements and less access to natural systems than the early Iroquois did), we can use the diet of earlier inhabitants to guide us as we develop new models of sustainable agriculture and sustainable food systems. As we seek to restore health and vitality to our forests through sustainable management, clean our waters through good stewardship, and manage wildlife ecologically, we should pay attention to what was once the diet native to our region.

Food Security

Community food security is a condition in which all community residents obtain a safe, culturally acceptable, nutritionally adequate diet through a sustainable food system that maximizes community self-reliance and social justice.

— Mike Hamm and Anne Bellows,
The Community Food Security Coalition

In November of 1996, the United Nations Food and Agriculture Organization released the Rome Declaration on World Food Security. This document "re-affirms the right of everyone to have access to safe and nutritious food, consistent with the right to adequate food and the fundamental right of everyone to be free from hunger."

Among the stated commitments the document makes is "the pursuit of sustainable food, agriculture, fisheries, forestry and rural development." While the signatory nations, through the international body, recognized "the need to adopt policies to achieve food security" it falls, as is too often the case, to individuals and non-governmental organizations to achieve this end.

At the dawn of the new century, the international trend is for globalization of the food system. In "The Argument for Local Food," Brian Halweil of the World Watch Institute reports that, due to advances in technology of food preservation, cheap fuel and transportation subsidies "the value of international food trade has tripled since 1960; the volume has quadrupled." He tells us that a food item will travel 2,500 to 4,000 miles from farm to plate. Globalization, unfortunately, leads to exploitative, energy-intensive systems and processes, which create and support international commodity markets at the expense of local food systems, local cultures and the environment. Government policies and research often promote large-scale centralized agriculture and agricultural industry special interests at the expense of small-scale local production and local markets. Competition with large-scale agriculture creates low food profits and limits access to markets. While a graduate student at Slippery Rock University, Heather House did research in this area and concluded that the government bias toward large industrial agriculture results from four aims:

- the desire to prevent disruptions of the food production (maintain status quo)
- the maintenance of existing farm and rural policies
- the promotion of cheap food to deter civil unrest
- the use of food as an economic and political weapon

These policies create social and economic barriers to small farmers while promoting large-scale industrial agriculture. And these policies come with costs, some more obvious than others. The costs begin with the erosion of rural social and cultural life. Money leaves local economies to pay for fertilizers and farm chemicals. Commodity prices kept low by the global trade system mean that profits are more likely to go to food processors and distributors than the farmers that grow the crops. The environmental costs are likely all too familiar to readers by now: soil erosion, synthetic pesticides in the air and water, monocultures that decimate biodiversity — to name but a few. The current high input-high output agriculture promoted by agribusiness and governments can only result in further displacement of traditional agro-ecosystems and farming methods with chemical dependence and environmental degradation.

Beyond food production, institutional food purchasing practices and preparation methods can be roadblocks to the use of local agricultural products. Also, consumer habits and lifestyles limit the use of fresh food. Single parent households and homes with two working parents combined with commuter lifestyles and busy schedules make it difficult for families to shop for local food or prepare meals with fresh food. Fast food consumption represents nearly half of the food dollars spent in the US. These foods tend to be high in undesirable fats and calories and low in other nutrients. Thus, a challenge to local food producers is to promote healthy eating while seeking to make products accessible to consumers. Access to local foods and consumer education can both be provided through well-planned promotion of local farmers markets, direct sales, and subscription produce programs such as Community Supported Agriculture (CSA).

Happily, beneath the wave of international trade there is a growing countercurrent of developing local food systems. Farmers are responding to consumer demands for fresh, safe food. Consumers are responding to the farmer's need for secure markets and customer support. Whether through an alliance of CSA farms, a marketing co-op or an organic certification group, farmers and consumers are coming together to discuss the need for regional food systems.

As urban sprawl consumes more and more farmland, states are beginning to enact conservation easements, land trusts are setting aside farmland and natural areas, and individual farm owners are working to keep a foothold on the land.

All these elements are part of the movement to build regional food security. A sensible food policy for a sustainable society promotes the production and processing of many agricultural products close to home. Regional agricultural systems should address the community's need for a safe, accessible and secure food system.

Planning for Food Systems

Professional planners have long overlooked the value of planning for food systems. This lack of planning on the part of public officials leaves local populations dependent on the global commodities trade, with all the associated environmental, economic and social costs. If, instead, farms were viewed in the context of the community they serve, diversified systems could be developed to meet specific, local needs. Integrating fresh food sales into local networks or creating local markets helps ensure the farm's place in the community.

Access to food for urban dwellers is a concern. Many inner city neighborhoods do not have sufficient access to grocery stores and farmers' markets. New trends in markets and urban farming are helping to solve many of these problems. Local governments and community organizations across the country are responding by promoting farm markets. This helps the small farmer by providing access to retail customers. Community gardens also allow urban dwellers access to fresh seasonal food by providing growing space. Subscription marketing brings weekly deliveries of fresh produce to customers while allowing farmers to receive retail prices for produce deliveries. Food banks have developed innovative relationships with nearby farmers, gleaning excess produce or using donated capital to purchase produce directly from farmers.

Farmland Protection

Local agriculture and local food systems depend on the preservation of farmland. The high value of land near urban and suburban areas creates economic difficulties for farms in the form of high property taxes. Pressure for development is eliminating farmland. The characteristics of good farmland — relatively level, well-drained sites — make it just as desirable for housing and business. Highway systems have promoted the conversion of farmland into commuter communities by allowing easier access from new suburbs to cities. This in turn forces the farmer out or onto marginal land. In many rural areas, farmland is subdivided into small lots as people seek homes in the "country."

Many states offer conservation easements, i.e., reduced property taxes, to help solve this problem. Other regional planning organizations seek to promote an understanding of the value of farmland preservation for the region. County and state agencies concerned with farmland preservation and food security offer farmer assistance programs to support these aims. These and similar national programs such as ATTRA and SARE are very important because farmers are often too busy with production and other jobs to effectively innovate with new methods and new markets on their own.

Search for Solutions

A number of state and national organizations are working hard to engage government agencies to shift support toward local food systems. Every year, more diverse organizations come to understand the issues of food security and join the effort to support local sustainable agriculture. The National Campaign for Sustainable Agriculture (NCSA) and the Organic Consumers Association both promote consumer education on issues of food security and safety and organic standards. The Sustainable Agriculture Research and Education Program (SARE) is a government-funded program promoting farmer and university research into sustainable production. Another government-funded program is Appropriate Technology Transfer to Rural Areas (ATTRA). ATTRA staff and researchers disseminate a wide range of detailed information on sustainable agriculture.

While it is not practical for regional farms to produce all the products consumed locally, a community benefits when it supports a strong regional agriculture network. Agencies and organizations can and are promoting local food systems by educating the consumer and the

Home-scale Food Systems: The Benek Homestead

The home food system is the first step in a sustainable food system. To best illustrate a home food system, I offer the example of my first garden mentors. Ellen and Bob Benek have been producing abundance on a small farm for 30 years. Their home food system is built around a seasonal cycle of animal husbandry, gardening and home processing.

Their rabbits, goats, chickens and riding horses have assured a plentiful supply of manure and bedding for creating compost. After 30 years of applying the farm's compost to the intensively planted raised beds, the soil is extremely productive. Vegetables are on their table 12 months of the year. Home food processing of fruits, sauces and vegetables is an annual tradition. Squash, cabbage, potatoes and apples are stored in the basement. Onions and garlic are stored on a cool side porch.

A small herd of goats provided milk for cheese, yogurt, custards and puddings as they raised their family of five children. Hens provide fresh eggs. Most summers, the family raises a few dozen meat chickens. These are processed and frozen, along with venison and goat meat. Some years they may raise a pig, also destined for the freezer or smoker.

The Beneks' garden begins literally at the doorstep. Each spring they set up a small propagation greenhouse on the south-facing deck just outside the kitchen door. Here they start many of their garden plants. In the summer they move the metal-framed greenhouse to the side, where it is then used for curing onions and garlic, to dry herbs and to cure seeds for the next year's garden. A picnic table replaces the greenhouse on the deck each summer. An adjoining terraced herb and flower ☞

Ellen Benek cold frame with endive

Ellen Benek's squash harvest

garden provides an intimate setting for outdoor dining and fresh herbs for the kitchen.

The main garden consists of about 4,000 square feet of raised beds. There are beds of asparagus and small fruits, many annual crop beds, and abundant flowers. The rich soil yields superior harvests. Carrots, given special treatment with rabbit manure compost, are huge, yet tender and juicy. Sweet corn ears are long and full. Cabbages, head lettuces and other greens also seem extra large and healthy here. Ellen is a sharp observer of her garden's biodiversity. She plants her cabbages, other brassicas and potatoes close to the sweet corn because corn attracts swarms of minute pirate bugs, which devour cabbage worms, Colorado potato beetles, aphids and other garden pests. Hoverfly, ladybugs, parasitic wasps and other beneficial insects are plentiful in the flower-filled landscape. Favorite herbs, like sage, chives and thyme are planted at the end of many garden beds. Patches of parsley, basil and cilantro, some newly sown and some in flower, are found interplanted with vegetables. Dill, borage, mache and annual red poppy self-sow and scatter themselves throughout the garden.

Other smaller gardens are located around the farmstead. These are planted in a regular rotation of main crops of sweet corn, potatoes, squash and other staples. The landscape includes many flowers, both perennial and annual, flowering shrubs as well as native trees and shrubs. Wild birds are fed in winter and are present in large numbers. Ellen reports that the maligned invader, the house sparrow, *Passer domesticus,* has been observed snatching the occasional cabbage worm and helps control pests both winter and summer.

Bob has constructed several small season extenders — cold frames, hot beds and a small plastic tunnel. These provide earlier crops in spring and summer and salads and cooking greens well into winter. The chicken house and pig pen are located at the far end of the garden.

Garden prunings, comfrey and weeds supplement purchased grain to feed the animals. Goats especially relish larger weeds, corn stalks and unusable produce. The gardens are fenced with "fake electric fence." Once trained to an electric fence, the horses avoid the highly visible wire strands of recycled romex and baler twine tied with ribbons. The horses are rotated through sections of the lawn to keep the grass neatly trimmed. The farm dog, the third in 30 years, keeps the gardens free of marauding groundhogs and rabbits. Goats are kept in their place behind heavy pasture fencing. Frequently, they are tied out to browse the edge of the surrounding woodland.

The landscape includes a lot of fruit. Apples, both homegrown and foraged from abandoned trees on neighboring farms, provide cider and sauce. (Ellen also makes her own vinegar from the apples.) Hardy kiwi provides heavy yields most years. A wild patch of blackberries has been encouraged and has, for three decades, provided more berries than they can pick. Red raspberries and blueberries are also harvested and devoured, with any excess frozen for later use. A family favorite is thawed berries with dried fruit.

The Beneks tap a dozen maple trees during thawing weather, between January and March, to make their year's supply of maple syrup. The sap is simmered into syrup on the wood cook stove. Firewood cut from the surrounding forest supplies the cook stove and the larger wood stove that heats the house.

The farm's old-time barn stores garden equipment and baled hay and houses goats and horses. Each year the family contracts with a neighborhood farmer to ☞

receive several wagonloads of hay. Family and friends share the sweaty but satisfying labor of filling the barn with scratchy hay.

Much of the rest of the household food is purchased in bulk through a local buying club or from natural food wholesalers. Whole grains, beans, flour and other staples are stored in containers in the basement. Most foods are made from scratch, with a focus on economy and healthful vitality.

The Benek household avails itself of the convenience of the local grocer regularly, but purchases are far fewer than for the average household. Dinner at the Beneks' is always a pleasure. A midsummer meal might begin with a salad of mixed lettuce, spinach and wild edibles, snap peas and early tomatoes. An appetizer may be slices of herbed goat cheese (chevre), or crumbly hot pepper goat feta. The main course is likely rich venison or goat stew, heavy with potatoes, onions and carrots with a side dish of steamed asparagus. Red raspberry pie for dessert and a cup of Ellen's Power Woman Tea (a personal blend of beneficial herbs) is a fitting end to a perfect meal.

Abundance is the word that comes most to mind at the Beneks' farm. Ellen's work as a midwife and Bob's construction contracting business keep them off the farm much of the time. Yet all this home food production is done with a rhythm and a schedule that does not seem to overtax the family; there is abundant time for family activities and travel. Boy Scouts, 4-H Club, horseback riding, cross-country skiing and just plain relaxing all somehow fit in.

Certainly, the entire family works hard. But the work is so integrated into their lives that a balance is achieved. Bob and Ellen and their teenage children all help with the spring garden rush. The animal stalls are cleaned regularly and the bedding is piled to compost. Beds are fertilized as needed. Many crops are mulched. Weeding the garden becomes feeding the goats and chickens. Harvesting and tending the garden are a natural part of daily life. The pleasures of tending the land and producing abundance and beauty hold many rewards. Less than five acres of pasture, gardens and bountiful landscape are managed to support a healthy and happy family.

farmer to issues of sustainable agriculture. The sustained health and welfare of our communities depend on their success.

Community Access to Local Foods

If we are going to create sustainable, low energy, healthful and dynamic local food systems, farms need to be close to customers. Prior to the 20th century, most cities were interconnected with agricultural landscapes. The geographic interweaving of the two landscapes, urban and agricultural, benefited both. People need access to nature; farmers need access to markets and the city's cultural activities.

To get a sense of how farmers once provided a city with its food, read F.H. King's description, in *Farmers of Forty Centuries,* of a visit over one hundred years ago to the farmers' markets of Shanghai. He saw over 60 varieties of fresh produce as well as nuts, poultry, fish, pork, tofu, oat and eggs. Goods were brought to the markets daily from the surrounding countryside, by barges through canals and then by wheelbarrow to the city markets.

The Parisian market gardens were so numerous, they nearly encircled the city just outside the old city walls and moat. The goods sold at those and similar markets all over Europe were, by definition, regional, leading to unique regional cuisines that reflected the local landscape and climate.

As cities expanded over the past century or so, market gardens and local farms have been largely displaced by other development. But as our energy supplies become dearer and local food systems develop, prime agricultural land should again be recognized as too valuable to develop for other uses. State and local zoning and tax policies are required to support the protection of agricultural lands and to provide incentives for farmers and gardeners. And, as already mentioned, many states do offer programs such as conservation easements and land trust designation.

In sustainable towns and cities, nature and agriculture would interact even in densely populated areas. Small public parks and community gardens within walking distance of each home can bring the city dweller into contact with the food system and nature. Private or community garden space is critical to the deep sense of well being that comes from a connection to natural processes. Many people do not have time or the inclination to garden, but every neighborhood needs some type of garden. Not everyone needs a garden, but everyone should be able to know a gardener. Providing accessible community garden space to those who want to tend a garden would be a municipal priority in a sustainable food system.

Landscape architect and urban planner Ian McHarg, in his landmark book *Design with Nature,* calls for the identification of ecologically and agriculturally important land in and around our towns and cities. These are the areas that have intrinsic values to nature and society that are not reflected in local property values: prime agricultural soils, wetlands, forests and woodlands, floodplain, aquifers and aquifer recharge areas. Biodiversity preservation, horticultural and agricultural productivity, recreational opportunities, and watershed functions of the regional landscape must inform the regional and local planning process.

Seeking to improve on the concept of green belts around cities, McHarg says:

> Rather than impose a blanket standard of open space, we wish to find discrete aspects of natural processes that carry their own values and prohibitions: it is from these that open space should be selected, it is these that should provide the pattern, not only of metropolitan space, but also the positive pattern of development.

Such is the mission of the permaculture designer: to study the land and learn from its form and function where and how we should or should not develop it in order to preserve intrinsic ecological values. On a local and regional scale, this is a political statement. Zoning has always been contentious but is deemed necessary by the community; to suggest that certain lands should not be developed deprives an owner of perceived rights and potential rewards. Although difficult to calculate, it is in society's best interest to compensate landowners with tax incentives or other rewards for appropriate land use. In the US, zoning regulations are set by counties, townships and communities and are, in general, supported by federal court precedents. Groups and individuals concerned about preserving and enhancing natural biodiversity and local agriculture should participate in the planning process as much as possible.

The prime challenge for preserving and building local and especially urban agriculture is the fact that the best land to build on — flat, well-drained land — is also the best agricultural soil. This pits economic development against agriculture. The smaller potential for income and jobs creation of agriculture generally makes it the loser. When agriculture does find a way to maintain a foothold on the edge of suburbia, other challenges appear. Complaints of farm odors and crowing roosters turn into restrictions on livestock. There are still many issues to address if the regional, suburban and urban farms are going to thrive, or even survive.

A Place in the World Economy

Small-scale intensive farms are playing an important role in revitalizing communities in the US and abroad. Harvest festivals, workshops and educational activities all contribute to the health of communities as much as the good food they produce. Ultimately, farms will thrive and community food security will develop if local consumers come to understand the many benefits these farms provide.

There is an inherent contradiction in calling for reliance on local food systems that needs to be acknowledged. Local and regional foods

Food Security Resources

Community Food Security Initiative (CFSI):

Helping nonprofit groups, faith-based organizations, state and local government agencies, tribes, and individual citizens fight hunger, improve nutrition, strengthen local food systems, and empower low-income families to move toward self-sufficiency.

The USDA's Community Food Security Initiative seeks to cut hunger in America in half by the year 2015 by creating and expanding grass-roots partnerships that build local food systems and reduce hunger. The USDA is joining with states, municipalities, nonprofit groups, and the private sector to strengthen local food systems by replicating best practices of existing efforts and by catalyzing new community commitments to fight hunger.

To review the USDA's text and get more information on CFSI, go to: www.attra.ncat.org/guide/a_m/cfsi.html.

The Community Food Security Coalition

Dedicated to building strong, sustainable, local and regional food systems that ensure access to affordable, nutritious, and culturally appropriate food for all people at all times. We seek to develop self-reliance among all communities in obtaining their food and to create a system of growing, manufacturing, processing, making available, and selling food that is regionally based and grounded in the principles of justice, democracy, and sustainability.

The Community Food Security Coalition, with over 325 members, is a network of organizations working to promote community food security throughout the US. Their website lists six principles of food security:

- Addressing the needs of low income communities
- Addressing the broader issues of the food systems, such as urban sprawl, access to food markets in the inner city, and environmental issues related to food production
- Community focus through the promotion of urban farms, farmers' markets, neighborhood grocery stores and community food processing
- Promotion of individual self-reliance and empowerment of communities to plan and implement programs to increase food security
- Development of local agricultural-based food systems that provide good food, equitable farm income, preserve farmland and build ties between producers and consumers
- Promoting food security through an "interdisciplinary, systems-oriented approach" that includes many agencies and organizations

For more information, go to www.foodsecurity.org/.

will never completely replace national and global trade. A certain amount of trade is good for all economies. For example, the ability of the US Midwest to supply grain to the world has done a great deal to help make food affordable worldwide. Local production benefits local food systems, but it won't do it to the exclusion of international trade. For one thing, we are all used to consuming products from all over the world. As a purveyor of fine farm products to upscale restaurants, I can appreciate the connoisseur's desire to partake of the good life. And yet, as a student of ecology, I understand the need for balance and restraint. The answer lies, I believe in the age-old wisdom: "All things in moderation."

Freshly brewed, freshly ground coffee is a prime example. The aroma alone can alert the senses and sharpen the weary mind. Chocolate too has the power to awaken and inspire. Combined in mocha cappuccino, they can be the perfect pick-me-up. I personally prefer the subtler stimulation of a cup of tea, but regardless of the choice of stimulant, coffee, tea and chocolate are probably never going to be local products in western Pennsylvania. Neither will vanilla, avocado, citrus, coconut, luscious mangos, or the many other tropical fruits we so love. And the wine some of us enjoy — French Beaujolais or California cabernet (or that occasional sip of tequila) — will never be produced locally.

One way to balance the ecological equation of global trade is to support environmentally and socially responsible sustainable agriculture when we purchase imported products. Because they increase the economic security of the peasant farmer, fair and just international trade is a valuable tool to promote world peace and stability. Fair trade coffee and chocolate continue to enjoy expanding sales. Other crops marketed as fair trade include tea, herbs, fruits, sugar and rice. Worldwide sales of fair trade products are at least one billion dollars annually.

Sustainable food systems, both local and global, will grow and evolve through complex interactions among marketplace, lifestyle and availability of local commodities. We cannot predict today what such a system will look like in a generation. What we can do is offer consumers new food choices that fit their lifestyle and continue to educate them on the value of buying local farm products. Consumers, in turn, can support fair trade at home by developing long-term relationships with regional farmers and paying a fair price for their products.

CHAPTER 3

Direct Marketing for a Small Farm

*Our focus is to bring about those changes in the human side of
the economy that can help preserve and reconstitute living
systems, to try and show for now and all time to come that
there is no true separation between how we support ourselves
economically and ecologically.*

— PAUL HAWKEN, AMORY LOVINS, AND L. HUNTER LOVINS,
NATURAL CAPITALISM

MARKET GARDENING IS A BUSINESS. The term "sustainable" must
also refer to sustainable economic viability of a farm. This chapter
offers a look at the concepts and practices that make a market garden
sustainable. These include business planning, sound business management, appropriate marketing strategies and related income-generating
enterprises.

When we began making plans for Three Sisters Farm, we had three
goals. We wanted to: (1) grow high-quality farm products; (2) provide a
place for others to learn about sustainable design; and (3) enhance the
local environment.

Similar objectives are described by Trauger Groh and Steve McFadden
in *Farms of Tomorrow Revisited.* According Groh and McFadden, creating farms with such objectives requires "a leading concept that concerns
itself with the wisdom that lies in nature and with the relationship of the
human being to nature." While they arrived at this proposition from the
study of biodynamic agriculture, we at Three Sisters Farm found this
"leading concept" in our study of permaculture.

For the first few years of operation, we focused on developing the
market garden business. Later we phased in more education outreach in
the form of tours, lectures and permaculture classes. This combination
of selling farm products and providing education services brings a good

balance of work and income to the farm. This chapter examines the first two of our three goals for our farm: growing high-quality products and providing a place for others to learn about sustainable design.

Preparation

We entered the market gardening business after a brief study of entrepreneurial ventures. An entrepreneur is anyone who organizes a business — and assumes associated risks — in pursuit of profit. We understood that the ability to pay our bills would make or break our business.

The plan for Three Sisters Farm, however, developed over a period of several years. We had been offered the use of a five-acre property (which we subsequently bought) for a permaculture project. The intention was to apply permaculture design to a market garden farm. The plan expanded to include a nursery business and teaching programs. We were willing to devote much effort, most of our savings and as much time as necessary to turn our ideas into our reality.

In that strange synchronicity that life sometimes manifests, just when we were ready to begin farm development, a request for grant proposals addressing energy in agriculture appeared in our mailbox. The Pennsylvania Energy Office was offering matching funds to agricultural projects that would demonstrate energy conservation and efficiency. The funding came from the US Department of Energy judgments against major oil companies in the US for excessive pricing of crude oil between 1975 and 1981. The Department of Energy was charged with distributing over $3 billion of oil overcharge funds to state agencies for energy conservation projects.

For the next month we prepared and submitted a grant proposal for building our bioshelter. When we were awarded the grant, we spent nine months preparing our site, building the bioshelter, and setting up business.

We began slowly and expanded our production and sales as we completed the bioshelter and developed our gardens.

Business Management

Managing a small business requires an understanding of the specific enterprise, the market, and basic business practices. We chose a partnership as the most reasonable business structure. This enables us to keep personal affairs somewhat separate from the farm activities while still retaining a simpler structure than incorporation would require. A partnership is the simplest form of business structure for two or more people to enter. While a partnership does not offer the liability protection of a corporation, it has less complex tax and management requirements. Linda and I are general partners; we plan and manage all aspects of the business.

We highly recommend that the would-be entrepreneur seek out qualified advice as to the best legal structure for his or her own needs. The complexities of tax codes, state and federal regulations and individual business needs are beyond the scope of this book. We found a knowledgeable tax accountant who helped us determine our best structure. She has continued to provide us good advice and service.

Many types of farm insurance are available, and appropriate protection should be obtained. A good insurance agent will help you plan for your particular needs. Liability insurance is a necessity, both for farm products and for occasional visitors. Institutional and restaurant customers require us to carry a certain amount of liability insurance to cover any litigation related to our products. We have never had to use this in all our years of business. The best insurance is to establish and follow basic safety and sanitation practices. However, in a litigious society it best to be fully insured.

Market Surveys

Once we had agreed upon a business structure, we began market research. As we completed our bioshelter and began to establish gardens, we distributed samples of our crops. Our initial focus on specialty crops, such as salad mix, herbs and arugula, was met with indifference by local chefs and grocers. However, our first attempts at marketing in Pittsburgh, 60 miles away, were immediately successful. Pittsburgh chefs were very receptive to buying fresh and local products. Several chefs took us into their coolers and showed us exactly what they were buying and began to instruct us on custom growing for their needs. We also found a ready market in a specialty food store in Pittsburgh. This store provided a good market for our salad and gourmet greens. Continuous feedback from early customers helped us refine our products and improve our production as we expanded our gardens.

Planning

A clear business plan that considers all areas of the operation is the first step in business development. These areas include production strategies, accounting, inventory management and marketing.

Our business plan was simple. We would develop our gardens over a period of several years, as our experience and sales grew. Key to the whole process was the availability of free manure for compost. Keeping a small dump truck parked at a nearby stable and bringing home a load of horse manure and bedding each week allowed us to keep up with the nutrient demands of an expanding operation. As our reputation for quality and service grew, so did the demand for our products.

Many customers pay on delivery. All pay within 30 days. We keep clear records of all sales and expenses with a sales ledger and a business checking account. Expenses and income recorded in the ledger and checkbooks are entered into a computer bookkeeping program (I use Microsoft Money), which allows us to generate various useful reports on sales and expenses. We hire a tax accountant each year to process our tax returns.

Cash Flow

Farm cash flow can be directly correlated to available sunlight. During the winter, harvested crops are fewer, but many costs are the same. In our own snowbelt climate, crop production can be slow from December 1 until January 15 due to short and cloudy winter days. This can be a trying time without careful planning. Happily, we've found that bioshelter production can provide winter income; during the darker days of winter, we sell eggs, plants, and holiday and seasonal crafts, such as herbal wreaths and pressed flower cards.

Increasing day length increases production from the bioshelter and the outdoor cold frames in late winter. The sales of produce subscriptions and plants in late winter and early spring provide a much-needed increase in farm income, allowing us to purchase the upcoming season's gardening supplies. In the fall, a farm needs to have adequate cash reserves to meet financial needs and obligations through the shorter days ahead. A farm also needs to enter the winter with enough funds to purchase seeds, plants and other supplies for the following year. Planning an annual budget that accounts for costs and income, with reserves for the year ahead is critical to sustaining a farm.

Inventory and Supply

Produce

We do not keep our products in inventory. Basically, we only harvest the amount of produce we need to fill orders plus a small amount for farm sale. We generally eat whatever produce is left after weekend sales, but we actually have very little unsold produce. Bolted lettuce and other crop residue are fed to the chickens.

Supplies and Maintenance

We do maintain an inventory of supplies. The garden manager keeps track of seed and plant material, fertilizer and potting soil, flats, pots, garden tools and related items. The farm manager (me) keeps track of general accounting, utility bills, and kitchen supplies such as packaging, boxes, labels and harvest tools. I also develop and maintain the facilities. Like the farmers of my grandfather's generation, I am, on various days, a mechanic, a plumber, an electrician, a carpenter, a designer, and a horticulturist. Each activity requires an inventory of tools and printed references. Tools are stored in the "barn," or tool shed, attached to the bioshelter. Service manuals and maintenance schedule are also stored there. Important references are kept in a small library in the bioshelter, or in the farm office.

Together, as farm manager and garden manager, we direct a small garden crew and interns.

Marketing Farm Products

Market gardening is two interwoven businesses; growing produce and selling produce. Writing in their book *Natural Capitalism*, Paul Hawken, Hunter Lovins and Amory Lovins, show that farmers' income represent less than one percent of the gross domestic product of the US. However, "those who sell to and buy from farmers ... have a share 14 times as large ... These interests tend to squeeze out small, independent, and diversified farms." Therefore, by selling farm products directly to retail customers, rather than to wholesale customers, a farmer can realize a better profit and develop a more secure position in the market.

The largest obstacles to profit in farming are the national and international policies that keep prices and wages low and resource-intensive production methods the norm.

Markets

Options for marketing produce include pick-your-own operations, on-farm markets, urban farmers' markets, subscription sales, selling to wholesalers and processors, and direct sales to grocers, restaurants and other consumers. As explained by Janet Bachmann, in ATTRA's *Market Gardening: A Startup Guide,* the direct marketing options of "farmers' markets, roadside stands, pick-your-own, subscription marketing arrangements and sales to restaurants [help farmers] "maximize their income [by] bypassing the middlemen."

Our sales strategy at Three Sisters Farm is a mix of several approaches. Our primary strategy has been direct sales and subscription sales; we also do some Saturday on-farm sales. In pursuit of the best price we could get for our produce, we set up a delivery system to pass over the wholesalers. Direct marketing also allows us better control over the quality of our produce. When our label reaches the customer, we want the product to be as fresh as possible.

Our marketing strategies reflect our gardening system and philosophy. Developing a rotation of diverse annual crops and establishing perennial plantings ensures a healthy garden and minimizes the effects of crop failures. We grow lettuces, herbs, flowers, various salad crops and other gourmet vegetables. The crop mix and planting schemes have evolved as we attempt to cater to customers' specific needs.

Our *mixed* sales strategies reflect an important permaculture principle: we have designed diversity into the system to create stability.

Direct Sales to Grocers and Restaurants

After a study of outlets for farm products, we concluded that selling directly to restaurants and grocers would be our best starting point. The local market was not large enough to support our sales goals. But driving 60 miles to Pittsburgh put us into a very large market. We began our sales promotion by calling chefs and arranging times to bring them samples. Face-to-face meetings allowed us to ask them directly what it was they needed and wanted.

Getting to know your customers' needs is very important in marketing. Anticipating new and interesting produce they may want and discussing this in advance of the growing season helps with planning and instills confidence in the customer. And understanding a restaurant's annual schedule of peak and slow times allows the grower to schedule plantings to avoid waste.

Each week during the main growing season, we produce over 40 pounds of our salad mix and cases of lettuce, herbs and other crops. We deliver to 12 to 15 regular customers, many of whom have standing orders. Weekly contact with customers allows for adjustments and special orders. Continuous discussion with customers allows us to plan our planting and adjust to changes in the market. This minimizes waste. Most of what we plant is pre-ordered, sometimes months in advance.

Produce delivery to chefs.

Personal interaction with customers is fulfilling. Both our customers and the farm benefit from discussion and feedback on gardening, cooking and sales. Several of our customers have been with us for over 20 years.

Customers benefit from the direct relationship with a farm in several ways. They receive top-quality produce within 36 hours of harvest at a reasonable price. Since our crops are pre-ordered and waste is minimal, our prices remain stable, even when other markets are fluctuating. While it was tempting to follow rising lettuce prices one spring, we felt that we should be fair to our customers; *our* costs did not rise because of flooding in California. In other seasons, our produce may be more expensive than California's, but our customers stay with us because our exceptional quality, freshness and dependability are worth a few extra dollars a week.

Our customer base is very stable from year to year. Each year, as we've expanded production, we've added one or two new restaurant customers. We occasionally drop a customer whose orders are only sporadic; sometimes we'll lose a customer if a restaurant changes ownership or staff. We generally seek new customers along our established delivery routes.

Demands and Costs of Delivery Service

There are certainly some drawbacks to our sales strategy. We absolutely must produce pre-ordered crops weekly, through the entire growing season, come dry spell or downpour. This requires precise planting schedules and a careful assessment of irrigation needs. Chefs and grocers want to receive top-quality produce — every week — for as long as the season lasts. The occasional crop failure is inevitable, but when they occur, customers are generally understanding. Restaurants also want their deliveries on time. And delivery is not something to take lightly. Delivery takes up to nine hours each delivery day, which means either time away from the farm or hiring someone else to do the deliveries.

There are significant costs associated with delivery: the delivery van, coolers, packaging, time on the road, and time spent interacting with customers. Additionally, special packaging and sorting procedures are necessary.

A drawback to restaurant sales is that restaurants often experience reduced business in the summer. Then, just as *our* season slows down, restaurant demand increases. Development of a seasonal subscription sales strategy helps to bridge this gap.

Be aware that direct selling is not for everyone; you have to enjoy dealing with people.

Subscription Marketing and CSAs

"Subscription marketing" is a system of providing a pre-ordered and pre-paid bag of produce to customers each week, for an agreed time period. This is different from the early concept of a Community Supported Agriculture (CSA) program. In its original meaning, Community Supported Agriculture was a process whereby the farmer is guaranteed funding for a fair wage and expenses by a "community" of consumers. The "community" households received a weekly share of the farm's products.

Many people mistakenly apply the name CSA to subscription produce programs. In a subscription program, the customer *is* guaranteed the value they pay for, but the farmer's income *is not* guaranteed. We point out the difference because we believe the CSA model is valid and important even though at Three Sisters Farm we have not attempted to find the committed community to support us with a CSA program. Instead, we have the subscription program. The subscriptions provide a reason to grow long-season crops such as cabbage, broccoli and sweet corn — vegetables that are needed for proper crop rotation in the garden beds. Our subscriptions tend to be "gourmet," and may include arugula, tea bouquets, heirloom tomatoes, shitake mushrooms, our salad mix, fingerling potatoes and other high-value crops. Generally, we do not sell many long-season crops to the restaurants. For restaurants, we focus on higher-value herbs, greens, salads and lettuces.

Subscription customers make three payments. The first is in early June, the second is due around August 1, and the last payment is due September 15. Weekly deliveries or pick-ups begin in the third week of June and continue for 16 weeks, into October. For $320, the customer receives $20 worth of produce each week. We charge a small per-week handling fee for deliveries to co-operative drop-off locations.

We offer a limited number of "spring greens" subscriptions, which is a lower-priced subscription that features salad, herbs, and cooking greens, plus a few other early crops. Many farms offer longer — even year-round — subscriptions. Increasingly, farms are joining together to form co-operative multi-farm subscription programs.

Most of the subscriptions we deliver are to private homes and small stores. Some customers pick up their produce at the farm. We try to keep in communication with subscribers with a newsletter every week or two. The newsletter provides recipes and updates on crop production. Each spring, a brochure describing the year's subscription program is sent to customers. New customers receive a brochure describing how fresh produce fits into a good diet and how to wash and prepare the main items in the weekly subscriptions.

Farm Sales

Produce and plants are also sold at the farm. Our weekend farm market is simple. We sell the produce left from our weekday harvest as well as products from the bioshelter kitchen. We have a core group of regular customers. Each year this market grows a little larger, solely by word of mouth. For our regulars, we'll often grow small amounts of certain crops, such as red celery. Fresh farm eggs are a big draw for local sales.

Customers are encouraged to call ahead and see what we have, because we often sell out fast. Some customers arrange specific pick-up times for pre-ordered produce.

LEO GLENN

Darrell Frey and Chris McHenry-Glenn,
with Ossian Glenn at spring plant sale.

Plant Sales

One of our original goals was the development of a useful plant nursery. The collection, testing and propagation of useful perennials is ongoing.

Many people stop by expecting to buy vegetable plants from our greenhouses. We explain that it is difficult to keep up with our own demands for seedlings. We never intended to develop a large vegetable plant enterprise because several farms in the area already had established spring plant businesses. We sell a few flats of vegetable plants from the farm each year, but this amounts to less than 2 percent of our gross sales. Instead, we found a market niche: our spring plant sales focus on heirloom tomato seedlings, native plants, herbs and flowers, unusual perennials and plants that are good for beneficial insect habitat. We divide many herbs and wildflowers and gather self-seeded plants from among our gardens, and pot them up for sale. We also start plants from seed we gather or buy locally. School gardens and other groups often pre-order specific plants.

Our native plant and useful plant nursery is a growing enterprise, with several part-time, limited partners involved each year.

Processing the Excess

Farm products not sold fresh can be processed into value-added products. We have made small market tests of crafts such as herbal wreaths and dried flowers, herbal vinegars, herbal body oils, and pesto. Each of these will require better production planning and marketing to be major contributors to the farm income. We would like to eventually develop these and similar products and sell them together as gift baskets. (Note: If you want to sell processed food from the farm, make sure you thoroughly investigate your state's regulations for the safe production of food products.)

Household Production vs. Growing For Market

Many people get started in market gardening by selling excess produce from their home garden. This is a reflection of the origin of traditional farming around the world. In traditional economies, the farm household produced many products for home use. Excess crops were sold or traded for other products and services. The opposite end of the marketing spectrum is growing only one or just a few crops for sale, with little or no home consumption. But this is the kind of commodity production that eventually leads to monocultures and using land as a site of production rather than within an integrated system. A permaculture farm, on the other hand, is a return to the traditional farm economy. Home processing and consumption is an important aspect of the sustainable farm. Producing posts, poles, fuel wood and other garden supplies and materials can reduce costs to the farmer and help balance the farm accounts.

A plan for feeding the farmer and a community of customers entails planning for a healthy diet and menu. So, creating a diverse food system for the farmer is as important as selling crops. We eat very well, with plenty of herbs and vegetables. Our diet varies with the seasons. Fruits include currants, raspberries, blueberries, strawberries, apples and pears. Asparagus and wild edibles such as cattails and fiddlehead ferns are used seasonally. Fish from the pond and sometimes the wild rabbits that threaten to eat our crops end up on our dinner table. Many farm products under trial are first used in family meals and work crew lunches.

Outreach and Education Programs

Integral to Three Sisters Farm's development was our initial inspiration to establish a permaculture farm as a teaching center. The demands of developing gardens, products, markets and a functional ecological landscape initially restricted our outreach effort. The first five or six years of operation, we were limited to giving tours and holding the occasional open house for local and regional customers. Otherwise, we focused on developing the farm.

In the mid 1990s, we began to market the information and experience we had been gaining by giving permaculture workshops and more tours. Since 1999 we have hosted a permaculture design course at the farm each year in August (when the restaurants and grocers

Farm based education.

are having their slow periods). Students camp on the farm and participate in two weeks of classes, field trips, farm work and design sessions. We use the design course to develop short- and long-range plans for further application of permaculture design to the farm. The permaculture design course and other workshops bring supplemental income to the farm.

Some crops are planted specifically to feed course participants — three meals a day for up to 12 people. The course fees pay the farm account for the produce. This aspect of the program — harvesting and eating fresh produce — greatly enhances the process for all involved.

Brand Recognition/Reputation

The name "Three Sisters Farm" is widely recognized in western Pennsylvania. Our farm business has maintained a reputation for high-quality produce and excellent service to our customers. Our reputation precedes us in our marketing and often leads new customers to seek out our products.

Our business logo is a colorful painting of the Three Sisters; corn, beans and squash. It is an easily recognized and attractive image. We use it for letterhead and business cards.

We are on the Phipps Conservatory's list of horticultural speakers in Pittsburgh. We give an average of four paid presentations to herb clubs in western Pennsylvania each year. This allows us to share our knowledge of ecological gardening and, at the same time, inform our audience about our various enterprises.

Our farm received a lot of publicity right from the start. Local and regional newspapers sought us out for interviews. We were featured in *New Farm Magazine,* and the book *Real Dirt.* Our farm is featured in a children's book about organic farming, *One Good Apple,* by Catherine

Paladino. In March of 2002, we were featured in *Organic Gardening Magazine* as an example of a permaculture farm. All this publicity has promoted both our sales and our outreach work.

Our Plans

One drawback to our marketing strategy is the need to drive to Pittsburgh. Although the Pittsburgh market allows us to sell all our weekly produce in one delivery, we would like to develop more local and on-farm sales strategies.

We view our farm as a small-scale agricultural enterprise incubator. Three Sisters Farm has developed to the point where several associated businesses could operate at the farm simultaneously. (See the sidebar for a list of some business possibilities we see for the farm.) Since 1998, through our permaculture design workshops, we have created basic design plans for a number of associated businesses. Some we have begun, some are in the planning stages. Some businesses are just waiting for the right person to do them. Implementation of most of these plans has been held up by the lack of capital. The uncertainty of markets and weather often makes profit margins slim. Consequently, available funds for labor limit the farm's ability to obtain and retain qualified workers. The cost of financing also limits extending farm enterprises.

Trials have already been made — with varying degrees of success. For example, we raised 50 broiler hens as a trial. They sold out well in advance, so we then developed a plan to raise 500 chickens in spring and again in the fall on rotational pastures and feed crop residue. We haven't found the time to do this yet, but we have the space and there is clearly a demand; the plan is described in more detail in Chapter 12.

We plan to greatly expand our fruit production in the coming years. Based on our trials, we will be developing a "forest garden" of fruits, perennial vegetables, herbs and flowers (the concept of "forest garden" is explained in Chapter 7). Many of the plants in the forest garden will be propagated on site. We plan to graft the best fruit varieties from our trial plantings and will divide and propagate herbs as we expand.

Each year, we make and freeze pesto and can salsa. We also freeze a lot of our berries. A long-term goal of the farm is to have a larger processing kitchen where we can make soups, sauces and otherwise process excess produce. We plan to begin this process by putting an addition on

Three Sisters Farm

TSF logo courtesy of Three Sisters Farm and Mary Grandelis, original design by Dawn Shiner.

the bioshelter. We will expand the current kitchen from 8 feet by 22 feet to 24 feet by 22 feet. This will give us a larger area to prepare produce, display products for sale and host classes. Above the expanded kitchen, we will build an office for the farm and our design and consultation business. The addition will improve our ability to attract local customers to our farm store.

Small-Scale Enterprises

The term "small-scale" is important in our development plans. We do not want to grow too big or hire a lot of workers. Rather, we see the small-scale intensive permaculture farm as a model of sustainable community development.

Allowing enterprising individuals to develop associated businesses at Three Sisters Farm accomplishes several goals. Each will contribute to the gross farm income to help take some financial burden off the market gardens. Some of the businesses are ideal for people seeking part-time and seasonal income. An expansion of products offered at the farm should bring more customers and so also increase our local produce sales.

The idea of providing consulting services on sustainable systems and organic gardening and farming is one we are currently exploring.

We share the enthusiasm of two ecological design pioneers, Nancy Jack Todd and John Todd. In their excellent book, *From Eco-Cities to Living Machines,* they are optimistic about the potential for urban enterprises surrounding a bioshelter: "The limits are as yet uncertain but such a local enterprise, drawing on the community for full- and part-time staff, has the potential to reverse the present agricultural equation."

New Partnerships

We have already entered into a partnership with associates who have worked with us to develop a native plant nursery at Three Sisters Farm. We grow about a thousand plants each year from seeds we collect or purchase. We start the native seeds in February to have them ready for sale in early spring. This is a month before we start our vegetable seeds, so it allows make use of our otherwise idle grow lights.

It has been great to see our permaculture workshops evolve into co-operative events. Each year, I coordinate a group of co-teachers and presenters to conduct class on harvest days and at other times. This allows me to do the behind-the-scenes management of food and facilities and gives the class a variety of qualified teachers to learn from.

A plan in which associated enterprises are combined into a whole system — managed by a co-operative group or non-profit corporation — is sound. Such a system would serve as a vital link between the local food system and the community.

Small-scale Agricultural Enterprise Incubator:

A Co-operative Sustainable Agriculture System

A small-scale permaculture farm can be an incubator for a number of enterprises. Combined into a whole system managed by a co-operative group or a non-profit corporation, these can provide seasonal and part-time incomes for those with a creative impulse.

Each of the enterprises and products listed below could be developed on a five-acre permaculture farm. No doubt others could be added.

- Processing facility for pesto, salsa, herbs, vinegars, jellies and other home-canned products.
- Weekend café with outdoor picnic areas.
- Storefront selling garden supplies, books, crafts, tools and farm products.
- Gift baskets containing farm products and crafts made from natural materials.
- Cut flowers, dried flowers and related crafts and products.
- Herbal teas and medicinal herbs.
- Bee products: honey, candles and bees wax.
- Organic poultry: meat and eggs.
- Aquaculture: pond-based fish farming and indoor fish tanks.
- Perennial nursery and plant sales and garden installations.
- Educational programs: tours, workshops and classes.
- Consultation services in sustainable systems and organic gardening and farming.
- Compost and potting soil sales.
- Vermiculture: sales of earthworms and vermiculture bins.
- Goat milk and cheese production.

CHAPTER 4

The Permaculture Farm: Design

The Design — 'A beneficial assembly of components
in their proper relationship'
— BILL MOLLISON,
PERMACULTURE: A DESIGNER'S MANUAL

PERMACULTURE HAS BEEN, from its very beginning, a design system. In *Permaculture: Principles and Pathways Beyond Sustainability*, David Holmgren defines permaculture design as the use of systems thinking and design principles that provide the organizing "framework for implementing … the vision of permanent agriculture and permanent (sustainable) culture." Bill Mollison, in *Permaculture: A Designer's Manual*, describes multiple approaches to design. These approaches range from the random scattering of seeds to detailed planning and management. This chapter examines several of these approaches and their application to Three Sisters Farm. The dedicated student of sustainable design is encouraged to seek out the references listed in the resources section for further study of systems of design and planning.

Farm Design and Planning

Design and planning of a permaculture farm includes planning for placement of buildings and gardens, energy conservation and green energy production, soil management, watershed management, pest control, crop production, arrangement of living spaces, and promotion of biodiversity. This chapter examines many these as they relate to farm design.

Design and planning for sustainability through energy efficiency, environmental integration, and positive social and economic impact is *site specific* and *situation specific*. In many cases, it means beginning

73

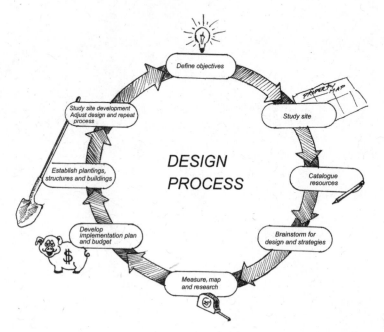

DESIGN
PROCESS

- Define objectives
- Study site
- Catalogue resources
- Brainstorm for design and strategies
- Measure, map and research
- Develop implementation plan and budget
- Establish plantings, structures and buildings
- Study site development Adjust design and repeat process

Sustainable design is an ongoing, circular process.

the process of transition to more eco-logical and sustainable practices. For a farm, this may mean converting to organic farming methods and includ-ing steps to promote biodiversity. Later, development of multiple crops and enterprises can be implemented in stages as funding and time allow.

To speak in simple terms, *designing* is identifying the farm's goals and studying its landscape, ecological and market niche, and available resources. *Planning* includes laying out a course to get there, developing a strategy of staged development, and deciding on a budget. It also includes deciding on management approaches, what tech-nology to use, and what skills might need to be learned.

Permaculture design begins with assessment of the site, its resources, and the goals of the developers. Objectives should be defined. The goals of the site developers, both long-term and near-term, need to be exam-ined. The skills, resources and experience of the developers are assessed. Strengths and limitations are listed.

Design sessions include researching and considering options for obtaining these goals. Options are compared with the needs of the developers and the needs of the land. What emerges is a preliminary design for the site. Preliminary plans lead to a new round of research and then design adjustments.

Site Analysis: The Land, the Community and the Environment

Soil types, water resources, landforms, geological features, existing spe-cies, and other aspects of the site are catalogued. Existing ecological relationships are noted. Special attention should be given to existing habitat for beneficial insects, birds and other vertebrates so that criti-cal wildlife habitat and wetlands can be preserved. Potential problems are noted (examples: a deer trail crosses your planned garden site; rabbits and voles are plentiful in the orchard site). This stage of site study should involve research into the potential for wild crafting and the use of site materials for construction and soil building.

The relationship of the site to the surrounding environment and bioregion should also be studied. *Sector analysis* (mentioned in Chapter 1 and discussed in more detail below) gives an overall view of the site in its relationship to the larger environment, e.g., climate, existing microclimates, solar access, the presence or absence of windbreaks, fire hazards, and floodplains. (Floodplain information is available from state and local officials.) Wildlife corridors, such as a tree line connecting woodlands, are noted.

Local and state zoning regulations are researched; each state, county and township may have different ordinances and codes. It is best to have a good relationship with the local officials and follow their guidelines for establishing new systems. You may find that local officials are aware of the need to develop alternative systems and are willing to work with you if you have done your homework and have prepared a good plan.

Many other legal issues relate to farm development. Business and tax laws, zoning rules, organic certification regulations, and other licenses need to be researched and addressed. Study of the property's deed and a title search will alert you to easements, right of ways and other restrictions on development.

Available local resources are catalogued. Sources of organic matter for preparing compost such as leaves, manure, straw and hay, food-processing wastes, sawdust or tree bark should be found. Sources of local building materials and soil amendments are also located during the planning stages.

Elders or mentors can be sought out for their knowledge. Whether gardener, farmer, beekeeper or builder, neighbors with years of experience in the local area can be of invaluable service. Many folks are willing to share their knowledge of the climate, soils and weather, as well as the local markets.

Potential markets are investigated and a community's needs assessed. As I already mentioned, when we started our farm, there were plenty of greenhouses offering bedding plants and vegetable seedlings. Rather than compete with them, we sought a new market niche as specialty produce growers. Talking with grocers, chefs and other produce buyers helped us plan what to grow. As our farm developed, we found more niches in the market for native plants and useful herbs.

Site Analysis: The Soil

The management of living, breathing soil is the essence of gardening. Understanding your soil's characteristics is the beginning of ecological management. The starting point is obtaining a soil survey report from the Natural Resources Conservation Service (NRCS). The NRCS's online *Web Soil Survey* tool will generate a detailed soil report with digital maps. The report provides general information on what characteristics to expect from the soil in any given location, assuming it has been undisturbed in the decades since the original survey was made.

Actual field conditions will vary from the report, so direct observation is vital to planning soil management and farm development in general. Placement of farm components should follow the best use of the site's soils and drainage patterns.

Soil Profile

Study of the soil profile reveals the depth and quality of the top soil, and subsoil, and also the site's drainage. Digging a post hole will reveal a number of things about the soil. A soil probe tool can also be used for a quick view of the soil's profile.

The soil layers are known as horizons. The O horizon is the top layer of organic matter in a forest, meadow or grassland. This layer includes leaf litter or old sod, animal wastes and dead creatures. Mulch serves as the O layer in the garden. The A horizon is the topsoil, with the most organic matter, air and life; depth ranges from a few inches to a foot. (In our area, it is usually about 6 inches.) The B horizon is the subsoil — more mineral than vegetable, and less lively, though still an active zone for roots and earthworms. It's enriched with nutrients leached from above, or decaying plant roots. This layer too can vary from a few inches to a foot or so, depending on the depth of the parent material. The C horizon is the underlying material that forms the mineral component of the soil. The A, B and C horizons are subdivided into layers, A1, A2, B1, B2 and so on, if clear differences are visible. Some locations have a D horizon, which is bedrock or other parent material near the surface.

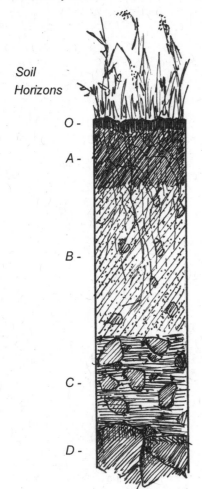

A simple post hole will allow a view the soil profile . The depth of various horizons reveals much about the condition of the soil.

Soil
Horizons

O -

A -

B -

C -

D -

All soils, presuming they drain sufficiently, can be improved with organic matter. The A horizon can be increased by incorporating organic matter into the B horizon. The question is how much and how long it will take? The answer is usually: a lot and a long time. One is much better off developing a farm on land with good healthy soil from the start. Many urban sites require soil to be restored or imported.

Determining Soil Type: Quart Jar Method for Finding Ratio of Sand, Silt and Clay

The relative size of the layers of sand, silt and clay is used to define the soil type. A soil that has too much sand will drain quickly and can be low in nutrients. A soil high in clay will compact, drain poorly and be low in air. To figure out what soil type(s) you are dealing with, fill a quart mason jar half full of the soil. Add water to within an inch of the top. Tighten the lid and shake until the soil is thoroughly mixed in the water. After about 30 minutes the soil will settle by weight into layers. The heaviest particles, small stones, and sand, will be on the bottom. Next will be a layer of silt, then the clay and then the organic matter.

Soil Testing

Chemical soil tests are a vital tool in farm planning and management. Testing kits range from inexpensive and possibly inaccurate to kits with very expensive equipment. Soil test kits give you what you pay for. The basic kits test for pH, nitrogen, phosphorus and potassium — the last three are known by their chemical symbols, N, P, and K.

A better test kit will range from $50 to $200 depending on the quantity of tests that can be done. Because the chemicals in the kit do not have long shelf lives, they need to be replaced each year, so buy only the size you need for the number of tests you want to conduct in a season. Keep accurate records of the areas tested to help manage the farm and to show to the organic inspector. You only need to test an area every few years if you know your inputs and follow a sound management plan. Testing will help you see the effects of crop rotations and inputs on fertility. Soil testing does have its limitations. Nutrients that become available through the biological activity in the soil during the growing season may not test accurately. But the tests will give a general idea of soil chemistry.

To determines the ratio of sand, silt, clay and organic matter in soil, put one cup of soil in a mason jar , add two or more cups of water, shake well and allow to settle.

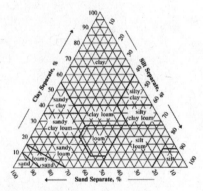

Soil triangle.

Testing for pH should be done every few years for all gardens. Wet and rainy years can leach out calcium and other water-soluble nutrients. Testing for pH helps to reveal lost soil minerals. As electrically charged mineral ions leach away, they are replaced by hydrogen, which increases acidity. Most crop plants cannot absorb minerals from acidic soil. On the other hand, adding too much lime will make the soil alkaline, which also limits a plant's ability to absorb nutrients. All application of lime should be informed by testing.

Testing for trace elements and the quality of farm-made compost requires special equipment. Commercial testing laboratories and state universities offer testing for a reasonable fee. The test results generally show what nutrients are in the soil, or what nutrients are needed to grow a specific crop, measured in pounds per acre. It is important to

Toxic Soils as a Design Element

Lead in the soil can be a significant problem in urban and suburban sites. Airborne lead from car exhaust and industry entered the soil for many decades in the 20th century. Soil near houses and on reclaimed vacant lots also can be high in lead from the paint used prior to the 1970s. State governments generally have programs addressing the problem and offer a lot of information on dealing with lead.

People, especially children, can be poisoned by soil-borne lead by breathing in dust from contaminated soils, from ingesting soil and, to a lesser extent, from eating plants grown in high-lead soils. Dust masks and gloves are required when working contaminated soils to prepare planting beds.

Soil pH, organic matter, moisture level, and plant health can all affect uptake of toxic minerals. Leafy greens are more likely to uptake lead than fruits, such as peppers, tomatoes, squash, tree crops and berries. Root crops grown in contaminated soil should be thoroughly washed or peeled.

Lead is relatively immobile in the soil, so levels can vary greatly across a site. A lot of testing may be required to determine the extent of the pollution. Initial testing from several locations on a site will give an indication of whether lead is present and in what concentration. Samples with more than 100 ppm (parts per million) are a problem. When results vary on a site, further testing can reveal the pattern of contamination.

Once a pattern of contamination is found, this information becomes an element to consider in the design process. The level of concentration, if present, can be a major influence on site design and development. It is possible that gardens can be located in clean soils, and areas with contaminated soils can be used for other purposes.

Soils with very high levels — over 300 ppm — may need to be removed and replaced. Some states require the removal of soils above a certain level of contamination. Gardens should not be developed in soils with lead levels higher than 100 ppm, although some agencies set the limit at 200–300 ppm. Some agencies recommend ☞

use a testing service that gives you the service you require. For example, we have our compost analyzed to help us gauge how much to apply to our crops. Be aware that many soil testing labs traditionally offer advice geared for chemical fertilizer applications, which must then be interpreted to the use of organic fertilizers.

Livestock feed crops and pasture forage can be tested with *leaf tissue testing*. This can help the farmer determine supplemental feed requirements and soil improvement strategies. Leaf tissue testing of green manure crops can also help determine the quality of soil fertility.

Site Analysis: The Watershed

Water management on the permaculture farm is approached by viewing the farm as a watershed and understanding its relationship to the larger

mixing contaminated soils with uncontaminated soil to reduce the levels present in any one site. Careful consideration should go into a decision to mix soils to determine expense, practicality and desirability of diluting the contamination. Other options should be ruled out first. Covering the soil with an impermeable barrier and importing new soil, or gardening in containers of imported soil mixes are options to consider when establishing gardens in contaminated sites.

Children, especially those under age 6, and pregnant women should have no contact with contaminated soil. It should be kept covered with sod or mulch. Even pathways should not be bare soil.

For soils with less than 100 ppm of lead, the following steps are recommended for reducing lead exposure: soils should not be tilled or dug when avoidable (mulch gardening is a good strategy for gardening without turning the soil); maintain a pH above 6.5; keep phosphorus levels high (this reduces lead uptake in plants); and maintain a high level of organic matter because it helps lock up lead in soil.

The presence of lead in the soil may be indicative of other contaminants. Copper, cadmium, zinc and arsenic can all be present in toxic levels. Residential sites may also have motor oil residue, asbestos and other contaminants. State public health agencies may have documentation of a site's potential for contamination from past industrial activities.

Phytoremediation of soil contamination is possible for some contaminants. This process involves planting crops known to absorb toxic elements. Harvesting and removing the plants can reduce levels of specific elements. Phytoremediation is still a relatively new area of research and the practical results are highly dependent on numerous factors. Cleanup may take many years and involves removal and careful disposal of the remediation crops. High levels of contamination may make the process impractical.

watershed. Issues of concern include catching, storing and using rainwater, sizing swales and developing rain gardens, wetlands and ponds. These last three are discussed in Chapter 13.

Nutrient management to keep excess nutrients from entering groundwater and streams is an important part of farm water management. Other concerns include planning for irrigation, water conservation, self-watering systems, riparian buffers, and preventing contamination of water with vehicle and equipment management.

The farm's relationship to the larger watershed is important in several ways. In many areas groundwater contamination from nearby pesticide use, excess fertilizers and industrial pollution can be a concern. Leaking gasoline tanks and industrial pollutants can send plumes of toxic substances moving through groundwater. It is important to research the area and the history of the area before planning to use well water in suburban and urban areas. The farm can affect the watershed if excess nutrients from on-farm composting, livestock, or fertilizer application leaches through the soil into the groundwater. Care should be taken to manage nutrients at the appropriate scale to prevent this. Soil testing is a valuable tool to help determine nutrient needs each year.

Streams passing through a farm may be quick to flood if the upstream watershed is not forested. A forest canopy and its understory reduce the impact of rainfall and help store it. Pastures and grasslands are more likely to drain quickly into streams. The riparian zone, or seasonal floodplain of a stream, is best kept in a cover of native trees and shrubs to prevent erosion. Cultivation too close to a stream allows excess nutrients to pollute the water. Allowing livestock uncontrolled access to streams breaks down the banks, leading to erosion, and also allows excess nutrients to enter the water. However, the riparian zone can be a valuable part of the farm landscape, providing woodland products, food and habitat for beneficial wildlife.

At our farm, we can expect to receive between 32 and 48 inches of precipitation distributed evenly throughout the year. Winter snows, averaging 50 inches each winter, tend to melt slowly and recharge groundwater. Summer storms bring a lot of rain — and potential for surface runoff. Storms bring 2.2 inches in 24 hours at least once a year, and 3.3 inches in 24 hours once every five years.

At least one summer growing season in ten, we have periods of six to eight weeks with little or no rain. Conversely, we have had 4 inches of rain in less than a half hour. That one storm flooded the pond 8 inches above the drainpipe, and the pond took nearly a day to get back to its normal level.

All these factors are of concern in the design process. Planning for irrigation in dry years, and proper drainage and catchment in wet years begins with calculations of both the watershed area (square footage of roof surface, pond site drainage acreage) and possible quantities of rain. Water needs are also assessed. Conservation and wise use of water is always important. It is commonly said that garden plants need an average of 1 inch of water each week. Our five-acre field will receive that much water three to four times each month in a normal spring and summer. Regular rain builds a reserve in the soil, especially under mulch.

For planning purposes, it is important to comprehend the volume of water needed. To cover one acre with 1 inch of water requires 27,154 gallons of water. Adding 1 inch of water to a 200 square foot garden bed requires 124 gallons of water (200 square feet = 28,800 square in./231 cubic inches per gallon = 124 gallons of water). One acre of beds this size need about 27,000 gallons per week.

Water needs for the 2,000 square feet of bioshelter growing space are an average of 1,240 gallons a week. In practice this varies through the year. On cloudy winter days, we do not water as much. Hot, sunny summers can require extra watering.

A polytunnel, measuring 45 feet by 15 feet has about 600 square

Keyline

When developing the permaculture concept, Holmgren and Mollison incorporated P.A. Yeomans's *keyline system* of water management as detailed in his book *Water for Every Farm*. The keyline system is a whole-farm watershed management approach used to direct water flow and maximize water retention on a farm. Keyline water management involves a close study of a property to take advantage of existing topography in the placement of ponds and swales to direct and retain water. Pastures, pathways and other farm elements can then be positioned to take advantage of the system of swales and ponds.

On sloping land, surface water moves downhill and into small valleys. The term *keyline* refers to a contour line system that catches water moving toward these small valleys and directs it along contour swales and furrows, horizontally across a hillside. It is a broad-scale strategy of contour plowing specifically designed to maximize water infiltration into the soil.

Keyline water management was developed for dry lands, but it has applications in all climates to help prevent soil erosion, build soil and manage rainwater. On our farm, we have incorporated the concept to both catch water for irrigation and to lead it away from our building to infiltration areas.

feet of beds and so will require about 372 gallons each week. Because a tunnel does not get rain and has higher temperatures, the tunnel needs to be watered heavily at least once each week. New seedlings and transplants in a polytunnel may need to be watered twice daily in sunny weather.

Water Conservation

One would hope for that average inch of rain to fall just after the weekly planting is in the ground and the weekend harvest is complete (on Friday, around noon please!). But, since nature generally has her own schemes, we must be prepared to catch, store and move rainwater where and when we need it. The huge amounts of water we need for our farm dictated that our primary strategy be pumping water from a well. Construction of cisterns to store rainwater would be extremely impractical on such a large scale.

Our farm, as already mentioned, is set on the surface of 60 feet of dense glacial sediment consisting of clay, sand, gravel, rock and boulders. This glacial sediment is a perfect filter and storage medium for water. We struck water at 32 feet, but took the well down to about 50 feet to allow for dry years and heavy use. This well provides household water, farm kitchen water and irrigation water. Our landscape collects rainwater in swales and contour beds to maximize the recharge of our groundwater.

Water conservation in the garden begins with garden design. Pathways are laid out as close as possible to level. These pathways become compacted with use. The beds are basically terrace gardens,

---

Water Facts

- A gallon of water weighs 8.34 pounds.
- 1 gallon of water equals 231 cubic inches.
- 1 cubic foot of water equals 7.5 gallons.
- 1 acre equals 43,560 square feet.
- 1 acre equals 6,272,640 square inches.
- 1 inch of rain provides .623 gallons per square feet of catchment area.

These are handy facts for figuring out water needed or water received.

Example 1:

To cover one acre with 1 inch of water requires 27,154 gallons of water:
(Calculated using the area in one acre [in square inches] divided by cubic inches in a gallon of water: 6,272,640 ÷ 231 = 27,154.29.)

Example 2:

A 1,000 square foot roof receiving 1 inch of water would yield 623.4 gallons of water:
(Calculated by converting the area of the roof in square feet to square inches [the multiplier is 144], then multiplying by the number of inches of water received, and then dividing by cubic inches in a gallon of water: 1,000 x 144 x 1 ÷ 231 = 623.4.)

gradually dropping downslope through the gardens. Whether from rain or a sprinkler, each bed is watered on the surface and excess runoff is caught in the pathways between beds. Because the beds are cultivated deeper than the pathways (and the pathways themselves are compacted and do not absorb the water) water collected in the paths seeps underneath the beds to help ensure deep watering.

Planting in blocks of four to six beds makes watering more efficient. The sprinkler covers an area up to 30 feet wide. Crops need water daily until they germinate or are established as transplants. Our solar-powered pump actually averages about 2.2 gallons per minute from morning until evening. With about 1,000 square feet of new plantings to irrigate each week, we need to run the pump for about five hours a day. Long-season crops that will be in the bed more than six weeks are mulched. Checking soil moisture, plant condition, germination and transplant vigor are a part of the daily routine.

Structures as Watershed

The amount of precipitation falling on a structure should be part of the calculations used to determine water management on the farm. For example, the south, glazed roof of our bioshelter measures 105 feet by

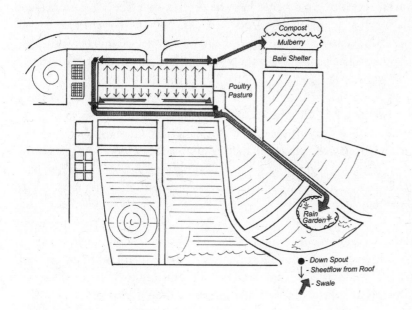

Rainwater running off the bioshelter is directed into swales. These wide shallow ditches allow water to collect and seep into the ground, recharging groundwater.

14 feet, or 1,470 square feet. Runoff from the roof is caught in rain gutters and directed into a swale located 10 feet from the south side of the cold frames.

Bioshelter south roof calculation:

1,470 square ft of roof catchment x 144 square inches per square feet = 211,680 square inches. One inch of rain falling on the roof is 211,680 cubic inches of water. Dividing that by the 231 cubic inches per gallon, means that 1 inch of water on the roof produces 916 gallons of water that can be directed where we want it.

The swale we constructed is 30 inches by 3 inches deep by 100 feet long and can hold 75 cubic feet of water, which is about 562 gallons. The swale continues downslope between the East Garden and the Southeast Garden through a space planted with currants, rugosa rose and goldenrod. The swale and any runoff from the bioshelter's foundation drain end up at a grassy uncultivated area.

The north roof of the bioshelter measures 105 feet by 26 feet, or 2,730 square feet, which equals 393,120 cubic inches. This divided by 231 cubic inches per gallon equals 1,701 gallons every time it rains 1 inch. The west half of the north roof drains through a swale along the west side of the bioshelter into the south swale. The other half enters a north side swale. This is directed past the poultry yard to the east, then to the row of mulberry trees and then dispersed into the yard. This swale also helps water the shaded native plant garden along the bioshelter's north wall.

Another rain catchment system is connected to the roof of our composting outhouse in the far southeast corner of the property. The roof measures 10 feet by 8 feet. The rain gutter is connected to a 50-gallon rain barrel. One inch of rain falling on this roof produces 49 gallons of water, so one night of hard rain will easily fill the barrel. This rain barrel is used for hand washing and occasionally to add water to speed up composting.

Roofing material is a concern when planning to catch and use rainwater from a roof. Acidic rain can leach potentially toxic minerals from treated wood shingles, zinc-plated roofing and zinc-impregnated asphalt shingles. Old and worn asphalt can leach toxic compounds and metals into the water. Slate, tile and metal roofs are preferred.

Nutrient Management and Water Quality

Farm watershed management includes maintaining the quality of water leaving the farm by keeping nutrients out of streams. Water leaving the farm with excess nutrients (or worse, eroded soil), is a pollutant and a loss to the farm ecosystem. As much as possible, water should be collected in the pathways and swales. In a heavy downpour or rapid snowmelt, some runoff is inevitable. We have located swales to direct heavy runoff drainage from compost piles toward trees, shrubs and berry plants. Buffer areas of native goldenrod between the gardens and ponds keep excess nutrients from entering our pond.

Riparian zones are the banks and floodplains of streams and wetlands. Riparian zones are best managed as mixed woodlands. Trees, shrubs and other perennials help stabilize stream banks and absorb nutrients before they can leach into the water.

Site Analysis: Mapping

After observing a site and cataloging resources, the next step is making a map of the property. The map is first used to help plan the placement of elements in the design. It is also a tool to convey and store information about the property. Drawing plans to scale on paper allows for testing and refinement of the plans before site work begins. Maps also serve to record species planted. Together with good garden records, maps aid in planning and documenting soil management and crop rotations. Organic certification procedures require such records and maps. A certification inspector or certifying agency will review these to assess the farm's management.

Mapmaking begins with gathering information on the site from existing maps, deeds and other property descriptions, and from site analysis. We were able to use topographic maps, soil survey maps, and a surveyor's map of our property to build our own base map.

The next task of mapping is putting it on paper. The scale of the map is determined by both the size of the area being mapped and the desired size of the finished map. An architect's ruler has inches divided into one-tenth-inch increments and is therefore very useful in mapping at scales of ten units per inch.

Standard notation and illustrations help to make a map look professional. Clearly marked scale and direction and legible labels are needed to make a map useful.

Obitz Road

Tree Line

Slopes East ⇨

Wet Area

Slopes Southwest ⇗

Level, Well-Drained Good Building Site Full Southern Exposure

Level Ground Slow to Drain

Slopes to South ⇩

South Slope Good Pond Site ⇩

Slopes Gradually To South ⇩

Slopes To Southeast ⇗

Tree Line

Driveway

N ↑

Bubble map of topography and main features.

For planning, transparent paper overlays are usually used; overlays are a good way to add information or make plans without cluttering a map. Final versions of overlays can be made on "acetate paper" or clear plastic with a fine-point permanent marker.

Other useful drawings include a *plan drawing,* which is an overhead view of the site or building and a *section drawing,* which is a side view, often a cross section of a particular part of a building or landscape.

In the early design phases it is good to identify key features of the landscape. Making a base map of the property is the place to start organizing the design process. A base map can begin as a rough sketch of the property's dimensions and key features.

As the process develops, better maps are needed. To make an accurate site plan map requires a lot of measuring and drawing. The effort is worthwhile if you use it as a development and planning tool. The process can be greatly shortened if you begin with a survey map of the site. Overhead aerial photographs are also very helpful in making a base map accurate, especially when the land in question covers several acres. When mapping individual gardens, the scale may be too small for aerial photographs to be useful.

Once you have a base map, you can use it as a template for making zone system analysis maps and sector analysis maps. The very process of creating the base map will help the designer understand site dynamics. More detailed field garden maps will be helpful for annual planting.

Soil types, water resources, landforms, geological features, existing species and other aspects of the site are catalogued. Soil survey maps from the US Department of Agriculture provide a wealth of information for a site. These maps provide information on the site's soil types, including a soil's depth and pH, drainage patterns, soil characteristics relating to construction, and a soil's suitability for various uses, i.e., ponds, driveways or agricultural development. Formerly, the soil survey maps were compiled into soil survey books by county. These books also provide information on local weather, including rainfall patterns, frost-free dates, and weather extremes. These resources can be found in the local library. Printed versions may no longer be available for purchase, but the county office of the Natural Resources Conservation Service usually has electronic versions available. The NRCS website also provides the online soil mapping services described above.

Digital Mapping

GIS (geographic information systems) mapping is a computer mapping system. The most widely used is the ESRI ArcView series of digital mapping tools. With GIS mapping software, various digital maps can be set to the same scale and coordinate system and then combined in layers to produce a wide range of maps. GIS maps can include digital links to site data, and be linked with Computer Aided Drafting (CAD) files. For example, on a GIS map we are developing of Three Sisters Farm, clicking on the data files on our buildings links to building plans and photograph files.

GIS tools include the use of a GPS (global positioning system) to provide coordinates for site mapping. GIS tools and programs cost thousands of dollars and require training and time to learn to use, but are powerful tools once mastered. While the average farm does not have the budget for a GIS program, the Natural Resources Conservation Service can provide GIS-based maps for site planning, and will include soil data and other information.

Mapping your Site

Measuring a garden or landscape for mapping is best done by two people. One person takes, and one person records the measurements. Begin at a known point, such as a corner of a building. From this known point,

Measuring a Site for Mapping

- Mapping tools:
  - compass
  - 250-, 100-, and 25-foot measuring tapes
- Contour study tools:
  - A-frame
  - water level
  - sight level
  - transit and measuring rod
  - field drawing board and pencils
- Drawing tools:
  - drawing table or board
  - T-square, engineers rule
  - pencils
  - assorted drawing triangles and curves

When measuring for a base map or laying out a building, knowing the dimensions of the right triangle improves accuracy. When the two sides of the 90 degree angle measure 3 feet and 4 feet, the hypotenuse will be 5 feet.

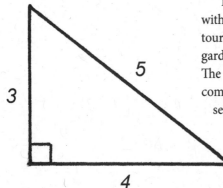

one line is measured parallel to the edge of the building and another line perpendicular to the building. All other notable points on the site are measured in relation to the two perpendicular lines.

At Three Sisters Farm, all maps use the bioshelter's corners or the property lines as a starting point. Because the building is aligned along an east-west axis and faces south, our maps are also aligned to the cardinal points. This is not actually necessary, but it works well with our solar-oriented planning.

To map the south garden, we laid out a 250-foot tape measure on the ground, beginning at the southwest corner of the building and stretching due south to the property line. This provides a square grid, with the building forming one edge and the tape running from the building to the south forming the second edge. We then marked the key locations of fencing, polytunnel, and garden paths on this grid. This is done by measuring with a 100-foot measuring tape the distance, at right angles, from the edges of the grid to the site feature.

Next we moved the tape, first to the middle of the south wall and then to southeast corner of the building, again marking key features along the line between the building and the property line.

This process, of starting with a known point and working at right angles to mark points on a grid is simple and straightforward. When laying out the tape on the edges of the grid you can check that the angle is 90 degrees by measuring the edges of a right triangle formed at the corner of the grid. Mark a point 3 feet along one axis and 4 feet along the other axis. A line connecting these points, the hypotenuse of the triangle, should be 5 feet. This is the 3, 4, 5 rule.

Laying out contour lines:

The scale of available contour maps is generally too large to help with detailed garden development. Several tools are used to lay out contour lines on the ground for building drainage, swales construction and garden pathways. These are the transit, the A-frame, and the water level. The latter two are low-tech tools and can be made at home. Transits come in many styles depending on use and can cost anywhere between several hundred and several thousand dollars.

A transit is the easiest tool to use. (You may be able to rent one.) At the simplest, a transit is a standard surveying, construction and engineering tool with a swiveling viewing scope and leveling knobs mounted on a tripod. Once the transit is set up level, the operator sights to a measuring rod held by an assistant. Keeping the scope level, you can move the measuring stake measure to find points of equal elevation and mark a contour line across the property with stakes. We also used a transit for laying level foundations and walls. Complex property surveys are made with a surveyor's transit.

For the low-tech and budget-conscious designer, an A-frame is a good tool for finding contour lines. An A-frame is generally made of wood and looks like a letter "A." Two legs, about 4 feet long and bolted together at top, are connected with a cross bar about a foot above the bottom of the legs. A weighted string is hung from the top of the "A." When both legs are set at the same elevation, the weighted string will point to the middle of the cross bar. To measure a contour line across a garden or field, simply "walk" the A-frame across the ground: start at one staked point and find the second level point. Then pivot the frame from the second point to the third level point. Set a new stake every few yards as you go, depending on how exact you want to be. For making a ditch or swale to drain water from one point to another, mark out a series of gradually descending points as you walk the A-frame across the landscape. A drawback of the A-frame is that it is difficult to use in the wind, and it can be hard to be accurate on rough ground, but it works very well on relatively level ground.

My favorite tool for finding contours is a water level. This is a 20–30 foot length of clear plastic tubing fastened to two six-foot measuring rods. (We made our own, but they can be purchased from building supply stores.) The rods are marked with ¼-inch increments. The stakes are held upright and the plastic tube is filled with water. When both rods are set on the same level and held upright, the water will be at the same height along the stake. The tubing is corked when being moved to keep water from spilling, and uncorked when they are in position.

Two people are required to use the water level. Each holds a rod. Beginning at one edge of a garden

The A-Frame is a low tech tool used to lay out contour lines.

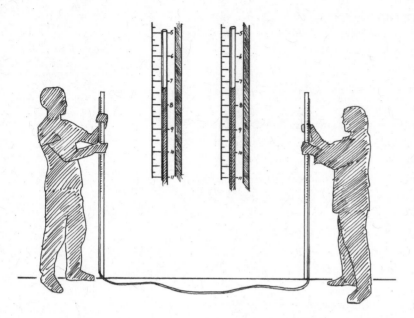

The water level, made from two measuring rods and clear plastic tubing, is used to measure relative elevations, lay out contour lines and to lay out level walls and foundations.

or field one person holds their rod at a starting point. The second person then sets the stake at another point and both check the water level. When the water is at the same level both points are at the same elevation, or "on contour." A little food coloring can help make the water easier to see. The first person moves past the second person to find the next point on the contour. Again, stakes are used to mark the contour line. The water level can also be used to measure changes in elevation for laying out foundations, leveling walls, and for laying out swales and drainage ditches. To measure elevation drop, plot a line downslope from the starting point and each time you move the rod record the difference in elevation between the rods. The water will always stay the same elevation, so the downhill rod will have water closer to the top of the rod.

During a permaculture workshop we had a race between teams using a transit, a water level and an A-frame. The transit team had laid out a 100-foot-long contour line by the time the water level had measured 50 feet, and the A-frame had measured 25 feet. So in that case, the transit was the most efficient tool to use. But with practice, both the A-frame and water level are effective and affordable tools.

The map is only a useful tool if it is used to plan and document work done on the farm. Accurate maps help plan the placement of gardens and structures in relation to windbreaks, pathways, perennial plantings and so forth. They are useful for communicating to others the plan for site development; they also serve as a record of the farm's assets.

Planning

Permaculture planning addresses farm components and activities as interconnected elements. Each facet of the farm — energy systems, water management, soil management, crops, enterprises, buildings, structures, and the landscape — should be reviewed for integrated design. After the infrastructure is in place, many microclimates can be developed and niches filled with useful species.

Brainstorming

Brainstorming is a planning technique in which you allow the free flow of ideas without critical filtering. Wild ideas, sensible ideas, random thoughts and logical conclusions can all add different perspectives. Pick a general topic such as land potential, development plans or staging schedules and write down ideas without making any judgment about the ideas' merits. Later, the ideas from the brainstorming session are reviewed for potential. Even impulsive and impractical ideas can lead to new, productive ways of thinking.

The Plan

Once the results of the brainstorming are reviewed and options are chosen, the development plan is created. This process begins with working out the details. The site should be measured and mapped accurately if this has not been done already. Again, ecological relationships are noted. Permaculture principles are applied to create ecological interactions and to conserve energy and resources. The zone system is used to place elements of the system in the right relationship to each other. Research is done on the plants, animals and structures needed for creating the design. The use of native plants and local and environmentally sound materials and practices are preferred throughout the development.

Budget planning is vital to the success of a project. The more accurate your projections, the more smoothly and successfully your site will develop. Establishing a permaculture farm requires a lot of hard work and a lot of time. Budgeting your time is as important as budgeting your money. Of course, there is often a difference between projections and reality. Equipment breaks down and holds up progress until repairs are made; there are weather delays; and generally life just happens.

Implementation of the design begins with staging the work in a logical progression. Early work includes soil building programs and gathering materials such as composts, mulches and plant materials. Perennials are added at a manageable rate. As trees and shrubs grow, they create niches for other plants, insects and vertebrates. Customers of the vegetables grown in the early years will eagerly purchase fruits when they become available.

Review and monitor progress and adjust to new observations and changing needs. Good record keeping of crop yields, weather patterns and worker productivity provides ongoing feedback for a process of annual design review and planning. Conducting trials of crops and markets allows you to adjust to what works best.

Phenology, the study of how plants and animals interact with the seasons, also provides continual insight into design development. When the spring peepers sing their spring song, it is time to prepare the beds to plant peas, spinach and onions. We try to be sensitive to the seasonal fluctuations and the cycles of the place. We are trying to develop the native knowledge that is lost to many modern farmers. Eventually the need to "design" is superseded by a cultural knowledge of land and resources. We must begin somewhere. Humility as to our own ignorance of nature's complexity will lead to the true permaculture farm. As a new traditional wisdom replaces design, we can find our way back to a deep knowing of the place as Indigenous cultures.

Season Extension

Season extension involves the creation and use of microclimates to expand the growing season. You will probably use some combination of the strategies listed here.

Windbreaks: Windbreaks help extend the season by cutting down the wind chill, allowing plants to get an early start or last longer into the year. A windbreak as small as a bale of hay can extend the harvest of greens for a week or two. Tree crops, berry bushes and other useful shrubs can be arranged to provide windbreaks for nearby plantings. Garden walls, fences, berry patches and tree lines can all afford some degree of protection until full winter sets in.

Mulch: Mulches can help extend the season on semi-hardy herbs like thyme and rosemary. Mulch minimizes the freezing of the soil and can extend a season by helping "store" a crop in the soil. Loosely mulched plants will stay fresh under leaves or hay until a heavy freeze. Cabbages, kale and other brassicas keep quite well this way. Mulch can also *slow* the warming of the soil in the spring.

Permaculture and Architecture

Buildings and other structures are an integral part of farm design. The Permaculture Institute, established by Bill Mollison and others in the 1980s through a series of International Permaculture Convergences, offers diplomas in ten areas of permaculture, one of which is Architecture and Building (see: www.tagari.com/courses/DiplomaOfPermacultureDesign). The permaculture books written by Bill Mollison and David Holmgren present only general recommendations for actual building design; although they present ecological design principles and make general suggestions, it is left up to the designer to apply permaculture principles to building design.

In *Permaculture Two,* Bill Mollison describes the elements of what he calls the reactive house — a house that reacts to its environment, thus minimizing the need for "external energy for climate control." Mollison lists essentials aspects of a reactive house:

- shelter from winds with windbreaks
- oriented for solar gain
- minimize windows on shaded sides
- well-insulated and sealed from drafts
- natural air conditioning with shade trees

    He also suggests:

- a greenhouse on south side
- a shade house on the north side
- high thermal mass masonry stoves

In *Permaculture: A Designer's Manual,* Mollison only devotes one page to home design. David Holmgren's *Permaculture: Principles and Pathways Beyond Sustainability* similarly provides little direction in actual home design. Rather, he stresses the need to design for energy efficiency and recommends the use of recycled and natural building materials and materials with low embodied energy (the energy use to create and transport the material). Holmgren also stresses designing for ease of maintenance and durability.

Both Mollison and Holmgren stress the importance of placing homes and settlements in the proper relationship to the landscape according to zone and sector analysis. Basically, sector analysis is concerned with harnessing or controlling the energy, such as wind, sunlight, or water flow, entering a site. Zone analysis is the layout of site components to create time- and energy-efficient relationships between components of the site. These two design tools aid in understanding the energy dynamic of a site and provide a basis for creating an integrated and sustainable system.

As permaculture has evolved, architecture, especially natural building, has become a prominent part of the movement. In the process, many teachers and proponents of permaculture have integrated regionally appropriate building design into their work. Straw-bale and clay-straw insulation and cob (clay reinforced with straw) walls are common in permaculture projects. Local lumber, stone and other materials are preferred. Similarly, green building concepts, such as energy efficiency, use of non-toxic natural materials, and ecologically responsible resource management are integrated into permaculture planning.

Green Design: Green building has much in common with the goals of permaculture design and provides an important tool for sustainable community development. The US Department of Energy (DOE) defines green building as an integrated approach to design used to create environmentally sound and resource-efficient buildings. Energy efficiency, resource conservation, renewable energy, and water conservation are important design ☞

Clay straw is a local natural building material.

considerations. Interior environments are planned to be healthy — lit and ventilated naturally when possible.

To quantify and standardize green design concepts, the LEED (a trademark of the US Green Building Council) system was developed by the US Green Building Council. Many project funders and state and local governments now give preferential consideration to LEED-certified projects.

LEED stands for Leadership in Energy and Environmental Design. It is a checklist system that provides measurable guidelines for green and sustainable building design and construction. Certified LEED consultants work with design and construction teams to meet goals in six areas of concern: sustainable site development,

water efficiency, energy conservation and air quality, material and resource use, indoor environmental quality, and innovation and design process. Some of the items that can qualify a project for LEED certification are reuse of old industrial sites (brownfield development), storm water management, reduced site disturbance, water use reduction, use of renewable energy, use of local materials, optimizing energy performance, management of construction waste, maximizing daylight in a building, reduction of indoor air pollution and innovative design. Projects can be LEED-certified Bronze, Silver, Gold and Platinum, depending on the total LEED score of a project.

There is a LEED checklist that gives a project a possible 69 points. Completed projects are rated according to the points earned in the review process. To be LEED certified, a project needs 26 to 32 points. A Silver LEED rating is given for projects with 33 to 38 points; Gold to projects with 39 to 51 points; and Platinum to projects with 52 to 69 points. The LEED standards are readily available online to serve as a guideline for builders (www.usgbc. org). The cost for the review process varies depending on the size of the building project. There are now LEED standards for private homes. Points are awarded based on assessment technologies that gauge indoor air quality, ventilation and efficiency. The location of the home is also a factor; points are given for homes located in areas of high-housing density.

Natural Building: Local and natural materials can be interesting to work with and provide unique character to a house. The low cost of the materials is, however, usually balanced by higher labor costs. When a building owner has more time than money and wishes to avoid excess debt, going local can be a good tradeoff. We highly recommend gaining hands-on experience with ☞

natural building before proceeding on one's own project. Many natural building workshops are offered each year throughout the US. Building codes and other regulations relating to natural building vary and should be thoroughly reviewed before proceeding with design and construction projects.

Including natural materials in a bioshelter design can and should be a primary design criteria. A well-designed building, insulated to conserve energy, will last for decades or longer. Such a building pays back the energy embodied in the construction with the heating and cooling energy saved. When choosing natural materials for a high humidity building such as a bioshelter, care must be taken to insure the long-term stability of the materials. Local lumber is preferred, though rough-cut lumber can be difficult to work with, and finished local lumber can be expensive.

Straw bales, used as insulation, are especially vulnerable to decay from moisture. Any use of straw in a bioshelter needs to be well protected. Cob and clay-straw are not as good at insulating a wall but may be more appropriate for bioshelters.

Row Covers: Row covers are synthetic or natural fiber translucent blankets that lightly cover a crop to break the wind and trap the heat of the ground and the sun. They are often made from spun polyester or cheesecloth. Row covers only provide a few degrees of protection, but they can add weeks to a growing season — on both ends. They allow spring crops to get a head start and protect established crops from light frost in the fall.

Cold Frames: Cold frames are small, unheated structures designed to create a microenvironment to protect plants in harsh weather. Cold frames come in many shapes and sizes. Many materials have been used in the past. The American Garden Calendar, published in 1806, mentions the use of paper painted with linseed oil as a cold frame covering. Oiled sheepskin and parchment have been used for the same purpose. Today, the best cold frames are freestanding boxes with hinged lids or, even better, structures attached to the south side of a building. But a sheet of plastic or an old window laid over a hay bale frame will provide some protection. (Be sure to fasten and support plastic sheets, as a heavy rain or snow will quickly collapse them. And old windows can collapse or break if not properly supported.)

Although glass would seem a more sustainable option than plastic; many factors need to be considered. Primarily, one must consider scale and cost of the structure. Smaller frames can be built of glass at reasonable cost. Larger structures glazed with glass may be prohibitively expensive and heavy. Whatever you use, make sure it is properly fastened; wind gusts can turn unfastened frames into flying health risks.

Hot Beds: A hot bed is a cold frame with heat. The ancient method is to bury composting manure under 6 to 10 inches of loose soil. The heat released by the microorganisms in the compost provide warmth to the soil and frame. Hot beds are discussed in detail in Chapter 11.

Polytunnels: A polytunnel is a single layer, unheated plastic greenhouse. Polytunnels can provide six to eight weeks of season extension for most crops. Tunnels should be large enough to walk into but small enough to allow the ground to provide thermal mass. A width of 15–24 feet allows for headroom and enough curve to shed rain and snow. A maximum length of 50 feet allows for natural ventilation through the end walls. For tunnels longer than 50 feet, rollup sides are needed to help ventilation. Clearance for rotary tilling equipment may be a design consideration, depending on the management plan. "High tunnel" is a name increasingly applied to polytunnels. High tunnel greenhouses, also known hoop houses, are becoming important components of agriculture in Pennsylvania and cold climates nationwide. Their relatively low cost and potential for extending the growing season provide a valuable tool for the market grower to increase productivity and profitability. Cultivation in high tunnels allows three crops to be grown in the same space in one

Poly Tunnel Cross Section Graphic Draft

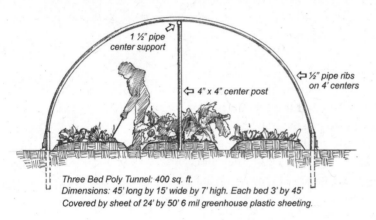

1 ½" pipe center support

⇦ ½" pipe ribs on 4' centers

⇦ 4" x 4" center post

Three Bed Poly Tunnel: 400 sq. ft.
Dimensions: 45' long by 15' wide by 7' high. Each bed 3' by 45'
Covered by sheet of 24' by 50' 6 mil greenhouse plastic sheeting.

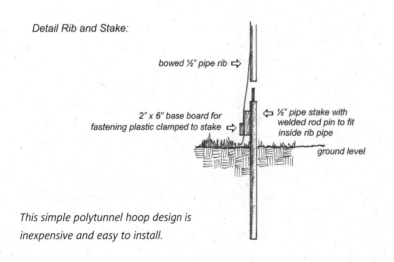

Detail Rib and Stake:

bowed ½" pipe rib ⇨

2" x 6" base board for fastening plastic clamped to stake ⇨

⇦ ½" pipe stake with welded rod pin to fit inside rib pipe

ground level

This simple polytunnel hoop design is inexpensive and easy to install.

extended growing season. For example, spring salad can be followed by summer eggplants and then a fall kale crop.

It is important to build strong end walls and to use sturdy frames. Otherwise the tunnel can collapse in a storm or under snow. Pipe frames are most common. Framing with local lumber is preferred when practical (if the wood is a renewable local resource).

Although an unheated polytunnel can add six to eight weeks to each end of the growing season for hardy crops, frost-tender crops will only get about four weeks of extension beyond frost dates. Using mulches and row covers on crops in the tunnel further protects plantings. Adding thermal mass such as stones, brick pathways and buckets of water to polytunnels can help stabilize temperature extremes and create microclimates.

Bale-shelters: A bale-shelter is a modified polytunnel. Bales are stacked inside the polytunnel to form the north wall and end walls. This provides insulation and so increases the performance of the structure. In the spring, the bales are taken down and used to mulch summer crops. Tunnels and bale-shelters can also be used in summer as poultry housing.

Greenhouse: A greenhouse is a more heavy-duty, double-layered polytunnel or some other type of glazed building with a heating system. A greenhouse is used for year-round production. The design, heating and ventilation are determined by the crop mix desired.

Bioshelter: A bioshelter is a well-built greenhouse managed as an indoor ecosystem. Bioshelters often contain some form of animal life (livestock, poultry, or people) for gas exchange and body heat. The plantings include insectary plants for beneficial insects. Chapter 9 and 10 provide a detailed study of bioshelters.

Integrated Design for Season Extension

The more beneficial combinations we make, the higher our overall productivity can be. For example, mulched crops in a cold frame protected by a windbreak will provide a high degree of protection. In Maine, Barbara Damrosch and Eliot Coleman of Four Season Farm use a system of cold

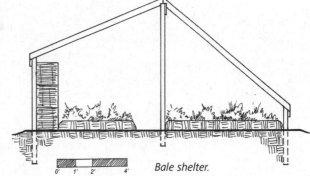

Bale shelter.

0' 1' 2' 4'

frames inside polytunnels to nurse crops through the harsh winters. Continued development of season-extension strategies is imperative as we seek more sustainable and regional food systems.

Case Study: Design of Three Sisters Farm

Before we started to develop our farm, we had the opportunity to study our site for several years. We were able to walk the tree line, field and surrounding woodlots to learn the terrain, note the practices of neighboring farmers, and study the surrounding natural systems. We were able to witness several seasons of important details: the patterns of the snowdrifts; the small, seasonal spring seeps at the base of the slopes; the wetland plants in places with poor drainage; and the annual floods of the nearby riparian zone of Mill Creek.

The property is a five-acre rectangle measuring 260 feet north to south and 842 feet east to west. The north and east property lines, adjoining a neighboring dairy farm's crop fields, had the beginnings of a young tree line. The west property line is a county road, and the south property line bordered our neighbor's five acres, which included his homestead, several acres of woodlot and forest riparian zone along Mill Creek.

Site Analysis

During planning stages, we conducted a more thorough study of the property. Understanding the environmental history and the nature of the human impact on the regional ecology is critical for long-range planning for any sustainable development. Sustainable land use on the farm benefits from this initial consideration of the region's natural history by informing the farm designer about the site's hydrology, soils, past use, native biodiversity, and most important, its relationship to the watershed and bioregion.

Climate

The climate of northwestern Pennsylvania is "humid continental." *Continental* refers to our location in the interior of the continent, which results in relatively stable seasonal weather patterns; *humid* refers to the average annual rainfall of 40 to 45 inches per year. The region is characterized by cold, snowy winters and warm, temperate summers. The primary influences on the regional weather are the latitude, atmosphere, terrain and the effects of Lake Erie. The latitude is around 42 degrees north of the equator; this brings both hot summer days, when the sun is as high as 72 degrees above the horizon at noon, and cold winter days when the sun is as low as 22 degrees above the horizon at noon.

The atmosphere in this region is subject to extremely variable daily weather because it receives the effects of the jet stream coming from the dry west, the arctic wind from the cold north and the warm, moist air coming from the Gulf of Mexico. The interaction of these systems provides variable and often stormy weather as the different air masses meet in the Midwest. Wind speeds can reach 60 miles per hour as storm fronts roll in. This is an important concern when designing buildings, polytunnels, cold frames, and other season extenders. During the early winter, cold wind crossing Lake Erie from the north and west picks up moisture; this can develop into "lake effect" snow over much of the region.

Throughout the year, the region has an average of 200 cloudy days and 64 clear days; the rest are partly cloudy. November through January, there are often only a few sunny days a month, although there are sometimes a few partly sunny days and an occasional long period of sunny weather. When Lake Erie freezes in winter, the lake effect is reduced, allowing more sun in February and early March.

The frost-free growing season ranges from 135 days in the northern section to 150 days in the southern section. Within two to five miles of Lake Erie (50 miles north of Three Sisters Farm), the lake moderates the climate, providing a growing season up to six weeks longer than the surrounding region.

Topography

We live in the Appalachian Plateau bioregion. Our farm is a rolling landscape, which is typical of the region. The field varies in elevation by about 10 feet from the original lowest point (now under the pond) to the highest ground. The farm property is above the highest flood level of the adjoining Mill Creek. The lowest part of the field, in the western half, drained a seasonal wetland and spring on a neighboring property to the north (making it a good candidate for our pond site). This area also serves as surface water drainage for several acres to the north of the property.

The northeast corner of the property contains a seasonal wetland covering several hundred square feet. Part of the headwaters of a spring the runs through a neighboring field, this wet spot is partly fed by a swale located along the edge of the north tree line.

Hydrology

The glacial deposits of the region help create extensive reserves of clean groundwater. The underlying sandstone in much of the region also contains major aquifers. These aquifers are recharged by rain and snow infiltrating porous soils, wetlands and streams. Numerous springs feed the many streams and wetlands.

Agricultural development during the 20th century drained many wetlands and soggy soils. Current federal regulations prohibit installation of new pipes to promote soil drainage and restrict repair of drains to those installed prior to the 1980s, unless permits are obtained. Some of the drained fields are reverting to wet soils as drain tiles clog in abandoned fields. Pastures often develop small sinkholes as soil above the old drain tiles erodes.

Soils

Soils in the glaciated region of northwest Pennsylvania are geologically young, having evolved in the 10,000 to 12,000 years since the glaciers retreated. The parent material from which the soil formed is generally weathered glacial till in the uplands and slopes and weathered glacial outwash and till along stream valleys and associated terraces. The formation of the present soil began with the complex prairie ecosystem that colonized the region as the ice retreated. Subsequently, the successive forest systems and related ecosystems built up the organic matter to feed the soil ecology.

At our site, the soil is primarily silt loam, underlain with a dense fragipan (a hard mix of clay, silt and stones that is poorly drained). The soil, while depleted when we started our farm, and seasonally wet, was still considered to be prime agricultural land for field crops. After the winter snowmelt drains away below the fragipan, the cultivated topsoil layer is reasonably well drained. This is especially true with our raised beds. Because the fragipan acts as a barrier to excess mineral leaching, when loosened with subsoiling tools and organic matter is added, the subsoil provides a good mineral base for expanding the topsoil layer.

Our soil is marginal for fruit trees because the high water table kills roots, leading to shallow root systems. This makes the trees susceptible to drought in dry years and to toppling over in wet, stormy years. Such trees must be planted in the better-drained portions of the property.

Sector Analysis

Solar Access: Sunlight access is extremely important in farm design. Because we had an open field, we were able to place the bioshelter and season extenders in full light. For us, the lack of shade was the problem. We have added trees and other plants and made use of the shade cast by the bioshelter to provide shady microclimates for shade crops, such as woodland plants and medicinal herbs.

Wind: Prevailing winds on our site are from the west and northwest. During certain weather conditions (usually only a few days each year), wind also can blow from the east. Even rarer is the approach of storms from the south, such as a hurricane's final downpours.

Control of the wind at the farm has been a priority. The existing tree line provided the beginnings of a good windbreak for the farm. A windbreak will filter the wind and slow it for a distance downwind. This distance is generally considered about ten times the height of the windbreak, so trees 50 feet tall will moderate the wind for 500 feet. A hedgerow of 6- to 8-foot-tall shrubs will reduce the wind speed for a distance of 60 to 80 feet.

When we began farm development, the young tree line was thin and needed to be expanded. We were concerned not only about the wind, but about stopping airborne herbicide spray from neighboring fields to the north and east from entering the farm. In the years since, we have allowed dogwood shrubs to expand 25 feet into our side of the tree lines, and added other species. This proved a very good strategy. When the neighbor decided to brush cut his side of the tree line in 2006, our side's shrub layer had become dense enough to act as a good barrier to the wind.

As we have developed the farm, we have been aware of the site's potential for wind-powered electrical generation. A tower would need to be located away from wind turbulence caused by the tree line. For this reason, a location on the eastern, leeward side of the pond is reserved for a windmill site.

Privacy: Privacy is a concern for a family farm. Living in a public space has many drawbacks. Visitors can show up anytime unexpectedly. (You should plan ahead if you care to visit any farm.) By locating our home in the northeast section of the property, behind the bioshelter, we allowed our family a sense of privacy.

Floodplain and Stream: The floodplain of Mill Creek is on the adjoining property to the south. This area is subject to inundation up to a foot deep after major storms every year. Beavers reside upstream a few hundred yards, and downstream a few hundred yards. While not part of our property, the wooded floodplain and stream are important habitat for many birds and other creatures that visit and live on the farm. Raccoons, skunks, opossums, and weasels live or travel through this area. Turtles and snakes pass back and forth between the riparian zone and the farm. Woodchucks, rabbits and many other creatures do not respect the property line either.

Along the south side of the driveway, on the neighboring property, a power line has an understory of dogwood shrubs and wild plants. These shrubs suppress tree growth and so are allowed by the power company.

Other Existing Features: The tree line species we inherited were mostly silver maple and black cherry, with a few shrubs. Over the years, birds have added the seeds of diversity. As we began to develop the farm, we allowed a gas pipeline to cross the northwest corner of the property. This pipeline became the location of the path to and through the west garden.

Overlaying the Farm onto the Landscape

During the six years we observed the property and made initial plans for our farm, we had time to walk the property and observe the changes occurring as the field began to develop. In 1983, we began to establish useful plant trials along the southern edge of the field. We left room for an access road at the field's edge and established Siberian pea shrubs (*Caragana arborescence*), rugosa roses, and honey locust trees just beyond the late-summer shade line of the woodlot to the south of the field. With the property owner, we also established a row of a dozen hazelnut bushes 30 feet farther into the field. A planting of 25 honey locust trees was established in the southeast corner of the field.

Bioshelter

When we finally had the opportunity to apply for funding to build the bioshelter, we had long known what we wanted to do with the site.

The first step was to locate the building. We wanted the site to be high, dry and level so foundation drains and the septic system could be easily placed. The property only offered one such space — in the north central area of the property. We wanted to set the building far enough in from the north property line to allow for the tree line to expand and for supply storage.

The next step was to locate the building's support structures, including a water well, buried electric and telephone lines, and the septic system. We put the well between the parking area and the west side of the bioshelter. This allows access for drilling equipment if the well ever needs to be serviced. Waterlines are buried running from the well to the bioshelter's west end. Electric and telephone lines also enter the bioshelter there. The sand mound septic system is near the west property line. Drainage lines for the bioshelter's foundation run southeast from the southeast corner of the building.

Farm map with basic structures outlined.

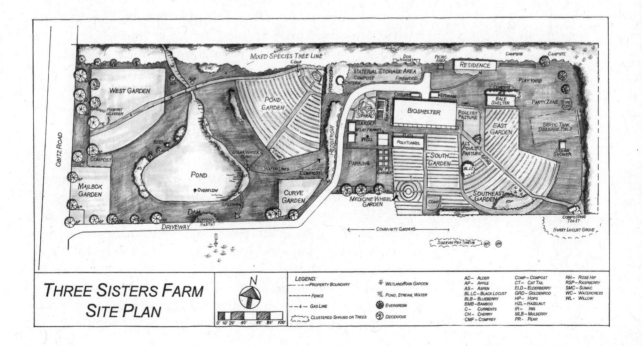

Other needs included a driveway, public and private parking areas, and access to the north side of the building. Here we allowed space for compost yards and storage areas for leaves, mulch, sand, and firewood.

We share the first part of the driveway with the neighbor to the south, who built his home overlooking Mill Creek. The driveway follows the property line along the south edge, then turns north in the center of the field toward the bioshelter. Parking for visitors is located just inside our property line southeast of the bioshelter. Private parking is to the north of the bioshelter.

We also allowed space in the plan for a future home for the farm managers. Currently, there is a renovated mobile home located along the tree line north of the bioshelter. This location allowed for connections to the phone and electric service and well water and to the septic system. We reserved an area of about one third acre for open space, variously a yard and meadow. This gave our children a play field and camping space. A fire circle near the east end of the yard provides a setting for social events.

Pond

Our pond is described in detail in Chapter 13. Located in the lowest spot of the field and naturally fed by a spring in the neighboring yard, the pond is a key feature in the landscape, providing irrigation, fish and habitat. Pond water is used to water newly transplanted crops via a sprinkler, and to irrigate mulched, long-season crops by flooding the pathways. This has greatly reduced our need for well water and associated energy use.

Building The Pond

During the initial fallow period, between 1981 and 1986, a small spring and wet meadow in the neighboring yard drained through our gently sloping field. A small patch of cattails and a young willow tree were establishing themselves. These were eliminated in 1986, when the entire field was plowed and planted with mixed clover hay. About the time we began to garden around the bioshelter, in 1988, a neighbor diverted the spring away from our field, drying the area. When we wanted to build the pond, we had to negotiate with the neighbor; we ended up burying a four-inch pipe through their yard to re-establish the original watercourse.

The pond has evolved from a newly bulldozed site to a diverse habitat (for more description and a list of species, see Chapter 13).

Wildlife at Three Sisters Farm: Many Neighbors

Our efforts to apply permaculture design to the farm include integrating the farm into the matrix of nature. Although, it would be highly arrogant of us to think we could actually design ecosystems, we believe that we can create a balanced ecological agricultural landscape by allowing nature to return to the land.

As mentioned earlier, the five-acre field that was to become Three Sisters Farm was in 1983 just sparse corn stubble and bare soil. The 1,200-foot tree line on the north edge of the property and the 300-foot tree line on the east edge had a minimal understory developing under a row of red maple and black cherry. Shrubs included scattered panicled dogwood, osier dogwood, and Juneberry.

While the tree line probably provided habitat for birds and other small creatures, the field was, ecologically, near death. From 1981 to 1986 the field was left fallow. Weeds such as Queen Anne's lace, yellow dock, dandelion, quackgrass and other herbaceous plants established themselves in the field. In 1986, a neighbor planted a mixed clover hay. In 1988, we took over full-time management of the farm when we built our bioshelter and began our market garden development. One of our primary goals was to create a farm that would enhance habitat and support diverse wildlife. The farm would, in addition to producing marketable products, restore a local ecology. As we worked to develop the landscape plan, nature moved in.

After several years, the farm was already quite diverse. I awoke at dawn one spring morning and went into the bioshelter to do the chores. On a whim I opened the window overlooking the pond garden. I was just in time to see a great horned owl lift off from a rabbit he had killed during his nightly patrol. Now, during the day, a red-tailed hawk patrols

Above: Garder Snake.
Below: Ladybug with larve.

the sky overhead. On rare occasions, the local bald eagle passes over as well. Often the small group of crows in the neighborhood will harass the hawks for fun. In the winter of 2006–2007, a pileated woodpecker began knocking about in the tree lines early in the morning.

The pond is visited each morning by a number of birds. Each dawn, spring through fall, you can see a great blue heron hunting in the pond, while Canada geese and wild ducks forage nearby. Mallard and crested merganser ducks feed in the pond. The green heron and kingfisher rotate between the pond and the nearby wooded stream. However, none of these water birds have been observed nesting here. This is probably due to the small size of the pond and the high level of activity near the pond (fishing and gardening) during the springtime. Some birds do nest on the farm. Baltimore orioles forage in our garden from their nest near the stream and killdeer have nested in gardens near the pond. On several occasions an osprey, which nests at a nearby lake, has been seen diving into our pond.

Redwing blackbirds nest in the cattails around the pond. Acrobatic barn swallows swoop and tumble, competing with dragonflies for mosquitoes. Bluebirds nest in boxes around the farm, as do English sparrows, swallows and wrens. Sleek grey catbirds mew in the bushes. Phoebes and robins nest in the eaves. Yellow warblers and thrushes flit and hop through the tree line and goldfinches hang on the weeds. Hummingbirds delight us as they sip from the scarlet runner beans. In the twilight, undetermined species of bats emerge to skim the pond for water and hunt for insects.

The deer seem to avoid our farm. For 15 years, our well-trained Australian shepherd/collie mix, Nick, assured that with his presence. Because of him, the wily woodchucks retreated to the edges of the tree line. Skunks and opossums pass through but do not dally. The raccoons stay close to the stream, avoiding the farm. Rabbits come and go in cycles,

- Changes in the five-acre field have contributed to increased biodiversity on Three Sisters Farm
  - the building of the pond
  - expansion of the shrub layer in the tree line
  - planting of habitat species and other trees
  - promotion of beneficial habitat with crop plantings and farm management
  - allowance for unmanaged fallow areas between gardens
  - state efforts to restore the bald eagle and osprey to nearby Goddard State Park
- Natural elements that promote species diversity
  - neighboring wood lots
  - neighboring stream floodplains
  - natural expansion of the tree line

but they, too, are wary of dogs. After Nick passed away, the population of woodchucks and rabbits began increasing. A new team of dogs now patrols our gardens — and again the wildlife is under control.

The farm cat helps out by hunting young rabbits in the spring, the moles that tunnel through soft garden beds hunting grubs, and the voles that follow them, hunting our crops.

Many species of animals use the tree lines for either homes or forage. In addition to the animals already mentioned, countless insects and birds either live there or visit in their foraging rounds. Juneberry, black cherry, nannyberry, sumac, elderberry, red ossier dogwood and panicled dogwood all provide fruit to birds and attract many types of insects that birds consume. Maple, oak and slippery elm also attract birds and insects with seeds, foliage and nest sites. We have expanded the north tree line with raspberries and blackberries and allowed native grasses and wildflowers to get established. We continue to add useful trees along our side of the tree line. Basswood for bee forage and edible leaves, heartnut, butternut, peach and sour cherry have all been planted there.

A multi-storied windbreak provides multiple crops, promotes biodiversity and enhances the effectiveness of the windbreak.

STACY DOMER

Bee balm: edible flower, tea herb and a favorite of hummingbirds.

Cardinal flower is a popular wildflower for damp soil.

DARRELL FREY

Promotion of Beneficial Habitat with Crop Plantings and Farm Management

Promoting biodiversity on the farm meant including plants that encourage beneficial insects to dwell in our gardens. Some of the plants we have used include tansy, flowering oregano, milkweed, ironweed, asters, goldenrod, Joe Pye weed, jewelweed, native grasses, clovers, motherwort, anise hyssop, bronze fennel. Most of these provide nectar and pollen for insects.

We manage our mowing and pruning with a sensitivity to the lifecycle of birds and insects, watching for nests and larvae. Herbs are allowed to flower to provide forage for a range of pollinators and beneficial insects. Dill, cilantro and flowering oregano blossoms all are valuable habitat for insects.

We watch for turtle nests near the pond as we work the gardens, and similarly try to avoid injury to snakes, frogs and toads when tilling. Piles of rocks near gardens encourage snakes and toads to take up residence as well.

Allowance For Unmanaged Fallow Areas between Gardens

When we began farming, the field was planted in a mix of clovers and alfalfa. After one season of laying fallow, self-seeded red clover establishes itself in our beds. The next year, aster and goldenrod moves in, and the third year goldenrod moves in. This is an ongoing process, anytime we quit tilling this process happens. As much as 25 percent of our property is dominated by goldenrod. This plant suppresses other plants by its dense growth and allelopathic properties; its roots exude chemicals to inhibit other seeds from germinating. Some sections of the farm have been in goldenrod for 15 years. The suppression keeps trees and brambles from establishing and probably dampens biodiversity. However, goldenrod does supply plentiful nectar in the fall and is known as a host to many beneficial insects. Meadow birds nest in it. Unfortunately, woodchucks like to hide and burrow in the goldenrod meadows.

We generally allow for short-term fallow periods in the gardens and for a lot of "weeds" in and among the gardens. Species such as thistles, burdock, dandelion, chickweed, purslane and lambsquarters all have high food value for wildlife. (For more information, see Martin et al., *American Wildlife and Plants*.) Bull thistles and burdock are allowed to grow in and around our tree lines and orchards. Burdock leaves, stalks and seeds are cut and fed to the chickens spring through fall.

Applied Observation

As an ecological farm, we strive to continually observe and remember the lives and habits of our many neighbors. The timing of the emergence of various species in the spring, the return of migratory species and their nesting and other reproductive habits are observed — and nurtured when possible. When we see metallic bees emerging from the pith of tansy stems in the spring, we know we should use care when cutting last year's dead stems; we place the stems where the bees will be unharmed. We remove any unwanted milkweed from the gardens before July 5th, when the monarch butterfly returns to lay eggs on them.

As the sparrow population has increased, bluebirds and swallows have declined on the farm. We say each year we will institute a more aggressive campaign to control the population of English sparrows, which compete with swallows and bluebirds for the birdhouses. But it is hard to keep up with them.

Through ongoing observation, we notice new plants and use them as indicators of microclimates and seasonal changes to help guide our land management. It also allows us to do selective weeding and mowing, which are important for habitat management.

Observation is a continuous activity at the farm. Each new species observed is identified with field guides. As best we can, we try to learn about its habits and lifestyle. Just 15 minutes spent observing the myriad syrphid flies, bees, wasps, and other tiny pollinators in our Spiral Garden on a hot summer day will convince any observer of the diversity of life on this plot of land. An hour observing the bird life will likewise testify to the health of the ecosystem. As the farm ecosystem has developed, our production and harvest of diverse products also increase. Yet, we still see room for expanding both productivity and wildlife habitat on the farm.

Wildlife Observed at Three Sisters Farm since 1988

- Mammals
  - Common muskrat, *Ondatra zibethicus*
  - Eastern mole, *Scalopus aquaticus*
  - Hairy-tailed mole, *Parascalops breweri*
  - Star-nosed mole, *Condylura cristata*
  - Meadow vole, *Microtus pennsylvanicus*
  - White-footed mouse, *Peromyscus leucopus*
  - Deer mouse, *Peromyscus maniculatus*
  - Norway rat, *Rattus norvegicus*
  - Eastern cottontail, *Sylvilagus floridanus*
  - Woodchuck, *Marmota monax*
  - Eastern chipmunk, *Tamias striatus*
  - Striped skunk, *Mephitis mephitis*
  - Common raccoon, *Procyon lotor*
  - Long-tailed weasel, *Mustela frenata*
  - Short-tailed weasel, *Mustela erminea*
- Rare: Transient visitors
  - White tail deer, *Odocoileus virginianus*
  - Common gray fox, *Urocyon cinereoargenteus* (presumed)
  - Red fox, *Vulpes vulpes (Vulpes fulva)* (presumed)
  - Virginia opossum, *Didelphis virginiana*
  - Common porcupine, *Erethizon dorsatum*
  - Coyote, *Canis latrans* (presumed)
- Amphibian and reptiles
  - Snapping turtle, *Chelydra serpentina*
  - Painted turtle, *Chrysemys picta marginata*
  - Gray treefrog, *Hyla versicolor*
  - Spring peeper, *Pseudacris crucifer*
  - American bullfrog, *Rana catesbeiana*
  - Northern leopard frog, *Rana pipiens*
  - American toad, *Bufo americanus*
  - Eastern milk snake, *Lampropeltis triangulum triangulum*
  - Smooth green snake, *Opheodrys vernalis (Liochlorophis vernalis)*
  - Eastern ribbon snake, *Thamnophis sauritus*
  - Common garter snake, *Thamnophis sirtalis*
  - Racer snake, *Coluber constrictor*
  - Northern two-lined salamander, *Eurycea bislineata*
  - Eastern red-backed salamander, *Plethodon cinereus*
- Fish
  - Bullhead catfish, *Ameiurus* spp.
  - Bluegill, *Lepomis macrochiru*
  - Smallmouth bass, *Micropterus dolomieu*
  - Largemouth bass, *Micropterus salmoides*
  - Yellow perch, *Perca flavescens*
- Birds
  - Ruby-throated hummingbird, *Archilochus colubris*
  - Redwinged blackbird, *Agelaius phoeniceus*
  - American woodcock, *Scolopax minor*
  - Mourning dove, *Zenaida macroura*
  - Common yellowthroat, *Geothlypis trichas*
  - Song sparrow, *Melospiza melodia*
  - Field sparrow, *Spizella pusilla*
  - Dark-eyed junco, *Junco hyemalis*
  - House sparrow, *Passer domesticus*
  - Indigo bunting, *Passerina cyanea*
  - Black capped chickadee, *Parus atricapillus*
  - House wren, *Troglodytes aedon*
  - Eastern bluebird, *Sialia sialis*
  - Eastern phoebe, *Sayornis phoebe*
  - Baltimore oriole, *Icterus galbula*
  - Cardinal, *Cardinalis cardinalis* ☞

- Blue Jay, *Cyanocitta cristata*
- Cedar waxwing, *Bombycilla cedrorum*
- Robin, *Turdus migratorius*
- Tufted titmouse, *Parus bicolor*
- White-breasted nuthatch, *Sitta carolinensis*
- Golden-crowned kinglet, *Regulus satrapa*
- American crow, *Corvus brachyrhynchos*
- Northern flicker, *Colaptes auratus*
- Red-bellied woodpecker, *Melanerpes carolinus*
- Downy woodpecker, *Picoides pubescens*
- American goldfinch, *Carduelis tristis*
- House finch, *Carpodacus mexicanus*
- Eastern meadowlark, *Sturnella magna*
- Common nighthawk ,*Chordeiles minor*
- Tree swallow, *Tachycineta bicolor*
- Barn swallow, *Hirundo rustica*
- Gray catbird, *Dumetella carolinensis*
- Brown-headed cowbird, *Molothrus ater*
- Killdeer, *Charadrius vociferus*
- Green heron, *Butorides virescens*
- Great blue heron, *Ardea herodias*
- Belted kingfisher, *Ceryle alcyon*
- Canada goose, *Branta canadensis*
- Hooded merganser, *Lophodytes cucullatus*
- Mallard, *Anas platyrhynchos*
- Great horned Owl, *Bubo virginianus*
- Red-tailed hawk, *Buteo jamaicensis*
- Osprey, *Pandion haliaetus*
- Bald eagle, *Haliaeetus leucocephalus*
- American kestrel, *Falco sparverius*
- Cooper's hawk, *Accipiter cooperii*
- Red-shouldered hawk, *Buteo lineatus*
- Sharp-shinned hawk, *Accipiter striatus*
- Turkey vulture, *Cathartes aura*
- Important beneficial insects
- Honeybee, *Apis mellifera*
- Orchard bees, *Osmia* spp.
- Eastern carpenter bee, *Xylocopa virginica*
- American bumblebee, *Bombus pennsylvanicus*
- Paper wasps, *Polites* spp.
- Potter wasps, *Eumenes fraternus*
- Eastern yellow jacket, *Vespula maculifrons*
- Braconid wasp, *Apanteles* spp.
- Black-and-yellow garden spider, *Argiope aurantia*
- Shamrock spider, *Araneus trifolium*
- Goldenrod spider, *Misumena vatia*
- Crab spider, *Misumenoides formosipe*
- Marbled orb weaver, *Araneus marmoreus*
- Jumping spiders, *Metaphidippus* spp.
- Ladybugs, *Coccinella, Adalia* and *Hippodamia* spp.
- Asian ladybug (Harlequin ladybird), *Harmonia axyridis*
- Pyralis firefly (Eastern firefly), *Photinus pyralis*
- Pennsylvania firefly, *Photuris pennsylvanica*
- Amercan hoverfly, *Metasyrphus americanus*
- Gall midge various types
- Preying mantis, *Mantis religiosa*
- Lacewing, *Chrysopa* spp.
- Brown lacewing, *Hemerobius* spp.
- Ground beetles, *Pterostichus* spp.
- Butterflies and moths
- Tiger swallowtail, *Papilio glaucus*
- Monarch, *Danaus plexippus*
- Viceroy, *Limenitis archippus* ☞

- Mourning cloak, *Nymphalis antiopa*
- Spangled fritillary, *Speyeria cybele*
- Black swallowtail (parsley), *Papilio polyxenes asterius*
- American painted lady, *Cynthia virginiensis*
- Red admiral, *Vanessa atalanta*
- Cabbage, *Pieris rapae*
- Sulfur, *Phoebis sennae*
- Questionmark, *Polygonia interrogationis*
- Hummingbird moth, *Hemaris thysbe*
- Woolly bear moth, *Isia Isabella*

CHAPTER 5

Energy Systems on the Farm

ANYONE READING THIS BOOK is already aware of global climate change and the urgent need to slow or stop global warming. And if you are reading this book, it is likely that you are interested in how a permaculture farm might contribute to that effort. This section looks closely at the issues related to farm energy use and presents strategies for reducing the use of fossil fuels on the farm. It is intended as a primer in solar energy design and the use of biothermal heat from compost and poultry.

Crop and livestock production in the US currently accounts for 6 percent of all fossil fuel consumption. Our food systems, including production, processing, storage and transportation are responsible for nearly one fifth of all energy use in the US. More sustainable agricultural operations — that do not rely on fossil fuel — reduce the release of greenhouse gases that are creating global warming. Moreover, agricultural operations can contribute to the search for alternatives to fossil fuels. An advantage that the market gardener has is the ability to tailor small-scale production systems to specific tasks. The non-industrial farmer can try various conservation strategies, rediscover traditional farming practices, and work with the cycles of nature. These systems can be much more efficient than electricity from the power grid.

Bioshelter in spring.

Approaches to Better Energy Use on the Farm

- Farm Facilities and Operations
  - Sustainable design is site specific.
  - Energy planning begins by observing and auditing use.
  - Upgrade to energy efficiency.
  - Explore options of alternative technologies.
  - Purchase green energy from utility companies.
  - Reduce use through conservation measures.
  - Observe energy use.
  - Develop appropriate alternatives that are site specific.
  - Conserve energy through sound design and construction.
  - Minimize operating costs with conservation and efficient use of energy.
  - Incorporate renewable energy when possible.
  - Use local materials and locally appropriate design measures.
  - Vote for representatives that support alternative energy policies and pay attention to political issues surrounding renewable energy.

(Based on ATTRA Publication #IP220 "Efficient Agricultural Buildings: An Overview.")

- Building Design Elements
  - Natural lighting (as much as possible).
  - Airtight construction.
  - Adequate and proper insulation.
  - Passive solar capture.
  - Summer shading.
  - Natural ventilation.
  - Energy- efficient doors and windows.
  - Efficient heating and cooling systems.
  - Windbreaks that minimize heat loss by reducing wind chill effect.
  - Earth berms that reduce heat loss.
  - Re-used or recycled materials (when appropriate).
  - Upgrade buildings and equipment to efficiency).
  - Building heat: dual fuel, wood and biofuel.
  - Transition equipment to biodiesel and ethanol when appropriate.
  - Reduce energy use by weatherizing buildings.
  - Build soil fertility with organic matter, compost and manures to add carbon to soil.
  - Combine trips to reduce fuel consumption.

Conservation

Energy conservation is the first step in reducing fossil fuel use. Performing an audit of your farm's energy use will help identify conservation strategies. To begin, make a list of all energy uses:

- Electrical motors (fans, pumps, refrigerators and coolers, etc.)
- Lighting
- Space heating
- Water heating
- Office equipment
- Farm equipment

Although your electric bill probably lists the total kilowatt hours (thousand watts used each hour) you use each month, a review of the energy demand for each machine, appliance, etc., can help guide your reduction of energy use. Once you have a clear picture of farm energy use, you can brainstorm strategies to reduce energy use and replace purchased energy with alternative sources.

Energy Strategies at Three Sisters Farm

We have designed many energy conservation measures into our bioshelter. These include the use of windbreaks and earth berms (discussed in detail in Chapter 7), heavy insulation and passive cooling.

When we built our bioshelter in 1988, we estimated a similar-sized greenhouse would require $3,000 per year to heat, using over 4,000 gallons of fuel oil. At 2010 prices, this would be nearly $12,000 per year. By incorporating energy-saving features into our design, most of our $80,000 initial construction cost has been paid back in fuel cost savings alone.

The design of our polytunnel greenhouses eliminates the need for cooling fans in the warmer months. These two "high tunnels" are 15 feet wide and 45 feet long, providing 600 square feet of protected growing space. Because they are oriented to run east to west, are under 50 feet long, and have wide doors on each end, these tunnels are cooled passively and do not require electric fans to ventilate. During the winter season, they are sealed on the ends and protected on the north side by a stack of straw bales, thus reducing heat loss and extending the season by several weeks in the fall and spring.

Cooling costs are reduced by limiting the use of our refrigerated walk-in cooler to two days per week. We only use it for harvests prior to delivery. No inventory is kept in the cooler between harvests.

Our primary tillage equipment consists of an 8-horsepower gasoline rototiller and simple hand tools. Mulching, interplanting and weeding with hoes minimize the use of the tiller and helps protect soil structure. We do as little tilling as is practical.

Human Energy

Human energy is a vital part of the permaculture farm. So, keeping the farmer healthy and able is important. Good food, interesting work and enough rest is essential. Good organization and the proper tools (and good weather) allow for production to flow smoothly without overtaxing the gardener. We have established daily and weekly cycles of tasks; a variety of tasks helps keep the mind fresh and free from daily monotony.

Much of our garden work takes place before 1 p.m. and after 4 p.m. each day. This allows for a long lunch break during the heat of the day. We generally provide our farm crew with a balanced lunch, including a lot of salad and other vegetables. The combination of good food with the mid-day break helps keep us productive and happy.

We have applied permaculture design concepts in our human energy planning as much as possible. Because of this, the tasks of tending the chickens, young seedlings and bioshelter plantings flow in an easy routine. Gardens are laid out so that those needing the most intensive management are closest to the bioshelter. Composts piles are located up slope from gardens so we do not have to haul wheelbarrows and cart loads uphill. Water storage tanks are placed so gravity can do the job of moving irrigation water down slope to the gardens. We have established harvest-day routines that everyone follows. This allows for a smooth process of harvesting, cleaning and storing produce.

Wood: A Renewable Resource

Burning wood for heat does not return fossil carbon dioxide to the atmosphere and, in theory, will not increase carbon levels (on a geological timeframe). However, large-scale burning of forests to convert them to agriculture does contribute to atmospheric carbon dioxide buildup and global climate change because forests that could absorb CO_2 are lost. Therefore, the use of wood as a fuel requires sustainable forest management to be considered a viable fuel option. The acreage of forest in Pennsylvania has increased in recent decades as farmland is abandoned and reverts to woodlands. Many foresters are advocating sustainable forest management practices. These considerations allow the proper use of firewood to be considered sustainable in Pennsylvania.

We obtain our firewood for the bioshelter from local sawmills. We buy mostly cut slabs that are a byproduct of milling lumber. We buy firewood in summer to allow it time to dry before we need it.

Efficient use of firewood, that is, gaining the most heat and producing the least pollution, requires two things: fast burning of the wood and storing heat in thermal mass. (Thermal mass is any substance that will absorb heat and then slowly release it. It is discussed in more detail below.)

Our bioshelter has two wood stoves. A cast iron stove heats the air when outside nighttime temperatures fall below freezing and days are cloudy. This stove is adjacent to a large mass of concrete and barrels of water, which absorb some of the heat and radiate it back to the building when the stove cools down.

The key to burning this stove efficiently is to get a load of wood burning very hot to burn off the wood's gases before turning the stove to a lower setting. Closing down the stove too soon and burning wood too slowly is inefficient and creates polluting gases and flammable creosote in the chimney.

Home Heating

Home heating is responsible for 62 percent of the energy used in the average household in the northeastern US. The average household uses 730 gallons of fuel oil a year. At approximately 134,000 Btus per gallon, this represents nearly 98,000,000 Btus per household. According to the US Department of Energy (DOE), using this much fuel oil releases over 15,000 pounds of fossil CO_2 into the atmosphere. Heating the average home with propane releases nearly 14,000 pounds of fossil CO_2. One should note that the use of electricity to heat a home generally requires 3 Btus of fuel to create 1 Btu of home heat. Therefore, use of electricity produced in a coal-fired power plant can release three times as much CO_2 as a home furnace. For the average home in the northeast US, that means 45,000 pounds of CO_2 are released annually.

Using firewood to provide the same amount of heat releases over 19,000 pounds of CO_2 into the atmosphere. However, this CO_2 does not add to the net increase of atmospheric CO_2 because the carbon released has not been stored geologically for many million years. Moreover, wood is a locally available resource, is renewable in a lifetime, and is considered "biogenic" or of biological origin (as opposed to "fossil" fuels). According to the DOE's Energy Information Administration (EIA) "Under international greenhouse gas accounting methods developed by the Intergovernmental Panel on Climate Change, biogenic carbon is part of the natural carbon balance and it will not add to atmospheric concentrations of carbon dioxide. Reporters may wish to use an emission factor of zero for wood, wood waste, and other biomass fuels in which the carbon is entirely biogenic." (Data is from eia.doe.gov.)

Regardless of its origin, it is important to reduce all CO_2 use. By using energy-efficient wood stoves, we've reduced the amount of wood we use on the farm to about 2½ cords of hardwood. That amount of wood provides us with approximately 60,000,000 Btus and releases approximately 11,700 pounds of biogenic CO_2.

Our second wood stove in the bioshelter is an aluminum "Snorkel Stove" (see it at snorkel.com) submerged in a 600-gallon water tank. This is used when outside nighttime temperatures fall below 20°F. (Temperature measurements throughout this book are Fahrenheit) A fast-burning fire transfer heat directly to the water, resulting in minimal heat loss and maximum efficiency. The stove will heat the 600 gallons of water 10° per hour. This is an output of 49,860 Btus per hour. The water tank then acts as a radiator, providing heat to the building through the cold night. The warmed water is also used for irrigating plants in the bioshelter. The water flows by gravity out of the tank through weep hoses, warming the soil and encouraging plant growth.

Together the two stoves use fewer than three cords of wood each year.

Biothermal Resource Recovery

Biothermal energy resources used at Three Sisters Farm include heat recovery from compost and the body heat of chickens. We first learned about this somewhat novel greenhouse heating system through the New Alchemy Institute. After visiting the Institute in 1988, we decided to integrate the system into our own bioshelter. We found support for the approach in other places as well. In *Permaculture One,* Mollison and Holmgren used the idea of combining a chicken house and greenhouse as an illustration of permaculture design. Anna Edey applied this idea — using poultry for supplemental heat and CO_2 — to her Solviva Bioshelter on Martha's Vineyard.

Biothermal Heat

The capture and use of heat from a compost pile is an ancient practice. But it is not used much these days. Although it is standard greenhouse practice to encourage seed germination by supplying bottom heat to germination trays, this is usually done using electric heating pads — which, of course, rely on networks of nuclear, coal and natural gas power plants. The use of biothermal heat from compost to provide bottom heat for germinating seedlings requires only a source of compost materials, a compost bin, a pitchfork and some human labor.

(The composting process also generates CO_2, water vapor and ammonia gas. Chapters 9 through 11 detail the use of compost for supplemental heat and CO_2 enrichment in the bioshelter.)

Chickens in the Greenhouse

Inspired by permaculture theory, we wanted to integrate egg-laying hens into our farm nutrient cycle and pest control strategies. The success of Anna Edey's Solviva Bioshelter encouraged us further, and we built a room for a flock of chickens in the bioshelter so we could "harvest" the chickens' body heat and CO_2 to aid plant growth. Following the lead of Solviva, we selected the northeast corner of the building. This location, away from the food processing side of the bioshelter, allowed for the development of an adjoining chicken yard.

The room is sealed from the rest of the bioshelter with a vapor barrier, so we can control the air exchange between the poultry room and the rest of the bioshelter. Fresh straw is placed on the poultry room floor daily to prevent odor and conserve nitrogen on the droppings. This bedding is allowed to accumulate during the winter; it becomes a source of heat and CO_2 when it is composted.

Our flock of 40 hens and two roosters has free access to the outdoors most of the year (except during periods of very cold weather, when they are kept inside). A compost chamber blower pulls the warm, CO_2-rich air out of the poultry room and sends it to rock storage areas beneath the raised growing beds which absorb the heat and act as thermal mass. Each bird produces around 720 Btus per day, so 40 birds produce about 28,000 Btus per day. It's a small amount of heat — equivalent to a gallon of fuel oil every five days, but in the three to four months of a heating season, the chickens provide enough body heat to save 20 to 25 gallons of conventional fuel.

Solar

Winter Heating: The Ideal

Just before dawn, the temperature outside is 25°; inside the bioshelter it is 50°. The day breaks bright in the cloudless sky. As the sun arcs low across the blue winter sky, its light and energy sweep through the solar building, penetrating the far corners, heating the air inside, recharging thermal mass throughout the building. Plants respond to the light and warmth, synthesizing sugar and processing nutrients, building leaf and bud.

Temperatures in the building rise slowly through the morning, from 50° to 60° by 11 a.m. The blowers are turned on to direct the extra heat so it can be stored in the rock and soil of the deep growing beds. At noon, the temperature on the second floor is 70°; by 2 p.m. it is 90° on the second floor and 75° on the first floor. At 3 p.m., as the temperature approaches 100° on the second floor, windows and a vent or two are opened to release excess heat. Soil surface temperatures rise several degrees by midday.

As the sun approaches the horizon and the temperature begins to drop, windows and vents are closed. The temperature drops slowly through the evening. By 9 p.m. the second floor is 75°; the first floor is 60°. The heat recovery blowers are turned off. By midnight, the building settles to a uniform 60°. In the pre-dawn darkness, the air temperature returns to 50° once again.

Winter Heating: Reality

The reality is that perfect days are spaced out, especially here in northwestern Pennsylvania. Most discouraging is a heavily overcast day, with virtually no solar gain, followed by a clear, starry night when outdoor temperatures plummet. Many factors affect building heating and cooling. Presuming a well-built structure, one can expect heat to be lost at a calculable rate through walls, glazing and air leaks. This rate increases as the temperature outside drops. Wind chill increases heat loss by, in effect, dropping outdoor temperatures.

Heat gain is dependent on the amount of sunlight entering the building, especially at midday, when the angle is most direct. On slightly overcast days some light, and thus heat, is gained. Charts used to estimate solar gain in a given region generally report *average gain,* that is, the average of sunlight typical for a given month. The numbers are given as Btus per square foot. This average presumes the normal pattern of sunny days and cloudy days in a region.

When heat loss from the solar building due to cold temperatures outside exceeds heat gain from the sun, supplemental heating is required to prevent the building from getting too cold. Thermal mass can be viewed as a battery. Sunlight charges the mass. During periods of extended cold, cloudy weather the "batteries" of thermal mass run down. Supplemental heating has to maintain the temperature until the sun shines again. Because our ground temperatures average 50°, the un-insulated floor acts as a heat source when the inside temperature is below 50°.

A sunny mid-winter day will give one to two days charge to the thermal mass — no supplemental heat is needed. After that, the wood stove is burned at night. We prefer a nighttime air temperature no lower than 45°, but 50° is preferable. Without supplemental heat, our building tends to stay about 20 degrees above outside temperatures through winter. So anytime the outside air temperature is below 25°, we burn the wood stove. When the outdoor temperature is below 15° we also heat the irrigation tank to at least 100°. This radiates heat through the cold night. When it is below zero outside, we burn both stoves very hot and use a fan to circulate the warm air from near the stoves to the east and west ends of the building.

Understanding Solar Energy

Passive solar design is designing to provide for the heating of a structure with direct sunlight. Active solar is the use of fans or water pumps to move the accumulated heat to storage. Good solar design begins with *passive* solar collection. Depending on climate and building design criteria, *actively* moving heat to storage can enhance building performance considerably.

The use of sunlight for heating a structure requires understanding a number of factors. These include:

How Hot Is It? British Thermal Units, Calories and Joules

The British thermal unit (Btu) is commonly used as a measure of heat energy in the US. *Heat* is energy that moves from a warmer object or area to a cooler object or area. In the equations used to measure energy transfer, heat is symbolized by the letter "Q." The physics of heat, also known as thermodynamics, is a complex science, and exact measurements vary slightly based on initial temperatures and other variable factors. This variance is not critical to solar design. Basically, a *Btu* is the amount of heat energy required to raise the temperature of one pound of water one degree Fahrenheit. A Btu is also equivalent to approximately 253 calories. A *calorie* is the measure of the amount of energy required to raise the temperature of one gram of water one degree Celsius. Most heating systems measure the number of Btus produced per hour. So a 150,000 Btu furnace has the capacity to produce 150,000 Btus each hour.

Outside of the US, the joule is the common measure of heat energy. A *joule* can be quantified as the measure of the energy required to raise the temperature of one gram of dry air one degree Celsius. A calorie is 4.184 joules. Roughly speaking, a Btu is equivalent to 156 joules.

- Orientation of the structure for maximum light and heating
- Angle of glazing relative to angle of sun above horizon
- Average solar gain at various times of the year
- Heat loss from the building
- Heating and cooling needs of the structure
- Heat storage capacity of the thermal mass materials

Average solar gain is measured in Btus per square foot. This is a direct measurement of the heat generated when sunlight strikes a given surface. Maps and charts of average solar gain for specific locations are found in publications on solar design and are available from most state governments. The term *average* means the measure takes into consideration the area's cloudy days. So it is not a measure of actual heat expected to be gained on a sunny day, but the daily average for the time of year. We used data in "The Pennsylvania Renewable Energy Assessment" (see Resource List) for our calculations. This assessment provided the average solar gain at the summer and winter solstices and the spring and fall equinoxes. (Note: All the figures, formulas and charts mentioned in this chapter are included in the appendix at the end of this book, "Greenhouse Heat Dynamics: Figuring Solar Gain, Solar Storage, and Heat Loss.")

Design for Extreme Weather

When designing a greenhouse or similar solar structure designed to let sunlight in, you need a plan that will allow you to maintain desired temperatures even during extremes of weather. This means planning for insulation, weatherization and possibly a backup heating system that will keep temperatures above freezing (at least) on the coldest day in winter. You also need a ventilation system that protects the plants and glazing system on the hottest day of summer.

Orientation

Besides insulation, orientation to the sun is the simplest way for any building to capture and retain energy. Maximum gain is achieved by facing the long, glazed side of a greenhouse toward true south. If obstructions prevent this, make sure to include the difference in your calculations. Variation to the east or west by up to 20° reduces solar gain

Finding the Right Direction in a Changing World

Solar south varies from magnetic south in most places. The compass alone is not an accurate tool for finding true solar south. This is because magnetic north varies from true solar north. This variation is known as *magnetic declination*. True solar south is determined by locating magnetic south and then adjusting for the compass's deviation.

The magnetic poles are the points where the earth's magnetic field passes through the planet's surface. A compass needle aligns with the lines of magnetic force connecting the northern and southern magnetic poles, therefore it points toward magnetic north and south.

As of 2008, magnetic north is located in Arctic Canada; it moves about 40 kilometers north each year. So, in any given location the difference between the physical North Pole and the magnetic north pole changes slightly each year.

To accurately plot solar south with a compass, one needs to know the current deviation between the North Pole and magnetic north for the site. NOAA's National Geophysical Data Center's website www.ngdc.noaa.gov/geomag/declination.shtml provides maps and tools for calculating the current magnetic declination for any location in the US.

At the time of this writing, in western Pennsylvania the magnetic deviation of south is about 9° west. This means that solar south is 9° west of magnetic south. So, for a greenhouse at this location, you would begin by placing a stake in the ground where the northwest corner will be located. Place a compass on the stake and sight a line leading 9° west of magnetic south. (A good compass will have 360° marked on the dial.) Place the stake along this line at the southwest corner of the building. The structure's south wall is then laid out perpendicular to this north-south baseline.

Other ways to find solar north and south include sighting a base line from the North Star at night or aligning it to the sun's position at solar noon.

by 5 percent. Variation by 30° reduces gain by 10 percent. The numbers seem small, but they can be significant relative to heat gain and plant growth, so any orientation over 30° away from true south is to be avoided. When a building is oriented 45° away from true south, solar gain is reduced by over 25 percent.

Because sustainable design is site specific, there may be locations and situations where such a variation is helpful. For example, if a location consistently has morning fog or cloud cover, orienting for the afternoon sun may provide the best gain. On steep land and in urban areas, the landform, vegetation, or surrounding buildings may restrict some solar gain in the morning or late afternoon. In these cases, the building should be oriented to receive the maximum amount of sunlight available.

When adding a greenhouse to existing buildings, careful consideration of solar restrictions should be made to determine potential performance.

Glazing — What's Your Angle?

The angle of the glazing is relevant to the quantity of sunlight entering a greenhouse and how far it penetrates into a building. Design of the glazing system (glazing material and glazing support components) is critical to the successful functioning of a greenhouse, in terms of both solar gain and plant growth. Maximum gain of light into a building is achieved when the glazing is perpendicular to the sun. But this ideal really only occurs for a few minutes each day, at noon. What you are planning for is optimum solar gain throughout the day.

You need ask a number of questions when planning your glazing angle: What thermal performance is really desired? What is the worst-case scenario of winter snow load? What type of glazing will work the best and be affordable?

There are many options for the designer to consider, depending on production goals. Glazing may be vertical or angled or it may arch completely over the structure. For glass to be angled, it must be tempered safety glass and have an appropriate (strong) support system. The cost of such a system is out of reach for most agricultural enterprises. However, glass is well suited to use on vertical walls. For angled glazing, rigid plastics such as acrylic, polycarbonate, fiberglass-reinforced plastic, and plastic films are often better choices.

Seasonal Sunlight

Solar angle: The path of the sun across the sky changes daily. The higher the latitude, the lower the sun is in the winter and summer sky. Three Sisters Farm is 42° north of the equator, so the angle of the sun above the horizon at noon is roughly 25° on December 22, 49° on March 22 and September 22, and 72° on June 22.

Solar azimuth: The azimuth is the angle of the sun relative to the south along the horizon, as viewed from above. In winter, these angles are smaller than in summer. Roughly speaking, in the US, the winter sun rises in the southeast and sets in the southwest. On the spring and fall equinoxes, the sun rises in the east and sets in the west; and in summer, the sun rises in the northeast and sets in the northwest. More light will enter glazed east and west walls from late spring through early fall; less will enter in winter. Because solar gain through the end walls is negligible in winter months, it is often better to reduce end wall glazing and add insulation for winter production.

Incidental Light and Direct Light

Direct light is sunlight that shines directly into a building. Incidental light is light diffused by the clouds and sky. In Pennsylvania, clouds often cover the sky 50 percent of the time, so knowing how much incidental light is available was an important design consideration. While heat from incidental light is negligible, plants appreciate any light they can get. The choice of how much to glaze a roof is a tradeoff between the need for light (incidental and otherwise) and the need to supplement with heating or ventilation. After nearly two decades of managing the Three Sisters bioshelter, we have concluded that the extra light we would gain from a partially glazed north roof would be worth the extra cost of firewood to heat the building (glazing loses more heat than insulated roof). Our planned renovation incorporates a continuous, glazed vent in the north roof that will provide this extra light. If our bioshelter were located 40 miles south — farther away from the lake effect snow and the cloud belt — we would probably not need the adjustment.

Solar Gain

As already mentioned, a greenhouse should be oriented for maximum exposure to the sun, generally due south.

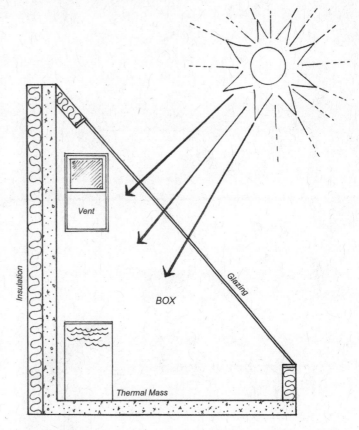

Whether designing a solarium, greenhouse, cold frame or water heater, all include the following components: a frame structure, glazing to allow light to enter, thermal mass to capture and store heat, and ventilation. Insulation helps improve the efficiency of a solar structure.

In summer, when the sun is higher in the sky, our bioshelter is partly shaded by the north roof, keeping building cooler.

Glazing should be angled to maximize light penetration into the building. Each surface the sunlight strikes in the building receives energy as heat. (See appendix for formulas.) To determine a building's solar gain for a given season, multiply the square footage of glazing by Btus per square foot. In northwestern Pennsylvania, we receive an average of 1,200 Btus per square foot in December. Our bioshelter has a south-facing glazed area of 2,000 square feet. So, in December we have an average daily solar gain of 2,400,000 Btus.

Thermal Mass

Thermal mass is material that acts to absorb and release heat. Different materials have different capacities to absorb and store heat (the thermal mass of the most common materials are listed in the appendix). In a sustainable greenhouse, thermal mass is used to store excess heat so it can be used later. Common storage media include water, stone, concrete, soil, wood and metal. Of these, water has the greatest capacity to absorb heat when the surrounding environment is warmer and then release it when it is cooler. Water's heat storage capacity is highest — by far — followed by wood. Soil, rock and concrete are much lower. Knowing the heat storage capacity in a greenhouse will help determine the size of the backup heating system.

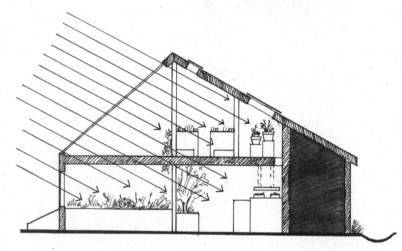

In winter, when the sun is lower in the sky, our bioshelter receives full sunlight throughout the growing areas.

Heat Absorption Capacity

As already mentioned, a Btu is the amount of heat needed to raise a pound of water one degree. You will see water listed in solar design publications as having a "specific heat capacity" of 1.00 Btu per pound per degree Fahrenheit (Btu/lb./degree F). That sets the *relative heat absorption capacity* of a pound of water at 1.00 (relative to the heat capacity of other material).

Compared to a pound of water's heat capacity of 1.00, a pound of wood is rated at 0.33, meaning it takes about three Btus of heat to raise the temperature of wood one degree. Concrete, soil, sand, brick and stone have rates between 0.20–0.23. Steel has a heat storage capacity of 0.12. So, pound for pound, it takes almost ten times as much heat to raise the temperature of steel one degree as it does to raise a pound of water one degree.

Density

The density of a material (measured as pounds per cubic foot) relates to its function as thermal mass. Water weighs 62.4 pounds per cubic foot (8.3 pounds per gallon). Concrete weighs 150 pounds per cubic foot; wood varies from 20 to 30 pounds per cubic foot; steel weighs almost 490 pounds per cubic foot. Therefore, a cubic foot of steel, while probably not cost effective, is nearly as good at storing heat as a cubic foot of water.

Concrete and stone each have about half the storage capacity of water per cubic foot. The other common heat storage media are soil, brick and sand, which have one third the heat storage capacity of water per cubic foot. However, a planter of soil, when properly moistened by daily watering, will have a higher heat storage capacity than dry soil, perhaps being twice as efficient at absorbing and releasing heat.

Heat Storage Capacity: The Simple Explanation

When calculating the performance of thermal mass, the density of the material (pounds per cubic foot) is multiplied by its heat absorption capacity to quantify its role in the solar heating system.

Specific heat capacity x density = heat storage
Btu/lb. per degree F x lb./cu. ft. = Btu/cu. ft. per degree F

Example: Water

Water weighs 62.4 pounds per cubic foot. If there is a one-degree temperature difference between surrounding air and a cubic foot of water, the water will absorb 62.4 Btus of heat. Conversely, if the air drops one degree, the cubic foot of water will release 62.4 Btus of heat.

1 x 1 degree x 62.4 = 62.4 Btus per cubic foot per degree F

Sizing Thermal Mass

The size of the thermal mass relates to its performance. For example, if hot air is actively blown onto a bed of 3- to 5-inch diameter stones, the air circulates around the stones and is easily absorbed. But a single large stone that weighed the same as all the smaller stones put together would have less surface area to absorb and release heat; therefore, it would be slower to heat and cool. Similarly, the size of pots, planters and water containers affects their efficiency at heat absorption. Larger tanks of water, like our 600-gallon tank, provide a lot of thermal stability but are slow to absorb and release heat over a period of days and weeks. Fifty-gallon barrels of water are more reactive; they can heat up and cool down over a period of days. Smaller containers, such as five-gallon buckets warm up quickly but can release absorbed heat before the night is over.

In a greenhouse, a mix of materials — water, concrete and soil — in different configurations provides stability. Large masses keep the average temperature from falling too low; medium and small masses act to moderate daily temperatures. True to permaculture principles, greenhouse design should incorporate the ability of thermal mass to serve multiple functions in a building. Examples of multifunctional mass are concrete foundation walls (insulated on the outside), concrete soil beds, massive masonry stoves, fish tanks, planter beds, potted plants and barrels of water used to support plant tables.

Because the heat storage capacity of the water vapor is high, humid air holds more heat than dry air. So the rocks under the deep beds that help drain them also capture and release the heat because when the warm, moist air passes through the soil, it condenses on the cooler rocks, and releases this heat. It's similar to what happens when warm air in the car fogs the cooler window.

In the Light

Location of the thermal mass in a greenhouse is important. Thermal mass is much more efficient as storage when sunlight shines directly on it. Direct sunlight on steel and plastic barrels heats the barrel, and the heat conducts through to the water. If the barrel is shaded, warm air will still warm the barrel and the water, but not as rapidly. However, while it is preferable to have mass in direct light, mass also needs to be evenly spaced around a building. And there are other factors to consider. For

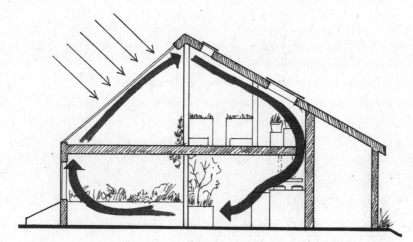

Warm air drawn from the second floor warms the first floor growing area. A thermal loop can form naturally in the well designed building, and can also be enhanced with fans or blowers.

example, we avoided placing our 400-pound barrels on the second floor. In addition to the weight, warm moisture condensing on the cool barrels would pool on the (wooden) floor and eventually cause decay. We limit most of the mass on the second floor to large planters with controlled drainage.

Because heat rises, a building will be much warmer near the ceiling. A thermal loop of circulating air can be created by allowing an air return space along the back wall of a greenhouse. Heated air will rise to the ceiling along the glazing and cooler air will fall down the back of the building. Actively moving air from the higher levels to lower levels of a building can help capture the extra heat and direct it to storage.

Thermal mass also can help *cool* a building in the summer. The mass in a ventilated building will cool at night and absorb heat in the day, stabilizing temperature extremes.

Building Performance

When the sun shines, the mass recharges. In the cool of nighttime, the heat radiates out from the mass. As already mentioned, storage capacity is the temperature increase in material from energy absorbed by the mass when the sunlight hits its surface. *Functional storage capacity* is the expected temperature range in the mass.

Calculations begin with the desired average temperature in mind. We want the soil temperature in the large planters and beds to be at least 50°, so our bioshelter temperatures need to average not less than 50° in the winter. Expected temperature range in the bioshelter is from 50° to 80°, or 30°. In other words, the base building temperature we desire is 50°, and the warmest our mass will get is about 80°.

To calculate storage capacity of a building in Btus:

- List the materials that make up the building's mass.
- Determine the total cubic foot volume of a given material.
- Consult the chart in the appendix for the material's heat storage capacity.
- Convert the volume to pounds per cubic foot using the chart's figure for the material.
- Multiply pounds by specific heat capacity of the material to get Btus per cubic foot per degree capacity.
- Multiply Btus per cubic foot by 30° functional capacity.
- Repeat for each material.
- Add together all calculations for total heat storage capacity.

Determining Building Heat Loss: Conduction, Convection and Infiltration

The next calculations required for determining a building's heating needs is the amount of heat lost by the building at night and on cold, cloudy days. Heat moves at measurable rates that increase as the temperature difference between the inside and the outside increases. In a well-sealed building, energy is lost mainly by conduction through the walls and roof, and convection of heat from the building's exterior surface. This convection loss increases with temperature difference and wind speed. The other main heat loss is from *infiltration*, which is the air exchanged between the building and the outdoors through general use of doors and all those small air leaks around windowsills, frames of doors and vents. (All buildings exchange air with the outside; by sealing leaks, the amount can be reduced but not eliminated, so infiltration should always be part of the calculation.) Total heat loss from the building is conduction loss plus the infiltration loss through air exchange.

Conduction

Heat loss from a building depends largely on the insulation value of materials and the temperature difference between inside and outside.

Anytime it is colder outside than inside, heat will conduct through the walls, roof and floor. Conduction, or heat transfer through a material is measured as a *U value*. The *R value* is the inverse of the *U* value; it is the measure of resistance to heat transfer, or insulation value. The rate of this conduction is determined with formulas provided in the appendix.

To determine and add heat transfer values from materials, calculate:

- Total square footage of glazing (probably R-1)
- Total area of insulated walls (R-19, or better, R-30)
- Total area of insulated roof (R-30 to R-40)

The total of these areas equals conduction loss from the building per degree per hour. To clarify, heat loss is a measure of the hourly loss *and* the heat loss increase as the temperature difference between the inside and outside increases. Therefore, the calculations for building heat loss are measured "per degree per hour."

Assuming the foundation's perimeter is insulated, the un-insulated floor may also be considered as a heat loss down to 50° (at our location), but this is not a significant factor. Ground temperatures vary depending on latitude. When the inside air temperature drops below 50°, the floor becomes a heat *source*. This can be significant; in fact, it is the basis for plant survival in unheated cold frames and polytunnels. The ground temperature also helps cool the building in summer.

Air Exchange Heat Loss

The next step in determining heating needs is to determine the average expected heat loss through normal air exchange (infiltration). Generally, it is assumed that all the air in a heavily insulated and weatherized building will change each hour. Air exchange heat loss is calculated as loss per degree of temperature difference between the inside and outside and per hour of the heating period.

See appendix for the formula used to determine the amount of heat stored in a cubic foot of air.

Adding It Up

Calculating the total heat loss, conduction (heat loss through the building's walls, roof and glazing) plus infiltration (heat loss through air exchange with outside) gives the total Btus of energy lost per degree of difference between the inside and the outside every hour. This total is used to figure out a building's heating needs per hour and then for a 14-hour winter night, the assumed heating period.

For example if the temperature is -10°, and you want a 50° inside temperature, you multiply the heat loss per degree of difference by 60°. This gives the heating requirement in Btus per hour.

Finally, compare solar gain, storage capacity and heat requirements to estimate performance. This can (and should) be repeated for other seasons for year-round performance estimates.

Determine Energy Needs

Once the heat gain and heat loss for a building is determined, you need to estimate how much supplemental heating and cooling you will need.

Heating: Compare the solar heat gain and the building's nighttime heat loss for the coldest expected day. For example, when the outside temperature is −10° and you want to maintain 50° inside, you need to balance heat loss from the building. First you determine the Btus lost per hour per degree from the building. Multiply this by the 14-hour winter nighttime heating period, then subtract the average daily heat gain in mid-winter. The difference is your heating needs for the worst-case night.

Example:

- Btus lost per hour: 100,000 x 14 hours = 1,400,000 Btus
- Daily solar heat gain: 800,000 Btus
- 1,400,000 minus 800,000 = 600,000 Btus of heat needed
- Dividing this number by the 14 hours you'll need the heat shows that you will need approximately 43,000 Btus per hour of supplemental heat.

A further consideration is a worst-case scenario of successive days of *no* gain due to clouds and snow. Obviously, you would need to increase supplemental heating capacity accordingly.

Although some greenhouse operations require hot water, we do not currently use hot water in the bioshelter. If we do install a hot water heater, we will use the sun to preheat the water. (We do, however, have a nifty solar shower for summer use. It consists of 300 feet of coiled plastic pipe set on top of a shower frame. Water from our well is pumped through the pipe and the sun warms the water in the pipe. With its 3-gallon-per-minute nozzle, it provides a long, hot shower in the afternoon on most summer days. We plan to upgrade this with a small holding tank to allow for evening showers as well.)

Cooling: The calculation of cooling needs is a similar process. Determine heat gain through glazing in the middle of June, when the days are the longest. Then determine how much volume of air needs to be moved out to keep temperatures at 90° or lower. Again, use the equation for heat storage capacity of a cubic foot of air multiplied by the cubic feet of building volume.

Deep beds and large planters will absorb heat during the day and release it at night. In the bioshelter, we generally leave the vents open 24 hours a day in the summer to aid the cooling process. On very hot summer days, we spray water on concrete planters and gravel walkways. Evaporating water absorbs heat from the mass and releases it through the ventilation system. Our bioshelter's passive cooling system draws outside air in through doors and windows on the first floor and sends it out by convection through second floor windows and vents. A large greenhouse ventilation fan turns on when temperatures exceed 95° on the second floor; this happens for about four hours each sunny day from June through early September. We plan to upgrade our passive roof ventilation system in the near future. The installation of two 30-foot by 4-foot continuous roof vents should eliminate the need for the ventilation fan.

Photovoltaic Applications

Photovoltaic (PV) power generation offers a way to reduce demand on the electric power grid and therefore reduce the air pollution and CO_2 production associated with conventional electric generation. Using PV power on the farm can help reduce peak demand on the power grid, especially in summer, when electricity is most in demand.

PV-generated power on the farm is best suited when sized for and applied to specific applications rather than providing general power for the whole farm. Seasonal irrigation and livestock watering are good examples of useful applications of solar power on the farm.

Basic Electrical Calculation

An important first step in controlling energy consumption is determining how much your equipment actually uses. The basic unit of measure for electricity is watts. Energy consumption is measured in kilowatt-hours (one kilowatt is equal to 1,000 watts). A kilowatt-hour (kWh) equals one kilowatt of electricity flowing for one hour. Ten 100-watt light bulbs, on for one hour, consume one kWh of electricity.

Grid tied solar array at Sally's Cider Press offsets electricity used in cider sterilization.

DARRELL FREY

Bioshelter Viewed from pond.

LEO GLENN

At Three Sisters Farm we use grid electricity for lighting, ventilation, water pumping and refrigeration. We have explored options for reducing purchased energy use in these areas. Planned improvements include installing larger passive greenhouse vents, building a super-insulated cooler, installing wind or solar electricity-generating systems, and using more low-energy lighting. We began with the installation of a photovoltaic irrigation system in 2004 and replacing light bulbs with compact florescent bulbs. The next phase in the application of PV to our farm will be to power the ventilation and circulation fans.

PV Irrigation

The original irrigation system at Three Sisters Farm consisted of a grid-powered ¾-horsepower pump, which sent pond water to sprinklers that watered newly planted crops. We also used a submersible well pump to water crops closer to harvest and for crops inside the bioshelter.

During the first 14 years of farm operation at Three Sisters Farm, using pond water to irrigate newly planted crops used an estimated 1,449 kWh per year. This estimate is based on an average of 35 hours per week of irrigation for 30 weeks of growing season, calculated as follows:

12 amp x 115 volt pump = 1,380 watts per hour x 35 hours irrigation per week = 48.3 kWh per week x 30 weeks = 1,449 kWh.

One kilowatt hour represents approximately 3,412 Btus of heat per hour. It takes an average of three Btus of fuel to make one Btu of electricity. Therefore, the original irrigation pump at Three Sisters Farm required approximately 14,832,000 Btus of fuel each year.

A kWh represents an average of 300 grams of coal. So, saving 1,449 kWh of electricity is comparable to saving of 958 pounds of coal.

In 2004, we replaced the ¾-horsepower pump with a PV system. System components include a 125-watt PV panel, a panel mounting system, a linear current booster to regulate electricity flow, a medium-flow 12-volt pump, two 1,000-gallon water storage tanks, and miscellaneous fittings and connectors for water lines.

Our 125-watt PV system powers a small, 3-gallon-per-minute pump. This system does not require batteries. While the sun shines, the system fills the irrigation tanks. Water is allowed to overflow the tank and is directed through a series of small pools and channels back to the pond or to crops as needed. Stored water is used mornings and evenings to water crops as needed. The system is sized to match irrigation use at the farm. The pump irrigates crops directly with a sprinkler as needed. This irrigation system is secure from power disruptions and substantially reduced farm energy usage from the power grid.

Other Energy Systems

There are many options to explore for increased energy efficiency. Innovative ways to conserve energy, produce energy, and eliminate energy use are awaiting our ingenuity. The full description and debate of the merits of various energy sources is beyond the scope of this book. But several topics are relevant to the immediate discussion of energy use on the farm.

Burning corn for fuel does not seem reasonable to me. In some bioregions, biofuels (from plant oils and ethanol) could and should be an important part of the sustainable future. But they can only be part of the sustainable energy equation when they can be made profitably and without the use of

Solar irrigation system at Three Sisters Farm.

LEO GLENN

petroleum. Large-scale production of biofuels to feed a wasteful lifestyle will only serve to raise food prices and degrade the larger environment. However, biofuel production *on a local scale* can cycle byproducts back to the land and be sensibly used to supplement livestock feed.

Methane: Methane digesters can be an important on-farm tool for both manure management and energy production. Methane gas is produced by anaerobic bacteria when water and organic materials are fermented in airtight tanks. The gas can then be used for fuel. Digesters are mainly suited to farms with livestock that are confined daily, such as dairy farms. They require engineered design, careful construction and a high amount of maintenance. Systems need to be scaled to the amount of manure available.

A wise farmer once told me that many dairy farm methane-production systems are based on the use of high-pressure water to wash manure into storage pits. His argument was that good compost materials are being mixed with clean water to create a problem: liquid manure lagoons. Because lagoon storage pits are a potential source of ground water pollution, and liquid manure application can result in excess nitrogen entering streams, state governments are pushing methane energy production as a way to handle what is essentially mismanagement of farm byproducts. However, with the right resources and expertise, heat and electricity from farm-produced methane can be sustainable. Some cheese factories, for example, produce methane from whey to generate heat and electricity that they use to power their operations. The choice between composting and methane production is a matter of scale, need, equipment and management capability.

Geothermal: Geothermal systems are referred to as *geoexchange* by the DOE to differentiate them from energy harvested from geysers and hot springs. A geothermal system uses electricity to circulate air or liquid that has been warmed by the earth through a heat pump to concentrate the heat energy for space heating. Geothermal systems require a specific volume of piping in the ground or amount of water pumped to heat a given space. A geothermal energy system may not be cost effective to heat a greenhouse that has high heat loss. However, geothermal systems are a cost effective and efficient way to heat well-insulated buildings and should be considered when viable. Geothermal can also provide cooling in summer by taking heat from a building and putting it back into the ground. For both heating and cooling, geothermal systems are most efficient when used in buildings designed to conserve energy use.

Wind and Water: Wind power and micro-hydroelectric systems are site specific. Small-scale hydroelectric production can be a good energy producer, but only if you have the right conditions and configuration of running water. When sufficient wind is available at the right times, wind-generated electricity and water pumping can be cost effective. But wind is capricious and may not be there when you want it. Battery systems or grid-tied systems are generally needed to make electricity generated from wind reliable.

Farm-scale energy systems can be tied together in innovative ways. For example, photovoltaic systems can run the agitators needed to promote methane digestion; they can run heat pumps for geothermal energy; or they can power components of other systems. Methane can be used to heat water already preheated in solar water tanks. Compost heat can warm soil for season extension. In Chinese-style pressure methane digesters, effluent exits through a pond built over the digester chamber. Duckweeds, water hyacinths, watercress and algae thrive in the nutrient-rich water and can be fed back into the digester or used as livestock feed or fertilizer. Such a system was designed and built in the early 1980s by Robert Hamburg of Orma, West Virginia (the Omega-Alpha Recycling System). He combined a digester, greenhouse and ponds into a symbiotically integrated organic recycling system providing methane gas, plants, and fertilizer. His methane tank was set on a south-facing slope. A greenhouse attached to the south side of the digester tank stabilized the digester temperatures to help assure year-round production. In turn, the digester tank provided thermal mass to stabilize the greenhouse environment. (For more information, visit omega-alpharecycling.com/index.html.)

Management of resources to make better use of biological and other natural energy sources is our challenge for the 21st century. The need to reduce fossil fuel use should be a concern of all inhabitants of the earth. In this regard, it is imperative that citizens of the US lead the way in the application of sustainable technologies.

Our recognition of these issues and a call to action was clear when we first visualized Three Sisters Farm over 20 years ago. As we have developed the farm — and as we plan future development — the conservation of energy and the use of sustainable energy technologies have been primary design goals.

CHAPTER 6

Beyond Integrated Pest Control: Insects, Disease, Weeds and Other Pests

The organiculturalist must realize that in him is placed a sacred trust, the task of producing food that will impart health to the people who consume it. As a patriotic duty he assumes an obligation to preserve the fertility of the soil, a precious heritage he must pass on, undefiled and even enriched, to subsequent generations.

— J.I. RODALE "THE ORGANICULTURIST'S CREED," *ORGANIC GARDENING MAGAZINE*, JANUARY, 1948

MANAGING A PERMACULTURE FARM begins with organic farming practices and expands on them by integrating habitat into the farm and garden landscape. Ecological design is an ongoing process. This chapter explores disease and pest control from an ecological perspective. I begin with a discussion of soil management as the basis of healthy ecosystems. This is followed by discussions of plant disease, beneficial insect ecology and control options for a wide range of pests.

Organic Methods

Whether or not a farm is *certified* organic, using recognized organic standards as a guide helps a farm's management to produce safe and healthful products. Organic pest control practices and material used in farm activities are defined by the USDA's National Organic Program (NOP). The national organic standards fill at least 400 pages. The NOP was established by The Organic Foods Production Act of 1990, and was a part of the 1990 US Farm Bill. The act authorized the secretary of agriculture to appoint the National Organic Standards Board (NOSB).

The 15-member board is made up of farmers and growers from four regions of the US; its members also include food processors, retailers, consumer advocates, scientists, environmentalists and organic certifiers. The board's mission is to assist the USDA in developing standards for organic production and to advise the USDA on implementing the national organic program.

Farm managers and workers should be familiar with the NOP standards that apply to farm crops and livestock. Because new practices and materials continue to be developed and organic markets are expanding, the standards change over time. The farmer is responsible for keeping up with current standards. Membership in a certification organization is the best way for a farmer to stay updated.

Materials for crop production, pest control, soil improvement, and animal feed are either *allowed, restricted,* or *prohibited* under organic standards. Associated standards also govern organic food processing. The Organic Materials Review Institute (OMRI) publishes the *OMRI Generic Materials List,* which lists 900 materials that are allowed, restricted, or prohibited.

Restricted materials may be used with certain precautions and prohibitions. Rotenone is an example of a restricted product. Once a staple organic insecticide because it is derived from plants (a botanical insecticide) and breaks down in a few days, rotenone is now restricted because it is toxic to earthworms, fish, insects, and mammals, including people. We do not use rotenone at Three Sisters Farm.

The shortcoming of the NOP is that the intent is to assure the public that organic products and their production meet *minimum* standards. Such minimum standards are indeed necessary to assure the integrity of the term *organic,* but they do not in themselves promote excellence. Organic farming and gardening, which began as a way to promote the health and vitality of the food system and the environment, is in danger of becoming simply a minimal threshold to cross into a more lucrative market.

Soil Management

The development and continuous care of healthy soil is a primary concern of the organic farmer. This is done through management of manure and composted farm waste and the use of green manure crops and mineral fertilizers to create a healthy nutrient cycle that will build the soil. Crop rotation is a primary strategy to maintain fertility and reduce pests and disease. Tillage methods that preserve soil structure and nutrients are encouraged. Author Ray Wolf described biodiversity as "a natural balance in the ecosystem." Biodiversity is a vital component of the farm for natural pest and disease control.

Mineral fertility is a primary concern for the market garden. The mineral nutrients phosphorus, potassium, nitrogen, carbon, magnesium, silica and oxygen are the main components of plants by volume and weight. Trace minerals, including calcium, iron, zinc, copper, manganese, chlorine, molybdenum, boron and sulfur are used by plants as biocatalysts. They are required in minute quantities for many metabolic processes including photosynthesis, growth, flower and seed formation, and disease resistance. Trace mineral deficiency can be identified through soil tests; it can also be observed directly in defects in plant growth, such as misshapen or discolored leaves. The chemistry of trace minerals is complex. Many minerals can be toxic when overabundant, and too much of some can limit the availability of others. For this reason, organic methods rely on unrefined rock minerals and biological sources of trace minerals. The nutrients in unrefined mineral products are made available as needed by the natural reactions of plants, fungi, and soil chemistry.

Biological sources of plant nutrients include manures, composted organic matter, accumulator plants, fish emulsion and seaweed. Again, the natural action of soil and plant chemistry releases minerals held in the soil and organic matter. Thus the standard organic dictum: Feed the soil, not the plants.

Whole-systems Planning

Permaculture builds on organic practices by attempting to design the integration of farm components into a whole system. Farm watershed management, use of accumulator plants, and integration of animals into the system promote on-farm nutrient cycles. This view of the permaculture farm as a whole — an integrated system — is strongly influenced by biodynamic agriculture. Rudolf Steiner's writings (and the work of others who contributed to the development of biodynamic methods) stress the view of the farm as an organism. The health and vitality of the farm and its products are the result of the total health of the farm system. The ideal biodynamic farm is self-contained, with minimal off-farm inputs.

Small-scale farms, especially in suburban and urban areas, are more dependent on off-farm inputs. Organic materials for compost, animal feed, hay and mulch may all need to be imported from elsewhere. Organic standards and practices help guide the use and management of off-farm resources.

Promoting Health

Health is something we can easily perceive. In many ways, our innate sense of natural beauty gives us the ability to recognize healthy landscapes. When a plant is healthy, we can see it: good color; strong, firm stems; a balance in proportion of leaves and flowers; and an overall appearance of vitality. A walk through the garden early on a summer evening, after a late afternoon rainstorm, when the rainbow has faded and the air is still, will reveal plants at their peak. The air is fresh and aromatic, the insects are calm, leaves are spread wide and are full of moisture. Flowers fluoresce as the sun sets. The smell of moist soil attracts robins and other birds hunting bugs in the garden paths. Sitting on a bench in the garden to rest a while, the evident health of the landscape has a therapeutic effect on the gardener as well.

Conversely, we can also recognize what is not healthy. Plants weakened and stunted by nutrient deficiency, competition with other plants, disease, insects, or drought are readily apparent.

The same general combination of factors that promote health in people and animals also promotes plant health. Good nutrition, clean water, the right amount of stress at the right time and a proper environment are all required for strong and productive plant growth. The new axiom in permaculture is that nutrient-dense soil produces nutrient-dense food. Plants need access to a steady supply of moisture, nutrients, proper pH, and, often, an association with soil bacteria and fungi. They also need a certain amount of air circulation, both for respiration and for building strong stems and root systems. The need for strengthening by stress is analogous to the need for adolescents to have sufficient exercise to build a skeleton strong enough to carry them through their next 60 to 80 years.

Unfortunately for gardeners, many other creatures also recognize and seek out a healthful food source. While insect pests are attracted to weak and stressed plants, that is not the whole story. No matter how healthy our soil or how well we manage plant growth to promote productivity and vigor, some living being will want to share our harvest. A tenet of ecology is that anytime a resource is available, somebody will use it. Spilled grain attracts mice and rats. Unprotected gardens attract woodchucks and rabbits. Plants attract aphids, slugs, flea beetles, thrips, whiteflies, and so on. So gardeners must defend their plants.

Ridding a farm of any living being presents a problem for a permaculture farmer who is also concerned with the health of the entire farm landscape. But using design and management to exploit the natural workings of an ecosystem can provide solutions. Control of pests begins with an understanding of a pest's habitat, life cycle, and the predators and diseases that affect them — and, in effect, using that understanding against them. For example, a particular pest may be minimized by reducing their available habitat or promoting their natural predators.

Masanobu Fukuoka, a Japanese pioneer in natural farming methods, has said, in his seminal book *One Straw Revolution,* that a different pest and predator relationship manifests each year. We have found this to be true, so pest control is an unfolding process of learning and adjusting. For example, spiders are abundant some years, and frogs and toads another. As weather conditions and field conditions vary, so do the dynamics of the farm ecosystem. As ecological farmers, we promote biodiversity and attempt to develop an awareness of the ongoing dynamics on the farm.

Spring peeper.

DARRELL FREY

Environmental Stress

Crop plants need to be suited to the soil and climate to be productive and healthy, and the pH of the soil needs to be suitable for the crop. Without these basic needs being met, plants are much less likely to flourish when they are stressed.

Weather is, of course, an important and variable environmental factor. A cool, wet fall can send a tomato or potato blight through a bioregion, causing a quick decline in yields in September. Hot, dry summers can cause lettuce to bolt faster than usual. Seed germination is greatly dependent on soil temperatures. Many seeds planted too early, in cool soils, will rot away rather than grow. (This is the main reason that seeds are sold with fungicidal treatment.) But cooler fall soils *are required* for the germination of self-seeded chickweed, chervil and mache.

After the shelter of a greenhouse and grow lights, full sun and wind can quickly damage plants. *Hardening off* is the process of gradually exposing seedlings to the outdoor environment before placing them into the soil. Protected seedling frames provide a transition for seedlings by protecting them from the wind and moderating the sunlight they receive. Use of row covers, polytunnels, cold frames and greenhouses for crop production mitigate some of the environmental stresses to crops by creating sheltered microclimates.

Water — either too much or too little — can damage or kill plants. New transplants and germinating seedlings are especially vulnerable to water stress. It is important to keep soil moist, but drained, until plants have established an adequate root system.

In a time of changing climate and variable weather, a wise farmer will grow multiple crops. A great year for cabbages and greens may be marginal for tomatoes and peppers. The next year, the situation may reverse. Growing many different crops for a CSA or a produce-subscription enterprise helps reduce the effects of extreme weather.

The Good Earth

Understanding your soil type and quality is the first step to promoting plant health. Just observing which plants are growing in an area can indicate whether the soil is acid or alkaline, wet or well-drained, fertile or deficient in nutrients. These observations can help you better allocate your soil testing budget by identifying the locations that require testing.

When the pH is low, plants that prefer acid soils such as docks and sorrels will grow. Wet and poorly drained soils will harbor milkweeds, Joe Pye weed, ironweed, smartweed, goldenrods and asters. We rely on field guides and the biodynamic classic *Weeds and What They Tell* by Ehrenfried Pfeiffer to help us understand our annual observations.

Do not be hasty in assuming that the presence of a particular plant always indicates a specific condition. We have good stands of sheep sorrel and milkweed in our well-tended gardens. That doesn't mean our soil is too acidic. But if widespread, or several indicator plants are present, you can make general deductions. Spring walks around the farm can reveal changing conditions.

Our woodland had a thin soil covered with mosses, ground pine, sheep sorrel and young wild blueberries. This indicated to us that years of heavy farming and grazing had eroded the soil and allowed calcium to wash away, leaving a poor, acid soil behind. Rich, balanced soil is also easy to read. Walking around our farm in the spring when green manure is growing fast, it is obvious which of our gardens have had the most improvement. The thickest growth is in the gardens we have used the most: the Pond Garden, and the South and East gardens (Zones 1 and 2). The Zone 3 gardens have had less compost over the past 20 years — and it is apparent to the observant gardener.

Gray tree frog at Three Sisters Farm.

Ecological Pest Control

Ecological pest control involves understanding the life cycle of the pest or disease and finding key points in the organism's life cycle to disrupt the spread of the problem. In a natural system, what we call pests are part of the natural balance. When they affect our ability to produce a marketable crop, we seek controls. Pest populations are controlled by predators, parasites and diseases. Finding the right natural control, and the best time to use it, helps keep pest problems minimal. Some crops may not be suitable in a given season or even in a given area due to pest pressure. On our farm, the prevalence of cucumber wilt (carried by cucumber beetles) makes it difficult for us to grow marketable cucumbers and squashes. Resistant varieties are our best option.

Not all pest damage is bad. A few nibbles can induce a plant's natural defenses and result in an increase of beneficial phytochemicals in the crop. The key question is whether a pest or disease is causing economic damage by reducing yields or marketability of the crop.

Many pests and diseases certainly do kill or stunt crops, but pests and disease are often a problem more for humans than the crop. Flea beetles can destroy young eggplant seedlings but only do cosmetic damage to arugula. Minor insect feeding on leaves may actually stimulate plants to grow stronger, but it is difficult to sell arugula with holes in the leaves. The scale of the problem also determines the response. A few flea beetles on potatoes may not be a problem, but in a dry year when leaf growth is slower, flea beetles can stunt early growth and reduce yields.

Control Methods

Using Insect Disease Against Them

Insect diseases are becoming important tools in pest control. *Bacillus thuringiensis* (BT) is used to control many types of caterpillars. (BT should be used with care to avoid non-target species.) Milky spore disease kills the larvae of Japanese beetles in the ground. *Beauveria bassiana* is one of a number of fungal diseases that kill insects. (It is discussed in detail later in this chapter, in "Insects in the Bioshelter.")

Botanical insecticides are often used — with restrictions — in organic crop production. Pyrethrum, rotenone, sabadilla, and neem tree oil are some examples. (These products all require precautions to protect the farmer who uses them.) They are of biological origin, so they break down in sunlight and in the soil. But they may persist longer in the greenhouse. We know organic farmers who rely on botanical insecticides, but we do not use them at Three Sisters Farm because we feel they affect beneficial insects.

Pheromone traps are useful for control of some insects. These traps emit subtle chemicals to attract insects. Insects are lead to believe they will find a mate, only to be trapped (sort of a "sting operation"). Care must be taken not to *attract* pests. They are best located away from the plants you want to protect. While Japanese beetle traps are very effective, we have found they work best for us when the neighbor puts them in their yard.

Pheromone traps are also used for monitoring and control of orchard pests. Monitoring the numbers of trapped pests helps determine when more controls are needed.

We have found that the dynamics of insect pests vary from year to year and farm to farm. We have never had trouble with bean beetles or Colorado potato beetles at the farm, but in a garden at our home, just five miles away, they were both present in high numbers.

Songbirds as Insect Control

The role of songbirds in the farm ecosystem is of vital importance in pest control. Each species of songbird has a role to play. The goldfinches eat thistle seeds, the Baltimore orioles patrol the orchard for caterpillars and the indigo buntings gather grasshoppers in the meadow. When they are raising their young in the spring, birds will gather an amazing amount of insects as food.

Certainly, birds also eat our fruit. Blueberries, especially, need protection from birds. The mulberries also attract a lot of birds competing with us for fruit. But on balance, we cannot do without the service songbirds provide. The aesthetic value of birds in the landscape alone is incentive enough for us to encourage them to visit our gardens and fields.

The book *American Wildlife and Plants: A Guide to Wildlife Food Habits* provides a wealth of data on the foods each type of bird and mammal eats in the US. The book is an important tool when designing a bird-friendly landscape.

Again, diversity is key. A varied landscape of meadows, shrubs, trees, hedgerows, gardens and buildings can provide nest sites and forage for numerous species. Adding a pond and lots of birdhouses further increases their presence.

Insects on the Farm
Bees and Wasps

When we observe the many varieties of insects visiting the flowers on our farm, we are astounded at their diversity. The order Hymenoptera, which includes wasps and bees as well as ants, is especially well represented on the farm. Bees and wasps are important predators in the insect world. They come in countless varieties, ranging from nearly invisible parasitic wasps to big, furry bees. Bumblebees and their furry kin may be all black, black with white stripes, black with yellow stripes,

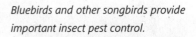

Bluebirds and other songbirds provide important insect pest control.

LEO GLENN

American Goldfinch nest at Three Sisters Farm.

Bees and Wasps at Three Sisters Farm

- Honeybee, *Apis mellifera*
- Orchard bees, *Osmia* spp.
- Eastern carpenter bee, *Xylocopa virginica*
- American bumblebee, *Bombus pennsylvanicus*
- Paper wasps, *Polistes* spp.
- Potter wasps, *Eumenes fraternus*
- Eastern yellow jacket, *Vespula maculifrons*
- Braconid wasp, *Apanteles* spp.
- Green metallic bees, *Augochlora* spp.
- Ground dwelling hornets
- Furry bees/tiny wasps (undetermined species)
- Other unidentified wasps and bees

or yellow with bands of black; some are tiny and some are quite large. Furry bees come in many shades, as do metallic bees. Wasps come in even greater variety. They can be shiny blue-black, brown and yellow, have narrow waists, or be full-bodied. Some have hair-like ovipositors longer than their bodies. Miniature wasps may be shades of black or brown, with red, white or yellow patterns and markings. Paper wasps are the most abundant. They can be found nesting in the polytunnel framing and practically any spot sheltered from the rain. Mud daubers are also abundant, though less so than the paper wasps.

Where the vast majority of these reside in the farm landscape is an ongoing mystery. Ground-dwelling hornets and bees are usually discovered by painful accident, while working in garden beds or shoveling aged compost. Other bees and wasps must find lodging in undiscovered hideaways in trees, shrubs, meadows, and the odd crevasse in a building. When we do identify a beneficial insect's residence, we try to preserve it. Since the fine spring day I observed tiny metallic bees emerging from a nest in the pith of the previous fall's tansy stem, we are more careful when pruning dead plant stalks. We leave a foot or two of stalk in a tansy patch, rather than trimming to the ground. Now, instead of using pithy plant stems to aerate the bottom of a compost pile, we often toss them along the edge of the tree line to provide the bees with more places to live.

Practically all wasps and bees are beneficial. In addition to pollinating plants, most wasps and hornets are predatory. Depending on the species, they may gather spiders, caterpillars, crickets and other insects. The wasps paralyze their prey, attach one of their own eggs to it and place it in a cell of a nest. The larval wasp consumes the live insect, starting with the non-vital parts.

Chalcid, braconid and ichneumon wasps, and the aphelinid whitefly predator wasp, *Encarsia formosa,* are parasitic wasps that deposit their eggs inside the bodies of other insects.

The 2,000 species of chalcid wasp in North America include the parasite of the tomato hornworm. Chalcids are among the smallest known insects, as small as .008 inches. They may lay their eggs in the *eggs* of other insects. Some species are hyperparasites, parasitizing other parasitic insects.

Braconid species number around 1,700 in North America. They lay eggs inside many types of insects and caterpillars, including aphids.

Braconids are commonly found anywhere aphids are found. They are especially active in the bioshelter. They came in on their own the first year we began to garden indoors. They have been present ever since, as have their host, the aphid. Infected aphids are easily recognized; they become mummified and are a shiny bronze color.

Ichneumons are the most abundant species in North America, with over 3,000 species. Most are internal parasites of other insects, as evidenced by their often-prominent ovipositors. They bore into wood and use the ovipositor to deposit their eggs in wood-eating larvae.

Some wasps' larvae feed mostly on plants, including the over 600 species of gall wasp, *Cynipid* spp. Gall wasps secrete hormones into a plant to induce gall formation. The developing larvae then live inside the gall, eating the gall walls. Gall wasps are primarily parasites of oak trees, but some species form their galls on roses, willow, daisies and other plants.

Other Beneficial Insects

A syrphid fly larva (family Syrphidae) will eat up to 400 aphids between hatching and pupating about ten days later. Most of the 900 species in North America resemble bees and wasps. The larvae resemble slender green slugs, with lovely yellow markings on their back; they can be seen searching for their prey on the underside of leaves. The adults hover and dart quickly, seeking nectar to fuel their flights.

Ladybugs, members of the Coccinellidae family, come in variations of red, yellow and orange, many with black spots. Both adults and larvae are predatory, consuming aphids, mites and other soft-bodied insects. A common species, the convergent ladybug, *Hippodamia convergens,* will live for several months and lay up to five hundred eggs. These orange eggs are usually in clusters of one to two dozen. Upon hatching, the larvae begin to hunt aphids, consuming many dozen before pupating. Adults winter over in leaves, under tree bark and in crevasses of building. Native ladybugs, such as the nine-spotted ladybug, have been heavily endangered by competition with non-native species.

Fireflies, including *Photinus pyralis, Photinus scintillans* and *Photuris pennsylvanica,* certainly enchant a summer evening with their aerial dance and flashing display. While the adults do not appear to eat anything, their larvae spend late summer through fall preying on slugs, snails and insects. (Firefly larvae, sometimes known as glow worms, are also luminous.) After resting through the winter in leaf litter and underground, they again feed on their prey for a few weeks before transforming into adults.

Praying mantis, *Mantis religiosa,* is a native of Europe introduced to North America early in the 20th century. After hatching from their brown, foam egg case in late spring, these voracious predators consume their kin. The fittest then proceed to patrol the landscape for prey, which includes caterpillars, moths, butterflies, bees and flies. Sitting perfectly still, praying mantis wait near leaf and flower, capturing and devouring any insect that comes within reach of their lightning-fast clutches.

Green lacewing, *Chrysopa* spp., larvae are predators of aphids, mites, caterpillars, and other soft insects and insect eggs. North America has dozens of species of these lovely lacy green insects. We have purchased green lacewing larvae for release in the bioshelter twice. Green lacewings, active in late spring and summer, lay their eggs on a silken stalk, which the books will tell you is done to prevent cannibalism as they hatch. But in the bioshelter, they have always disappeared, presumably having consumed one another.

Brown lacewing larvae, *Hemerobius* spp., also prey on aphids, scale, mites and other soft insects. While plainer than their green cousins, the dozen or so species of brown lacewings are important predators in the garden and the orchard. Both green and brown lacewings are attracted to bright lights at night. Because of this, we try to limit our use of outdoor lighting to avoid distracting all insects from their natural routines.

Minute pirate bugs, *Orius* spp., are an often-overlooked predator, because of their size, just 1/8 inch long. They have an easily recognized brown and tan diamond-patterned body. Nymphs and adults feed on a wide range of prey, including caterpillars, thrips,

Green lacewing, a predator of aphids, thrips and whiteflies, on bronze fennel.

DARRELL FREY

mites, aphids, and their eggs. Like a pirate, they stab their prey with a sword-like beak, after which they suck out the prey's juices. Minute pirate bugs are most abundant in a diverse garden landscape. They are attracted to legumes, corn and pollen-producing composite flowers. From these they move about, following their prey.

Dragonflies and damselflies, both of the order Odonata, are busy predators throughout their lives. More than 1,000 species are found in the Upper Midwest of North America alone. All spend their nymph stage as aquatic predators, devouring mosquito and other larvae in ponds and other standing water. Adults of damselflies and dragonflies range far from the water in search of prey. Dragonflies catch flies and mosquitoes in flight. Damselflies are known to eat almost any insect they find.

Thrips.

Predatory thrips, *Scolothrips* spp., are usually abundant in the summer garden, but you have to know where to look. They are one of the reasons pest thrips and spider mites are less of a problem outdoors than indoors. Predatory thrips resemble the pest thrip but are darker and have jagged or hairy sides, as opposed to the smooth-sided pest thrip. They consume many pests and their eggs, including spider mites, thrips, aphids and whitefly. Like other thrips, they are very small, about 1/20th of an inch. We find them on daisies, where they are drawn to the pollen as an alternate food source. It is likely that any composite flower will host them.

Aphids.

Aphid midges, *Aphidoletes aphidimyza,* are an easily overlooked beneficial insect. Adults look like small mosquitoes. They eat the honeydew secreted by aphids, whiteflies and scale. Eggs are laid on plants near aphid colonies. Aphid midge larvae, small, red and maggot-like, are voracious predators of aphids. After consuming dozens of aphids in about two weeks, aphid midge larvae pupate in the soil. They have up to six generations a year outside. Inside the bioshelter, they begin to appear in April and continue to patrol for aphids until late October.

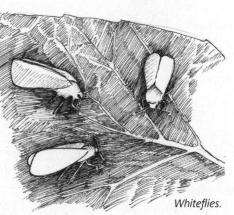

Other beneficial insects are found on our farm in smaller populations. Soldier beetles pierce and consume prey, much like a minute pirate bug. Assassin bugs, robber flies, tachnid flies and predatory mites all find a niche in the ecologically managed garden.

Whiteflies.

Thrips, aphids and whiteflies are major greenhouse pests.

Spiders on the Farm

The diversity of spiders on a permaculture farm can rival the diversity of insects. Spiders are an important part of the farm's pest control. Those we have identified are described below, but dozens of others spin webs and crawl about everywhere on the farm. All are predators that eat some type of insect. Luckily, we do not have the black widow or (we hope) the brown recluse spider in our area. Generally, when a spider *is* forced to bite a human, the result is not much worse than a mosquito bite. Hundreds of American house spiders (*Achaearanea tepidariorum*) live inside our bioshelter. These diminutive (¼ inch) spiders string their irregular strands around the window frames, among the grow lights, and in odd corners. Small colonies of house spiders weave erratic networks of silk, which become festooned with winged aphids, fungus gnats, flies, cabbage butterflies and other insects drawn to the window's light. Because cobwebs are unsightly, we clear out older webs every now and then, especially before tours. We try to allow the spiders time to retreat to the corner, or drop to the sill. They soon return.

Crab spiders include around 200 species in North America. Rather than weave a web, they sit on a flower and wait for their prey to land. White crab spiders will sit on a white rose, yellow on a goldenrod flower, and pink on a red rose. Some species are reported to adapt to the color of the host flower. The goldenrod spider is white with lovely pink lines; it will turn yellow on a yellow flower. Other types of crab spiders are shades of green or brown, depending on their chosen habitat.

Wolf spiders (family Lycosidae) are often seen prowling for prey on the ground while carrying their egg sac. The more than 200 species are fast-moving nighttime predators of ground-dwelling insects. Like crab spiders, they do not weave webs, preferring to live in mulch, leaf litter, small tunnels, under rocks or in buildings.

Jumping spiders are common in the bioshelter and around the farm. The two most common are the pretty *Metaphidippus* spider, which is furry black with green spots on its head, and the daring jumping spider (*Phidippus audax*), which smaller and gray and black. Funnel spiders were an endless source of summer entertainment to me and my childhood friends, much to the dismay of the flies we fed into the wide funnel webs. Similar in appearance to wolf spiders, their ground-hugging webs are built in grass and weeds and alongside rocks, planters and other

Garden spiders

- Black-and-yellow argiope, *Argiope aurantia*
- Shamrock spider, *Araneus trifolium*
- Goldenrod spider, *Misumena vatia*
- Crab spider, *Misumenoides formosipe*
- Marbled orb weaver, *Araneus marmoreus*
- Jumping spider, *Metaphidippus* spp.
- Wolf spider family, Lycosidae

Bioshelter spiders

- Jumping spider, *Metaphidippus* spp.
- Wolf spider family, Lycosidae
- American house spider, *Achaearanea tepidariorum*

shelter. They are quick to rush out and grasp any insect that falls into their web.

More conspicuous web builders on the farm include the shamrock spider, the marbled orb weaver and the black-and-yellow argiope, or garden spider. The argiope female grows to over an inch long — plus its long legs — and builds conspicuous webs in prominent locations. The smaller, male argiope lives on the edge of a female's web and makes a zigzag line when it moves across the web. The bright yellow marbled orb weaver and the brown shamrock spider are nearly as large, up to an inch long, but with shorter legs than the argiope. They like to nest along buildings, arbors and in shrubs and brambles. All three make an egg sack in the fall. Young spiders overwinter in the sack by the hundreds and emerge in the spring. They wait until the wind is right, and then drift away on a web strand to be scattered randomly across the landscape.

Daddy longlegs are arachnids but are not considered true spiders. The over 200 species in North America include both predators and scavengers. Most are nocturnal hunters of small insects, but they will also eat decaying organic matter. Some are said to prey on slugs and earthworms as well as spiders and flies.

Many other spiders fill every niche on the farm, weaving webs of various sizes, cruising the underbrush, and generally hanging out waiting to catch unwary prey. Very few spiders here are poisonous, though many can give a nasty, stinging bite. All are beneficial to the farm ecosystem. In addition to catching insects, spider webs are an important material for hummingbird nest construction.

Orb weaver.

Design for Ecological Control

With so many players "in the field," design for beneficial insect habitat is best approached with a broad strategy of providing a wide range of nectar and pollen plants throughout the growing season. Inside the bioshelter, with its reduced diversity, more specific strategies can be employed. The list of plants we use in the bioshelter and outdoor gardens gets longer as we observe insect behavior and add new flowering species. Some insect species have such short adult life spans or live in such small populations that we would only find them if we could devote all our time to observation and research. More practically, we try to provide suitable habitat and trust in nature to find a balance.

Plant-Insect Interaction

The biochemical interactions between insects and plants are both mysterious and fascinating. Many insects inject plants with chemical compounds that encourage the plant to form galls or release volatile chemicals. Predatory insects are drawn to plants besieged with insect pests by chemical clues that some plants release when distressed. Similarly, male gall wasps find their mates by following chemical clues emitted by infected plants.

The complex chemistry of plants and the interactions of plant-produced compounds with soil, insects and other plants are likely to never be fully understood. From the permaculture designer's point of view, these complex interactions are yet another reason to approach ecological design with humility and to support native diversity in our landscape.

Good garden sanitation will reduce many pest populations. (Sanitation measures include removing and composting crop residue or tilling it into the soil in a timely manner.) However, some beneficial insects overwinter under mulch and in soil. Uncultivated and fallow areas in and near gardens will act as refuge and reservoir for repopulating the garden ecosystem after cleanup and tillage. When clearing crop residue, we look for praying mantis egg cases, spider eggs sacks and dormant wasp nests. Moving these to a safe spot helps maintain populations.

As the diversity of trees and other plants increased in our tree lines and windbreaks, biodiversity increased. Certainly, the wide range of native pollinators helps keep our fruit trees, shrubs and brambles productive. The panicle and osier dogwood shrubs and viburnum species all provide forage for countless varieties of insects and many birds. The many other plants and trees each add some forage for insects and birds. Each flowering plant offers some food and shelter for insects. Leaf litter, mulch, dead fallen branches and tree bark all offer shelter and brooding places. As with most aspects of ecological design, we must acknowledge our limited information and allow for a wide range of species to fill in the gaps of our limited knowledge.

Below is a description of the beneficial habitat plants we use in our gardens, as well as those we allow to grow in the uncultivated areas between our gardens.

The monarch caterpillar feeds only on milkweed species. Note the caterpillar in the flower.

DARRELL FREY

Beneficial Habitat Plants in Outdoor Gardens

Tansy, *Phacelia tanacetifolia,* tops our list of beneficial insect habitat plants. We have established plantings of tansy near all of our gardens and in central locations between them. Doubling as a cut flower for drying, tansy never fails to host ladybugs and their larvae. Other beneficial insects on tansy leaves and flowers include soldier beetles, assassin bugs, robber flies and syrphid flies. Tansy persists where it is planted, so be careful where you establish it. It spreads easily by runners. We put it along the fence line of the South Garden and in a swale between the East and South gardens, as well on the edge of the Pond Garden and West Garden. Tansy grows at least 4 feet tall and tends to fall over in late summer and fall. Most of our patches are located where we can tie them to the fence line with cotton string. Other patches are surrounded with several stakes and tied, or simply left to fall over where they are out of the way.

DARRELL FREY

Flowering oregano is a preferred habitat for the beneficial syrphid fly.

Mints and their relatives have found many niches on the farm. (Mints will spread and can be hard to contain, so think before you plant.) All provide nectar for honeybees, metallic bees, and small wasps. The many members of the genus *Mentha,* like most of the family Lamiaceae, produce abundant, small flowers over a long period of time. These flowers are readily accessible to even the tiniest insects. Most produce abundant nectar, as well as oils and medicinal compounds. We have established patches of various mints and other Lamiaceae throughout the farm. In the West Garden orchard, ajuga, peppermint, *Mentha x piperita var. piperita,* and spearmint, *Mentha spicata,* are planted between trees. Peppermint and water horehound, *Marrubium vulgare,* are found along the pond's edge. Permanent patches of spearmint allow us to do a weekly rotation of harvest sites. English pennyroyal, *Mentha pulegium,* has found a home in a wet swale surrounding the fire circle. American pennyroyal, *Hedeoma pulegioides,* gathered from an old farmyard, has become well established along the swale that drains the north side of the bioshelter, and around the summer kitchen. Pennyroyal hosts gatherings of tiny bees and wasps, big wasps, syrphid flies, and lots of types of ichneumon wasps.

To promote insect diversity, provide:

- diverse pollen and nectar sources in each garden
- uncultivated wildflower areas between gardens
- plant diversity in tree lines and windbreaks

Spearmint and peppermint are also planted in the farm's kitchen garden and in our polytunnel. Several patches of spearmint are tended inside the bioshelter. These are rotated every few years.

Ajuga, *Ajuga reptans* 'Atropurpurea,' or common blue bugle, flowers in late spring. While we collected it primarily for its striking purple leaves and bright blue flower spikes, it is relished by the early generations of beneficial insects.

Purple flowering oregano, *Oreganum vulgare*, is an excellent habitat plant. This spreading plant produces its clusters of dark pink blossoms from late June through fall, with a few clusters lingering until a hard frost. All summer, they are busily worked by syrphid flies, bumblebees, metallic bees and other small bees. Less aromatic than culinary oregano, this plant is mainly used as a habitat and ornamental planting; we also often include it in our herbal bouquets. Our initial planting at the door to the bioshelter kitchen has spread around the north side of the building to fill an area of about 30 square feet. Other patches are established in our other gardens among the shrubs and along the fences.

Other oreganos we grow include the aromatic Greek oregano and the ornamental culinary golden oregano. Neither of these stands out as insect habitat, mostly because we discourage flowering by regular harvest.

Motherwort, *Leonurus cardiaca*, nurtures a great diversity of tiny wasps all summer long. Honeybees love to visit the ever-rising spire of fuzzy pale lilac flowers, each a small vial of nectar. Until killed back by frost, this herbaceous perennial member of the mint family will continue to produce its lovely flowers around its slow-growing stalk. Motherwort is allowed space in the middle or on the edge of our gardens and orchards. It spreads slowly by division and possibly by seed. As the name implies, motherwort is a traditional medicinal plant for women.

Catnip, *Nepeta cataria*, like motherwort, is a medicinal plant. It too has established itself in our gardens. Growing to over 6 feet tall on the edge of our chicken yard, catnip thrives in the rich, moist soil. It flowers continuously all summer and helps give the chickens shade from the hot mid-day sun in July and August. Catnip will spread by seed, though not as aggressively as the root-running spearmint and peppermint. The clusters of white and pink flowers feed honeybees, metallic bees, bumblebees and many tiny wasps. We occasionally have allowed catnip to live in the bioshelter, but it is not a high-value, marketable culinary plant, so we only keep a few small, self-seeded plants, which are allowed to flower in summer.

Catmint, *Nepeta mussinii,* is a small, delicate relative of catnip. It grows in clusters of 6- to 8-inch-long prostrate stems with lovely leaves and blue flowers. We mainly grow catmint for our herbal bouquets. It grows well in partial shade under our kiwi arbor and in the bioshelter beneath the lemon verbena.

Anise hyssop, *Agastache foeniculum,* flowers from early summer until stopped by a heavy frost. These dark purple flower spikes attract honeybees, metallic bees and many tiny wasps. Beautiful in the landscape, delicious in tea, edible and medicinal, anise hyssop has a prominent place in our bioshelter and herb gardens. It is an herbaceous perennial, though perhaps not long lived, though it readily self-seeds. Inside the bioshelter, excess anise hyssop seedlings are cut for the spring salad mix. The fresh flowers look good and are long lasting in an herbal bouquet. Picked with care, anise hyssop will branch densely and produce flowers all summer.

Thyme's tiny florets grow in clusters and attract many wasps and bees. We grow several varieties for the herb and edible flowers. German thyme, *Thymus vulgaris,* has a very good flavor and grows upright for easy harvest. Creeping thyme, *Thymus serpyllum,* also has good flavor and flowers profusely. It grows in low, self-mulching mats. Creeping thyme lives up to its name by creeping into the yard, where it competes well with grasses. When mowed regularly, it is fragrant and soft to walk or lie on (not that we have the time to lie in the thyme). We also grow silver thyme, *Thymus vulgaris* 'Argenteus,' for bouquets, flowers and culinary use. Lemon thyme, *Thymus x citriodorus,* is established in the bioshelter and several gardens. It is known to be a mosquito repellant. We do not seem to have many mosquitoes, but we do keep some lemon thyme handy in the Zone 1 areas of the farm. A large patch in the Spiral Garden is readily accessible for workers and visitors relaxing in the kiwi arbor. We simply pick a handful of lemon thyme stems, rub them in our hands and then rub the oil on our shoulders and neck to ward off irritating mosquito attacks.

Lavender, *Lavandula angustifolia,* flowers for only a short period of time, so we harvest the flower stalks for everlasting bouquets before they fully open. When not harvested, lavender flowers over several weeks and nurtures insects in early summer, when its cousins are just beginning to flower. Bumblebees and other tiny bees are attracted to the fragrant flowers.

Cilantro, *Coriandrum sativum*, and dill, *Anethum graveolens*, both attract large numbers of tiny bees, wasps and syrphid flies. We plant them every other week through the growing season to allow a steady harvest of their leaves. Many of these plantings are allowed to flower and set seed, for both culinary use and to promote biodiversity.

Bronze fennel, *Foeniculum vulgare dulce* 'Rubrum,' is given a prominent place in our Spiral Garden and is planted in several other gardens as well. When flowering, it attracts a steady stream of small bees, syrphid flies and wasps of all sizes. Bronze fennel is a perennial. We gather the self-seeded seedlings each spring for our plant sales. Through the summer we add the dark, ferny foliage to our salad mix, stripping the narrow, anise-flavored leaves from the stalks. The two-inch-wide umbels of yellow flowers are popular as an edible dessert garnish.

Lamb's ear, *Stachys byzantina*, attracts bumblebees and tiny wasps. Grown primarily for cut flowers and dried arrangements, the silvery, furry leaves and lavender flowers are quite lovely. We grow it in several locations to promote the population of wasps.

Wildflowers

Cultivated and unmanaged wildflowers, shrubs and vines all provide some degree of habitat and food for beneficial insects. We are still observing and learning about the dynamic relationships between plants and insects.

Queen Anne's lace, *Daucus carota*, the wild sister of the garden carrot, attracts syrphid flies, tiny bees and wasps. The delicate, flat white umbels bloom over a long season mid to late summer. They are an excellent craft material when dried. Because they are a biennial taproot, they are also important in maintaining an aerated soil.

Milkweed, *Asclepias* spp., is important summer forage for many insects. We have observed honeybees, bumblebees, butterflies, tiny wasps, small bees, wasps and hummingbirds all foraging on milkweed's fragrant pink flowers. Because milkweed is the sole food of the caterpillar of monarch butteries, we give it special privileges on the farm. Small, permanent patches of milkweed are protected from mowing on the edges of our gardens. Random milkweed plants are allowed to grow in our gardens most years. Because the first monarchs appear in early July here, we are very careful about pulling out milkweed plants after July 1. Any unwanted plant is inspected for eggs and young larvae. These are relocated to other milkweed patches as needed. Swamp milkweed has also been established in the Pond Garden and near the pond. This more delicate looking plant hosts the same insects as regular milkweed.

Wild yarrow, *Achillea millefolium,* self-seeds all over the farm and is left standing anywhere it is not in our way. Syrphid flies, metallic bees, other small bees and tiny wasps make steady use of its easily accessible flowers. Yarrow has established itself in sections of our yard as well. Mowed yarrow makes a soft, ferny carpet that spreads by underground runners, never getting a chance to flower. Some of our yard areas are thick with mowed yarrow.

Goldenrods provide important habitat for beneficial insects. With over 80 species native to North America, it is a common perennial. Canadian goldenrod, *Solidago canadensis,* is the most abundant at our farm, but at least a dozen species are found near the farm. We allow large patches of goldenrod to grow between and around our gardens. Goldenrods begin blooming in August and continue through October. They are an important honeybee forage plant, and they also support a wide range of beneficial insects. Several species are good for dried flower arrangements and cut flowers. Goldenrods are allelopaths, which means they have a capacity to suppress the germination of other plants' seeds.

Stinging nettle, *Urtica dioica,* has inconspicuous flowers but is known to host a wide range of beneficial insects. Nettles are valuable as a spring cooking green. Nettle tea, because it provides soluble nutrients, is used as a foliar spray to promote healthy plants.

Ox-eye daisy flowers, *Chrysanthemum leucanthemum,* host predatory thrips. If you look closely at the sunny yellow disc of a daisy, you can often see the tiny, slender, black insect crawling among the florets. Predatory thrips are distinguished from the smooth-bodied, plant-eating thrips by having a distinctly jagged appearance. They consume pest thrips, aphids, and other soft insects.

Asters and daisy fleabane are also common among our gardens. Daisy fleabane, *Erigeron annuus,* begins to bloom in June and can continue to flower until fall, feeding honeybees and other bees with pollen and nectar. Many asters, including the New England aster, *Aster novae-angliae*, calico, and crooked-stem asters are found in our meadows, tree line and orchard. Most flower in the fall and are important to honeybees and the last generations of bees and wasps in the fall.

Burdock, *Arctium lappa*, and bull thistles *Cirsium vulgare*, are favorites of the honeybee and bumblebee. Both are biennial tap roots and so help loosen and enrich the soil. Bull thistle is allowed space on the edge of our gardens, where it often supports large orb weaver spider webs.

Honeybees in the Gardens

Honeybees, *Apis mellifera*, deserve special mention. Their economic importance is considerable. They pollinate tens of millions of dollars worth of fruit and vegetable crops each year. As I write, the mystery of colony collapse disorder is unsolved. This sudden die-off of many bee colonies across North America is the latest disorder to plague honeybees. The varroa mites and tracheal mites that appeared and spread through bee colonies in the last part of the 20th century have already made beekeeping a more difficult enterprise. It is likely that the colony collapse disorder is the result of multiple stresses. Together, all the challenges of beekeeping raise the cost of honey production and the cost of pollination services bees provide. They have also led to a general decline in beekeeping as a hobby.

Certainly the troubles of the honeybee confirm the need to provide food and habitat for the native pollinators. Nonetheless, the products bees provide, honey, wax, propolis (described below) and pollen make the effort to maintain bees worthwhile. Rather than provide a detailed study of beekeeping, which is the topic of many good books, here we will focus on the placement and management of bees and bee plants in the permaculture landscape.

Bees are best located away from high traffic areas. They also need regular tending to be productive. This places them in a Zone 3 management area. Some farms keep a hive or two of bees but do not tend or harvest them. This places them in Zone 4.

At Three Sisters Farm, our beehives are located along the north-west edge of the pond, well away from the main work areas. The hives are set on concrete blocks at the base of a small slope where they are somewhat protected from the prevailing wind. (We are planting a hedge of useful shrubs to the north and west of the hives to provide additional wind protection.) The hives receive full sunlight, and the additional light reflected from the pond helps warm them early in the morning. Ready access to water is important to the bees.

When beekeeping in suburban and urban areas, bees should be located where they will not bother anyone. Bee stings are a minor irritation to most people but can be deadly to those with allergies to bee toxins. Rooftops and secluded back yards are the most reasonable locations. In rural areas, bears and raccoons will break into a hive. Raccoons can be deterred by placing a heavy stone or concrete block on top of the hive. Bears may be deterred by a heavy-duty or electric fence.

Honeybee Forage

Literally hundreds of wildflowers are foraged by bees for pollen and nectar. While most of these are only minor nectar sources, producing little if any excess honey, they are nonetheless important to the hive for building strong colonies in the spring and sustaining them between major nectar flows. Major honey plants vary with the bioregion. White clover, sweet clover and buckwheat are well known for their distinctive honey. Raspberry, blackberry, thistles, asters, goldenrod and milkweed also provide abundant bee forage. Willow, redbud, tulip poplar, catalpa, black locust, basswood, sumac, and sourwood are all important nectar-producing trees. Willow produces abundant pollen early in the year, providing an important food for the hive as it expands its population in the springtime.

We have a theory that bees collect and use a wide range of nectars and pollens to keep healthy. Just as we require a diverse diet, bees also, we suspect, need a wide range of plants to stay vigorous. The use of essential oils, including peppermint and wintergreen, rosemary and thyme, by beekeepers to kill and suppress tracheal mites and varroa mites, seems to support our pet theory. Formic acid, part of the sting of stinging nettles, also kills these mites while not harming the bees. For this reason it is good to have stinging nettle and peppermint, as well as other aromatic herbs, growing close to beehives.

Many plants should be observed for bee forage potential. For example, cowslip, *Calthus palustris*, a potherb found in sunny swamps and springs, is one of the earliest flowers each year. Whether it is worked by bees or not, I cannot say, but it flowers at a time when bees really need a good food supply. Any flowering plant might be used by bees from time to time, in any given location.

The distinction between major and minor bee forage is rarely absolute. The degree to which bees utilize a nectar source can also vary from year to year and place to place. When a minor plant is especially abundant, it can become a major source — as can a minor spring source, given an exceptionally warm, dry spring and a strong hive. A major nectar source might be a poor producer if the weather is unfavorable when it blooms. At such a time, bees will fall back on flowers they otherwise might ignore.

In our area, two species of viburnum, nannyberry (wild raisin) and arrowwood, and a shrub called ninebark, *Physocarpus opulifolius,* inhabit sunny, open wetlands and stream banks. Each alone is only minor bee forage. However, they bloom in a succession. First nannyberry (a fruit-bearing shrub that grows to 25 feet) flowers, followed by the blooms of arrowwood (a smaller shrub growing to 8 feet high) that are similar flat-topped umbels of white florets. After about a week, the arrowwood blossoms fade but the ninebark branches are loaded with lovely snowballs of florets 1½ inches in diameter for another week. These three shrubs together provide the colony with three weeks of forage, leading them to the major nectar flows in flowers of such trees as tulip, basswood and catalpa. If there is an abundance of blackberries, these shrubs may be largely ignored by the bees, who will work the high-nectar bramble blossoms instead.

Bees are attracted to flowers by color and scent. Bees see the colors blue, blue-green, yellow and ultraviolet. Strongly scented flowers generally have a lot of nectar. Pollen from a wide range of plants is gathered to raise young bees, especially in the spring. Plants also produce resins used by bees to make propolis, a glue-like substance that has been used in the production of antibiotics. Some plants also produce *extra-floral* nectars. It is believed that extra-floral nectar is secreted to attract ants. Ants then protect the plants from would-be foragers. Bees sometimes collect "honeydew" secreted by aphids. This is considered by some to make poor-grade honey due to its generally strong flavor.

Beekeeping for marketable production requires a good deal of care at the same time that the small farm requires lots of crop production.

Spring is the time to monitor hives and expand production by dividing hives. Maintenance and regular harvest after honey flows can conflict with gardening schedules. For this reason it can be difficult to do both well. Commercial-scale beekeeping should be viewed as one of the seasonal businesses that can be developed by enterprising partners on a small farm.

Insects in the Bioshelter (and Other Creatures)

Because the bioshelter is a semi-enclosed environment, the insect dynamics are a little more complex to manage. Three of the biggest greenhouse pests — aphids, mites and thrips — are rarely a problem outdoors. In the gardens, they are kept under control because they have to contend with environmental stresses (rain, humidity, direct sunlight, etc.) and a wide range of natural predators. Inside the bioshelter, these pests can quickly expand in numbers when predator numbers are low.

Whiteflies have fewer natural predators in our area and when allowed to leave the bioshelter on infected plants, they can become a summer pest. Inside the bioshelter, they are more subject to control. When we have whiteflies on our plants, we make sure to treat them with soap spray and do a close inspection before they are taken out to the gardens.

Habitat Management in the Bioshelter

Habitat management is at the heart of bioshelter management. Flowering plants that provide insect habitat are kept in the bioshelter all year: alyssum, scented geranium, blue basil, thyme, oregano, marjoram, lavender, and Johnny jump-ups are some examples. Alyssum is especially important. This low-growing plant flowers all year. It never needs deadheading, and it grows well in small planters and in the holes of the 6-inch concrete block walls that form our deep beds. Alyssum also looks good. Available in both white and violet varieties, the masses of flowers and green foliage hanging over the edge of the concrete block is quite lovely — as it provides pollen and nectar to adult parasitic wasps.

Pest control inside a bioshelter relies heavily on the greenhouse manager to keep track of populations of both

The Cast of Characters:

The Bad Guys	The Good Guys
• aphids	• labybugs
• thrips	• *Encarsia formosa*
• whiteflies	• syrphid flies
• spider mites	• aphid wasps
• pill bugs	• gall midge larvae
• ants	• spiders
• scale insects	• predatory thrips
• slugs	• predatory mites
• cabbage moth	• ground beetles
• leaf miner	• lacewings
• spittlebugs	• bumblebees

Other Players are

• toads
• tree frogs
• soil organisms

pests and their predators, and to step in with other organic control measures when necessary. Below are some of the players in the bioshelter ecosystem. Chapter 10 provides detailed discussion of bioshelter pest management.

Insect Control Methods

Insecticidal soap

Insecticidal soap is fast, easy and effective at killing insects. To work, the spray must cover the insect. Most die within a few minutes (sometimes as quickly as 30 seconds) as their cell walls dissolve. Made from potassium salts and fatty acids, insecticidal soap is non-toxic on skin contact for humans and wildlife. It can damage leaves, though, so it should be rinsed off within a day of application. It is best to apply soap spray in the evening, when whiteflies are less active.

We use distilled water for mixing the soap for application. Minerals in our water (which is hard) tend to reduce the soap's effect. When we do not have distilled water handy, we just add a little more soap than the directions call for. As with any product used on the farm, be sure that the brand you use is acceptable for organic certification. Some products contain pesticides or unnecessary additives. A small hand-held sprayer is appropriate for greenhouse use. In the gardens and for dealing with bees and wasps, it is better to have a backpack sprayer with a longer spray wand.

When bees and wasps build their homes in the wrong location, soap spray will eliminate them. It can take patience to stand near a nest long enough to spray every wasp as it enters or exits a nest, but it is preferable to applying pesticides.

Spider mites, thrips, aphids, scale, mealybugs and whiteflies are all effectively controlled with soap spray. But soap will also kill beneficial insect predators, so it is best to use soap sparingly in seasons when they are active.

Sticky traps and Other Methods

Thrips and whiteflies are attracted to color, especially yellow. Coating yellow paper with a clear, sticky substance makes an effective trap. (Thrips are also attracted to light blue traps.) For many years we bought the Tangle-Trap brand of trap coating. It is a petroleum-based product, but it is approved for organic production. Coated traps remain sticky for days or weeks and so continue to trap insects for a long time.

However, sticky traps also attract insect predators, so we use these traps only when we have major infestations. In recent years, we have begun to use cooking oil as a coating for paper traps. The oil is effective only for a day, but is easier to apply than Tangle-Trap coating. Because it is only sticky for a short time, the effect on predators is less severe. We have begun to experiment with adding essential oils to the cooking oil. Lavender oil has proven extremely effective. We fold a 14-inch by 24-inch piece of paper over a clothes hanger and cover both sides with the scented oil. Infested plants are shaken with the trap held a foot or two away. Whiteflies are drawn like magic and stick to the oil.

Occasionally, we have resorted to vacuuming up pests with a portable hand-held vacuum cleaner. When a major outbreak occurs, this is an effective way to catch whiteflies and thrips.

For controlling caterpillars, slugs and larger insects, picking them off by hand is sometimes the best solution. On a garden scale, though, this is impractical. Sometimes, pests can be knocked off plants with water. Aphids and other small pests infesting flats of seedling or potted plants can be rinsed off with a watering wand or rubbed off by hand.

Beauveria bassiana

Our need to import predators was greatly reduced by the appearance of the insect-consuming fungus *Beauveria bassiana* in the bioshelter and surrounding gardens. This entomopathogenic fungus spends the warm months breaking down organic matter in the soil. When the air temperature cools in the fall, we find it infecting aphid colonies in the gardens. The fungus becomes active in the bioshelter in November and early December, just as the insect predators are becoming less active. With our assistance, the *Beauveria* fungus virtually eliminates aphids, whiteflies, spider mites and thrips until March, when it becomes too warm inside for the fungus to infect insects. By then, the predators are active again. In a sunny winter, the fungus is not as effective, due to higher temperatures and lower humidity.

This fungus occurs naturally in organic soils when conditions are right, that is, when the humidity is high and temperatures are cool. Every November, it first appears on aphids that are feeding on the underside of

Aphids.

low-lying leaves of tatsoi mustard, spinach and borage. Infected aphids first appear stiff and yellowish, then become white, with a bloom of spores. As the building gets cooler in December, the fungus can rapidly spread among colonies of aphids and quickly eliminate them.

Beauveria bassiana spores grow into the cuticle of susceptible insects, including whiteflies, thrips, aphids and spider mites. Soil-dwelling insects tend to be immune to it, but many leaf-dwelling insects succumb to various strains of this fungus and other entomopathogenic fungi. Because the spores are heavier than air, they need assistance getting to the insect host. Washing aphids off of plants and onto the soil may bring them in contact with spores. A spray application of the fungus is more effective.

Beauveria bassiana is available from commercial sources. It is labeled as effective for specific pests at specific humidity and temperatures. However, I suspect finding and using your own local strain is a better alternative. We collect fresh, infected aphids and place them in a clean plastic deli tray. We add a little distilled water and place them in a cool, shaded spot for a few days to spore. With proper moisture and temperature, each infected aphid will produce millions of spores. We then add more water to the trays and shake them vigorously to suspend the spores in the solution. We use a spray applicator to spray the solution onto our seedlings and infected plants. We apply the fungus on a Friday or Saturday after a harvest and plant clean-up; a cloudy cool day is preferable. Infected aphids do not readily wash off, so infested leaves need pruned off and fed to the chickens (after we have collected the specimens we want for spore culture). After one or two applications, insect populations are so reduced as to not be a problem for a while. Spring seedlings are sprayed weekly to prevent leaf-dwelling insects from colonizing.

When not infecting insects, the spores can lie dormant in the soil and also work as decomposers of decaying organic matter.

The US EPA lists *Beauveria bassiana* as non-toxic and non-infective to mammals, noting that it grows at temperatures well below human body temperature. It is approved for outdoor use and indoor use in greenhouses. However, there is a slight possibility it could cause an allergic reaction in susceptible individuals. It is best to apply the spores in solution to avoid creating excess airborne spores. A dust mask is always a reasonable protection for application of any spray. Also, follow any guidelines on commercially available strains. The EPA also warns of potential to infect honeybees. For more information, visit www.epa. gov/pesticides/biopesticides/ingredients/index.htm.

Predatory Mites

We purchase predatory mites as needed to help reduce populations of thrips and two-spotted spider mites. These microscopic relatives of spiders consume the eggs, nymphs and adults of their prey. Spider mites apparently are also killed by the *Beauveria* fungus, but as the bioshelter warms in the spring, the fungus is less active.

When we see populations of spider mites in the bioshelter, we order the predatory mite *Phytoseiulus persimilis* from a commercial insectary. It is best to release these predators in late winter, when the bioshelter is neither too hot nor too cold for the predator to be active. After placing the order, we spray several applications of insecticidal soap to reduce the populations. After that, the persimilis mite is very effective in controlling the two-spotted spider mites.

We sometimes release the predatory mite *Amblyseius cucumeris* in spring to help control the population increase of thrips in the bioshelter. Thrips are especially active consumers of nasturtiums. Again, we first make several applications of soap spray to reduce populations of thrips before releasing the predator.

Weed Control

A weed, to repeat the old saying, is just a plant in the wrong place. Many plants some gardeners consider as weeds are valuable wild edibles to us. These, of course, can get out of control and suppress the main crop. Bioshelter weeds include snapdragon and nicotiana, which both self-seed profusely.

Ehrenfried Pfeiffer, the biodynamic farmer and writer, outlines three types of biological weed management. First is planting the crop seeds thick enough to outgrow and choke out weeds. This can be a good strategy for planting lettuce in warmer months. In colder months, a dense planting can cause disease in the lettuce.

The second strategy is to plant a crop that has allelopathic properties to suppress weeds. Rye grass is a good winter crop for suppressing winter and spring weeds, and clover can suppress weeds in the summer.

The third biological weed control Pfeiffer offers is to improve the soil with compost, manure, lime and drainage. Soil improvement is an ongoing process, though, and we wonder if this really only changes the type of weeds that grow.

To Pfeiffer's measures, we add that weed control begins with seed control. Undesired weeds should not be allowed to seed. When weeds go to seed, those seeds can continue to germinate for many years. Each seed has a slightly different chemistry for breaking dormancy. Sunlight, moisture and time are all elements.

If a farm has sufficient space to produce its own mulch, weed seeds will not be as big of a problem. It's *imported* mulch and manure that will bring in weed seeds. This is where our field got its Canada thistle and probably its red dead nettle. Canada thistle is a noxious weed by legal definition. It spreads by seed and perennial roots. We strive to eradicate it by mowing and cutting it before it can flower. When patches do flower, we burn them with a torch before cutting the stalks and burning them in the fire pit. Admittedly, each weed has some virtue: Canada thistle is a wonderful bee and butterfly forage plant. But that does not make it a desirable plant.

Disease Control

When a disease appears in the garden, the problem is often environmental. Weather, soil type, fertility, and other factors can stress plants and allow them to become diseased. A farmer should keep references on hand to identify plant disease. State plant inspectors are happy to help identify problems.

Mosaic Viruses

There are tomato, tobacco, lettuce, and cucumber mosaic viruses. All are similar — making it hard for the average gardener to identify which one is the problem. All are spread by infected seeds or carried from plant to plant by aphids and other sucking insects or the gardener's hands when managing diseased plants. The virus can survive up to two years in infected plant residue.

Cucumber mosaic virus, especially, is carried by many weeds and their seeds. Aphids get it from infected plants and transfer it to other plants. As with many pathogens, it is likely that mosaic virus has some benefit to the aphid. The weakened plants do draw other aphids. Cucumber beetles quickly spread the disease among cucumbers and squashes. Planting resistant varieties is often the only way to prevent crop loss.

There are three times when you have a chance to control these viruses: when buying seeds; when weeding out infected plants; and when choosing how to control vector insects.

Control in the bioshelter begins with the use of mosaic-tested lettuce seeds (MT0). These seeds are tested and declared to be virus-free by the seed company. Using only mosaic-tested seeds in the greenhouse will help reduce or prevent lettuce mosaic virus.

Because mosaic virus is present in the outdoor gardens, it eventually is carried inside a greenhouse by aphids and other sucking insects. The virus may be present is a healthy looking plants and not appear until the plant is stressed. In winter, when the soil is cooler and plant defenses are low, the virus' effect appears. The misshapen leaves and stunted plants are easily recognized. Infected plants should be gathered and fed to the chickens as soon as they are found.

Many plants in the bioshelter are susceptible. Our self-seeded wild edibles, chickweed, wood sorrel, purslane, lambsquarters and amaranth can all carry cucumber mosaic virus. (Because these all self-seed, infected plants can be hidden on the floor, under planters, and in odd corners of our polyculture beds.) Our collection of ornamental oxalis is prone to mosaic virus. During the summer, plants will appear fine, but in the middle of winter the leaves will be misshapen and stunted. Self-seeded snapdragons also are a common host. When they germinate and begin to grow in the winter greenhouse, obviously infected plants are removed, and healthy looking plants are allowed to grow. It is best to remove all infected plants as they appear.

Aphids are most able to spread the virus when they begin to grow wings in the middle of winter.

Powdery Mildew

Powdery mildew is often a problem in the bioshelter, primarily in the cooler months. Mildew doesn't usually overwinter outside in cold climates such as ours, but it returns with the summer rain. Outdoors, it appears in early fall on red clover, plantain, and other plants. Inside the bioshelter and cold frames it can grow as a faint white powder on the leaves of pansy, columbine, tomatoes, calendula, rosemary and a few other crops. The mildew can spread under cool, moist conditions and will kill a plant if unchecked. The first control for powdery mildew is environmental: *Do not get leaves wet when watering.* Water sitting on leaves in cool, cloudy weather will provide the conditions for powdery mildew to get established. Keeping leaves dry is not easy when watering. Poor air circulation in a closed building makes the problem worse. Also, there is a trade-off, because wetting leaves can help reduce other pests, such as whiteflies, thrips and especially spider mites. This is probably because wetting helps promote parasitic fungus and other insect disease. So a decision must be made whether to wet leaves or not. This is often a major source of disagreement between bioshelter managers at our farm. As with all agricultural practices, timing is everything. If spider mites and thrips are present and mildew is not, wetting the leaves will help control the pests.

When plants do become infected with powdery mildew, it can be cured by dusting the leaves with powdered sulfur. The sulfur will kill the mildew, but also will kill the leaf. So the treatment must be timely to save the plant. We have noticed that plants treated with the sulfur powder often grow back looking healthier; we suspect powdery mildew is related to sulfur deficiency.

Slugs and Snails

Terrestrial mollusks, slugs and snails can be the bane of the market gardener. Continuous battle is necessary some years to control their numbers. Our worst pest is the grey field slug, *Deroceras reticulatum*. This native of Europe has followed gardeners all over the planet. We also do battle with the garden arion slug, *Arion hortensis*. This is also a European immigrant that is partly identified by its orange mucus trail. We began to see a small snail in our gardens in 2003. While not over-abundant, it has spread throughout the farm. The appearance of snails can indicate an increase in available calcium, which is necessary for snail shell formation.

Slugs and snails are a favored food of fireflies. With all the thousands of fireflies we see each summer, I shudder to think how many more slugs we would have if there were no fireflies. Snakes also eat slugs, as do some songbirds.

The population of slugs in our gardens has risen and crashed several times in the past 20 years. Slugs are the worst in wet years and after mild winters. Old, damp straw and hay can harbor slugs and their eggs, so mulching with straw is often a big mistake. Slugs eat a wide range of crops and seedlings, including most salad and cooking greens.

When our children were young, we paid them a penny a piece to collect slugs. Wearing latex gloves, they picked slugs off plants by hand. Consequently, each spring, the kids had plenty of money for bike rides to the local ice cream store. Slugs come out in the early evening and can be collected in buckets of water. They are eagerly consumed by chickens. Since the children have grown, I have inherited the job of collecting both slugs and snails. Beer traps are effective and can collect dozens of slugs in a night when infestations are severe. I fill a tray half-full of beer and set it down into the soil. Slugs crawl into the beer and drown. Beer traps need to be emptied right away though, because slugs smell horrendous after they are dead a few days. It is best to feed them to the chickens each morning when they are freshly "pickled."

Diatomaceous earth, with its sharp, jagged texture, will deter slugs, but our gardens are far too large for its use to be practical. We do treat our outdoor seedling frames with diatomaceous earth. Slugs like to hide under the flats and come out at night to devour young seedlings. Between collection and the diatomaceous earth, we have been able to minimize slugs' effect on our seedlings.

In recent years, we have reduced our use of imported straw mulch and increased our use of leaf mulch. This may be responsible for the major crash of slug populations in 2006.

Mammals

White-tail Deer

The most common question we receive when presenting garden workshops is "How do you control deer?" I suggest a double fence system and a good dog. We have had almost no problem with deer on our rural farm in 20 years. I attribute this to the presence of our farm dogs and the local hunters who keep populations under control. Occasionally, a deer will sneak in to eat fallen apples, but they generally avoid us.

Many suburban communities struggle with the "Bambi phenomenon" — heavy resistance to reducing deer populations through hunting.

This is the result of an unfortunate misunderstanding of ecology. The natural processes of life require a balance of predators and their prey. Deer populations in nature are controlled by heavy predation. When we remove predators from the system, we must work to restore balance. Choosing *not* to control the population of deer and other game animals turns them into pests.

Groundhogs

You say groundhog, I say woodchuck. We both mean *Marmota monax*. This wily rodent, most closely related to the squirrel, is the bane of gardeners across North America. Woodchucks can destroy a case or more of lettuce in a morning and will take a bite out of each tomato that begins to ripen. They will climb over a fence or bite a hole through chicken wire to get at your garden. The damage they can cause goes beyond the loss of crops. They can even undermine building foundations.

To control woodchucks, we must first examine their role in nature. As a strict vegetarian, they are a primary consumer in the food chain. They, themselves, are food for wolves, coyotes, dogs (and people). Young woodchucks are among the prey of foxes, owls and hawks. Their habit of digging extensive burrows does have the positive effect of bringing mineral-rich subsoil to the surface and air into the soil. When they abandon their dens, skunks, foxes, rabbits and other creatures use them for shelter. In their natural environment, woodchucks are an edge species, preferring to forage the meadow from a burrow on the forest edge; they are a key species in the ecological balance wherever they live.

Woodchucks begin to breed in their second year and can produce a litter of up to six young. They actually do arise from three or four months of hibernation in late winter and their young are born in early spring.

Spring is a good time to inventory burrows and mark their location. We like to fill in the borrow entrance and then come back to see if the burrow is active. Later in spring and summer, we are alert for new burrows as the young strike out on their own to seek new pasture. They are elusive — up early and out late — though they may forage at any time of day. They like to stand guard over their burrows and watch for intruders. The only sign you may have of their presence in the underbrush is their shrill whistle of warning as they flee to their burrow.

Our primary control method has been to have a good-sized farm dog. As discussed elsewhere, setting up predator-prey relationships on the farm is an important part of ecological design. A young and energetic dog can kill its own weight in woodchucks every week. Older dogs that have been bit once or twice will get wary and may need assistance from the farmer. As our first farm dog aged, woodchucks and rabbits began to move into our tree lines and gardens again. When he was not as good at chasing them down, we had to help. One way was to flood out the burrow with the irrigation hose. First, all exits but one are filled in. Then the hose is turned on. When the animal comes out of the burrow, the dog can catch it.

We once had two dogs, ours and a neighbor's, that developed a tag-team approach. One would guard the burrow, while the other chased the foraging woodchuck toward him.

If you have no dog, a .22 caliber bullet or two usually is sufficient to kill a woodchuck. But be aware that there may be more than one woodchuck per burrow.

You can also buy sulfur dioxide smoke bombs that will kill them by asphyxiation. If you use these, be sure to cover all other holes and then the hole with the smoke bomb. Follow all instructions and do not breathe the smoke.

Woodchucks are easily caught in a large-sized live trap. Apples are common bait. Once you have the animal in the trap, you should either kill it or take it at least ten miles away. They are particularly nasty when cornered and will bite, so use all necessary precautions when handling, or otherwise controlling them (or any wild animal). I personally do not think it is fair to give someone else my problem, but perhaps you can take it to a state game land or other place where it will not be a pest.

Once a woodchuck is dead, we usually bury it back in the hole it came from. We do have friends who find the meat of the woodchuck to be fine for roasting. It is indeed tender when the animal is young; it is similar to roast beef. They are best in July, when they weigh about five pounds. But we do not usually eat them. If you chose to add them to your diet, be sure to follow all precautions in handling wild game. Woodchucks have several musk glands that need to be removed. Woodchuck fur is too coarse to be of value, although Native Americans used their hides for leather goods.

Rabbits

The Eastern cottontail, *Sylvilagus floridanus,* is the rabbit that we share the land with. They, too, are primary consumers, and are even more important to the diet of predators than the woodchuck. Rabbit populations rise and fall on a ten-year cycle, in ecological step with the populations of foxes and other predators. Rabbits can produce three new generations each year. About 25 percent of these survive to reproduce the next year. Fox, weasels, coyotes, dogs, cats, owls, hawks and snakes all prey on rabbits.

Rabbits can cause considerable damage to crops. They especially enjoy grazing on seedlings of lettuce and other greens. They can quickly girdle and kill a young fruit tree in the winter by eating the bark. Fencing for rabbits should be at least 2 feet high and bent flat to the ground at the base to prevent them digging underneath.

Rabbit control includes predation of young rabbits by the farm cat. We also hunt and eat them. During the summer month, rabbits often have warbles, a fly larva parasite that lives under a rabbit's skin, on the back of its neck. So we usually hunt them in the early spring and fall.

Rabbits can also be caught in a live trap and released off the farm. Almost as soon as one is removed from the farm, though, another moves in to take its place. The combination of hunting, fencing, and predation helps keep them from becoming too much of a problem.

Small Rodents

A vole looks like a large mouse. The meadow vole, *Microtus pennsylvanicus,* is a common and constant garden pest. Voles are often mistaken for moles. Moles also burrow in lawns and gardens, but they do not eat vegetables; they prefer worms, slugs, snails and even small mice. Moles actually are beneficial because they loosen and aerate soil while patrolling for their prey. Voles also burrow through garden beds and loosen soil. But they eat our crops. They can quickly ruin a crop of potatoes, bite into ripening tomatoes and consume many root crops. They especially love mulched beds.

Voles are controlled by predators as well. Weasels and foxes hunt them. Cats and dogs will catch them when they can. Dogs can learn to locate and dig them out of the garden beds during winter months. In the summer, when dogs are excluded from the gardens, we catch them in rat traps baited with cherry tomatoes or root crops. Regular rototilling between crops also seems to discourage them.

Voles occasionally move into the bioshelter's deep beds. When we see their tunnels, we set traps to catch and eliminate them.

Various types of mice can be a problem in the garden and greenhouse. They will eat seeds and young seedlings of lettuces, peas, squash and other crops. The farm cat and natural predators, including snakes and weasels, keep down the outdoor population. Inside the bioshelter, we regularly trap them to control the population. Seeds are stored in sealed containers.

The Farm Dog

A good dog or two is essential for farm pest control. First of all, the dog must be friendly and like children. They need to be almost a year old to really begin training. Dogs, contrary to what many people think, like to have a space of their own to hang out in. So being chained to a doghouse is not so bad from dog's point of view — so long as there is an adequate length of chain and access to fresh water. What is needed is daily exercise. Training to follow commands, including come, stop, walk, run, and sit is required. Never try to train a dog if it is not on a leash. The best reward for most dogs is a good rub behind the ears and words of praise, though dog food treats can also help with training.

The dog should be trained to walk the farm's pathways and to avoid garden beds. Daily walks around the farm's perimeter create a routine and makes the dog familiar with his territory. Dogs need to be trained to relieve themselves in appropriate places; in our case, that means along certain parts of the tree line and under the utility line, well away from foot traffic and harvest areas. Once the dog knows the property, and it is trained to stay on the farm, you can begin training it to hunt for pests. Most dogs have an instinct to kill woodchucks, chase rabbits and, sometimes, dig for moles. If the dog you are working with does not have that instinct, consider finding it another home and get another dog. Once they know the game, dogs will spend a lot of time finding a woodchuck's pathways and holes and chasing them down.

The Farm Cat

Cats are territorial hunters. They will patrol an area of about four to five acres. Young cats will climb trees and can catch birds. But as they get older, cats seem to learn that ground-dwelling rodents are easier prey.

In the spring, they help control the population of young rabbits. Cats are necessary in a landscape to control voles, mice and young rats. Overfed cats are less likely to hunt as frequently, but this varies from cat to cat.

The ecological view of cats in the landscape is based on carrying capacity. One cat hunting on a five-acre farm is a good balance. Even the occasional bird lost to a cat may be good for the ongoing evolution of songbirds. But adding more cats can quickly overburden an ecosystem. The classic food pyramid shows the population of predators as the small triangle at the top of the food chain. Predators are by nature territorial to keep in balance with their prey populations. As food farm managers, it is important to keep the number of cats in the neighborhood to a minimum. When we do get a new cat at the farm, we always are sure to schedule a veterinarian appointment for all shots and birth prevention procedures.

Cats should not be allowed to use the garden for their litter box. We prepared a space in the tree line with sand and cat litter and encourage our cat to use these facilities. Cats can carry the protozoan parasite *Toxoplasma gondii,* which causes toxoplasmosis in humans. Cats get the parasite from eating rodents and birds. The parasite breeds in a cat's intestines for several weeks and can spread to people through contact with cat feces. The parasite can live up to 18 months on the soil. While the effects are usually mild in adults, pregnant women should avoid contact with cat feces. Toxoplasmosis can cause severe complications for a developing fetus.

Pests and Permaculture

As with other aspects of permaculture design, ecological pest control is based on an understanding of ecological systems, continuous observation and experimentation. The diversity and balances that have developed in our five-acre field are a small step, but they are a good indication of the potential of permaculture design.

CHAPTER 7

The Market Garden Farm: Management

Apart from the character and condition of the soil, the location of the farm and the crops to be grown, two of the most important factors that influence profit and loss are the "lay" of the land and the "lay-out" of the land. Lay always refers to natural features: lay-out to the artificial ones. … Each of these has a bearing on the character of the farming to be conducted.

— M.G. KAINS, *FIVE ACRES AND INDEPENDENCE*

THIS CHAPTER CONTINUES THE STUDY of farm design with a look at the layout of Three Sisters Farm, our crops, and other important aspects of the farm.

Our gardens are treated as separate, but interacting, elements within the farm system. Their different aspects (variously sloping to the south, west and east), microclimates and drainage patterns, as well as their locations relative to the bioshelter, determine their role in the farm management and development plan. Annual records are kept of the crops grown in every garden and are referenced when planning crop rotations. The annual garden plan takes into account each garden's history and unique character.

Each garden is composed of raised beds laid out on contour. Pathways between beds collect rainwater or can be flooded with water pumped from the pond. Wide central paths, mulched with cardboard and sawdust, allow access to the contour paths. The four main gardens have fences, birdhouses, compost areas, and irrigation hoses.

As we develop gardens, we allow uncultivated areas within or near them. These biological islands, together with herbs and flowers and other insectary plants, provide a healthy balance of predatory creatures.

Birdhouses, rock piles and perennial plantings of fruit, nuts, berries, vines, shrubs and flowers are added each year. Plantings are placed to create windbreaks and shade zones. Large planters are used to store stones collected from the garden beds. These function as hose divots and also provide summer shelter for snakes and toads. In the spring and fall, the stones are used to hold down floating row covers.

Planting begins in the gardens closest to the bioshelter and we work outward to the other gardens as the season progresses. First, we plant the East and South gardens because they are sheltered from the winds by the bioshelter itself and slope to the southeast and south. When the weather is more stable we begin planting in the Pond Garden and Southeast Garden. Next we put summer crops in the West Garden and several smaller gardens around the farm. Culinary herbs are more likely to be planted in the gardens closer to the bioshelter, in Zones 1 and 2, where they can be better tended.

Zone 1 Landscape:

The area within 70 feet of the bioshelter's west side, including our 300-square-foot Spiral Garden, seedling frames, kiwi arbor and other plantings, is a good demonstration of the application of permaculture design. A series of windbreaks slows and deflects ground-level wind from the west and northwest side of the bioshelter. A hardy kiwi arbor provides a shaded break room and relaxation space. The Spiral Garden allows for easy harvest of herbs and flowers, promotes biodiversity, and creates an aesthetically pleasing setting for greeting visitors. Flower and herb gardens, habitat plantings, and frames for growing and displaying seedlings are all located here for ease of watering, tending and harvest.

The area immediately west of the bioshelter is a "zone one" or daily use area, where garden transplants and plants for sale are tended.

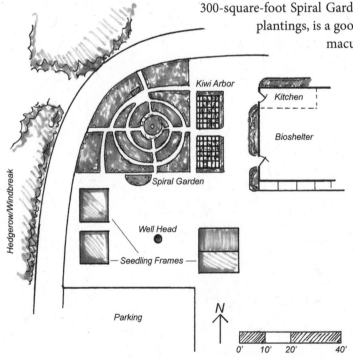

The Spiral Garden

Designing the Spiral Garden was an important process for Linda and me. We wanted to work closely on an important project to hone our skills as a design team. Our goal was to have an inviting and functional centerpiece of the farm landscape in the space between the bioshelter and the driveway. The first impression a visitor receives can be very important.

We spent an evening looking at the space and discussing several possibilities. We sketched out the area to scale. The curve of the driveway and the need for a wide pathway adjoining the bioshelter helped define the shape and area. We drew out two possible layouts. One was a medicine wheel of concentric circles of beds and footpaths, with paths to the four directions. The second design had a path beginning directly opposite the bioshelter's kitchen door spiraling into the center, with wide beds in between. Connecting pathways to the four directions would allow easy access from all sides. The spiral design spoke to us both. It added an element of mystery — a curving path — inviting one to wander through and linger. The medicine wheel design was very nice and simple but the spiral was more interesting. (We later used the medicine wheel layout for our medicinal herb garden.)

Linda would have primary responsibility for the management of the Spiral Garden. She laid out the beds at a width she could work with and had free reign to develop the plantings.

We allowed space on the outer edge for shrubs and left enough room for an arbor between the bioshelter and the Spiral Garden. The first years of the Spiral Garden, Linda planted mostly annuals, such as basils, edible flowers

1. American Black Currant
2. Bee Balm
3. Black Currant
4. Black-Eyed Susan
5. Blackberry
6. Bronze Fennel
7. Butterfly Bush
8. Chives
9. Columbine
10. Comfrey
11. Daffodil
12. Day Lily
13. Dolgo Crabapple
14. Edible Blue Honeysuckle
15. Forget-Me-Nots
16. Garlic Chives
17. Gladiolus
18. Golden Oregano
19. Good King Henry
20. Gooseberry
21. Juneberry
22. Lamb's Ear
23. Lamium
24. Lavender
25. Lemon Thyme
26. Lilac
27. Milkweed
28. Nettle
29. Oregano
30. Peony
31. Purple Flowering Oregano
32. Rose
33. Sage
34. Sedum
35. Strawberry
36. Tarragon
37. Thyme
38. Trellised Honeysuckle
39. Valerian
40. Violets
41. White Bergamot
42. Wild Bergamot
43. Yarrow

Both ornamental and functional, the spiral garden is evolving into a forest garden as the apple tree and shrubs mature.

and cut flowers. Every year she added more perennials; as the shrubs grew, she added an understory of herbaceous perennials. Over a period of 18 years, Linda has transformed this from a garden of annual plants to a complex, multi-level garden in the process of becoming a forest garden.

Each spring the path is hoed level and mulched with a few layers of newspaper and covered with a few inches of sawdust. The beds are cleared of grass and unwanted plants. Self-seeded herbs and flowers, such as garlic chives, bronze fennel, forget-me-nots and bee balm, are potted as needed. After thorough weeding, the beds are mulched with shredded tree bark.

The Spiral Garden is the Zone 1 garden for the bioshelter kitchen. It serves multiple functions:

- Allows quick access to small herbs for orders and harvest supplement.
- Supports a thriving biodiversity of insects, at the center of the farm.
- Adds beauty and interest to the visitor entrance.
- Creates an educational space for herb and flower identification, insect study, etc.
- Provides a quiet and beautiful sanctuary.
- Serves as a windbreak.

Young Terra Frey enjoys August raspberries in the driveway windbreak.

CATHERINE PALADINO

Windbreak

The outer edge of the Spiral Garden is planted with shrubs. The shrubs shelter the garden from wind, block dust from the adjoining driveway and create an intimate, enclosed setting. The shrubs include purple lilac, white lilac, Juneberry, edible honeysuckle, a trellised honeysuckle vine, blackberry and rugosa rose. Interplanted with these are foxglove, iris, lupine, peony, daylily, white bergamot, nettle, lamb's ear, rudbeckia, chives, and assorted groundcover plants. A stone bench in the north bed, among the Juneberry, bergamot and honeysuckle, allows one or two people to sit and observe the garden. The center bed is planted with a flowering and fruiting Dolgo crabapple with a perennial understory of valerian, columbine, violets, thyme and variegated lamium. (The Dolgo crabapples make a beautiful and deliciously tart applesauce.) Other shrubs in this garden include American black currant, black currant, and gooseberry.

Most beds in this garden have established perennial flowers and herbs interplanted with annuals each year. Plants include lemon thyme, creeping thyme, silver thyme, German thyme, dianthus, golden sage, tricolor sage, green sage, purple sage, clary sage, violets, lupine, daylily, peony, red bergamot, purple bergamot, lamb's ear, black-eyed Susan, stinging nettles, Good King Henry, garlic chives, chives, ox-eye daisy, bronze fennel, yarrow, gladiolus, iris, catmint, lavender, sheep sorrel, variegated lamium, and for-get-me-not. Potted rosemary, white sage, Mexican sage, blue sage, pineapple sage, calla lily and other tender perennials are placed in this garden most years.

Heavy stones placed at bed corners help frame and define the beds. They also serve as hose divots to prevent us from dragging the hose over plants when we are watering.

Summer harvest treats, red currants, black currants, and red and golden raspberries.

Kiwi Arbor

Between the bioshelter and the Spiral Garden, we built an arbor to support several hardy kiwi vines (*Actinidia arguta*). The arbor is actually two arbors, each 10 feet long and 6 feet wide. Eight posts, measuring 4 inches by 6 inches and 7 feet tall, are set on concrete piers. Angle bracing and rafters, made of 2-inch by 6-inch lumber, complete the framing. Trellises made of 2-inch-wide oak strips are placed on two sides of each arbor, creating an intimate setting and increasing the windbreak function of the arbor. Chairs and benches under the north half the arbor provide a lovely setting to visit with guests and enjoy the Spiral Garden.

Hardy Kiwi.

The south half of the arbor is used to store shade-loving wildflower seedlings. The arbor is framed on the north, east and south sides with stone-edged beds. These are planted with one male and two female kiwis and a variety of herbs. Sweet woodruff and wild ginger are on the north, shaded side, and yarrow and thyme are on the sunny, south side. We also plant an annual flowering vine on the trellis. In years past, we've planted morning glory, peas and scarlet runner beans.

Seedling Areas

Several covered frames and a greenhouse are located between the Spiral Garden and the parking area. These are used for holding seedlings and potted plants for the gardens and for sale. Every day, from spring through fall, the plants in the seedling frames, seedling greenhouse and under the arbor are watered from a garden hose on the west wall of the bioshelter. This hose is rolled up and kept out of traffic every day.

The earthen berm on the west side of the bioshelter is planted each year. A local red grape climbs a trellis over the main kitchen door, providing shade from the late afternoon sun as well as edible leaves and fruit. The vine is pruned each spring and puts out abundant new growth each year. Excess shoots are pruned and leaves are harvested for use in our kitchen. The rest of the berm has had a number of different plantings over the years. Lemon balm, dwarf evening primrose and an ornamental perennial geranium have persisted on the south end for 15 years. Wild oregano has grown on the north edge and around to the north end of the berm. The rest of the bed has been, at times, a strawberry bed, a couch of chamomile and thyme, and a cutting garden of rosemary and other herbs.

The north side of the bioshelter also has an earthen berm along the 20-foot-long kitchen wall. This is planted with a wildflower garden that has ostrich fern, trillium, wild ginger, sweet woodruff, wild clematis, Virginia creeper, snowdrops, wild oregano and a persistent wild blackberry. This, and a second wildflower garden on the opposite end of the bioshelter's north wall, take advantage of the shade cast by the building. The resulting microclimate is adequate for many wildflowers.

A wildflower bed on the east side is watered by a swale draining the driveway. This garden includes the purple-flowering raspberry, red currants, and a number of wildflowers: Dutchman's breeches, wild bleeding heart, trillium, black cohosh, wild columbine, Virginia waterleaf and others.

Tansy

A large patch of tansy, 5 feet in diameter, promotes beneficial insects in this Zone 1 area. The patch is watered by the swale between the bioshelter's west wall and the kiwi arbor. Because tansy grows tall and has a tendency to fall over, we tie it to a fencepost for support.

Square-foot Beds

Square foot gardening is a method of small-scale intensive gardening popularized by Mel Bartholomew in his book *Square Foot Gardening*. This system of gardening in wood-framed beds is good for small spaces in urban and backyard gardens. Most of our other gardens are large and sprawling and may not inform the home-scale gardener. Square-foot beds are more organized and can easily demonstrate intensive-gardening techniques. When visitors arrive, they are greeted with a simple display garden of square-foot beds. The beds are 5 feet on each side and are framed with 1- by 6-inch lumber. They are interplanted with herbs and edible flowers. One square is a "Three Sisters Garden" (with squash, beans and corn).

A pathway leads from the parking lot to the bioshelter, past the gardens and seedling frames. A small garden surrounds the water well casing in the lawn between the Spiral Garden and parking area. A small (but spreading) section of herbal lawn has developed along the south edge of the Spiral Garden. Creeping thyme and yarrow spread through and compete well with the grass. Regular mowing keeps this fragrant lawn soft. The open area between the Spiral Garden and the parking lot allows a direct view of the Spiral Garden as visitors arrive. This space serves as a staging area for farm tours; from there, we can point to most of the farm's systems and explain what we do.

Flower beds adjoin the square-foot beds and separate the parking space from the South Garden. These provide edible flowers, cut flowers and more material for herbal bouquets. The flowers provide an attractive first view of the farm.

The South Garden

Measuring approximately 10,000 square feet, the South Garden is our prime growing area. Beds are laid out on contour running east to west. The garden gradually drops about 3 feet in elevation from the bioshelter to the south property line.

The South Garden includes the cold frames (discussed in Chapter 4) that are attached to the bioshelter. A wide path separates the cold frames and bioshelter from the South Garden. This path leads past the bioshelter, along a swale that drains the building's south roof, through the East Garden and then to the far southeast corner of the farm. A polytunnel garden is located in the northwest corner of the South Garden, near the bioshelter. The rest of the South Garden is rotated with annual crops.

The South Garden is the first garden planted each year and the last one to be harvested in the fall. This is because it is closest to the bioshelter, so most convenient. Also, the south-facing aspect of this garden warms the soil earlier in the spring.

Three beds in the southeast corner of the South Garden are partially shaded by a row of filberts along the property line. These beds are often planted with cool-weather crops in summer to take advantage of the cooler microclimate in the shade zone.

On a few occasions, when grasses invaded the beds on the northeast quarter of the South Garden, we fenced in that section and let the chickens forage it for a season. We left the beds fallow overwinter and then put them back into the garden rotation.

A wide path along the western side of the South Garden follows the course of a buried electric line. By locating the

Lettuce Harvest in the south garden at Three Sisters Farm.

TOM GETTINGS

path over the line, we keep track of its location so the phone or electric company can find it if they ever need to.

The East Garden

The East Garden falls gently away from the bioshelter, sloping slightly to the east and southeast. This garden is protected from wind by the bioshelter and the bale-shelter to the west and north. It is protected from the occasional east wind by a mature tree line on the east. As with the South Garden, early spring and late fall crops are planted here to take advantage of the sheltered site and easy access to irrigation.

The lower portion of the East Garden is treated as a Zone 3 area. Perennial beds of chives, mint, sheep sorrel and wild coltsfoot are all harvested from there.

Linda Frey and Pam Owens harvest head lettuce.

The Southeast Garden

The Southeast Garden is a triangular wedge of land between the East and South gardens. Actually all three gardens are enclosed within a continuous fence, but the separate designations help us plan and manage them. The upper portion of the Southeast Garden includes a small grove of black locust trees with a comfrey understory. This area is difficult to rototill because of its small size and shape. It is fenced in to be a part of the poultry's rotational pasture. After a fall, winter or spring turn as poultry forage, the beds are shaped by hand and planted with hills of corn, beans and squash, or other mulched crops. The lower portion of the Southeast Garden borders the South Garden fence; it is planted in golden raspberries, a local variant of the black-capped raspberries that we discovered in the wild and have been propagating to increase production. The black-capped raspberry grew well on its own here, so we took note and added more, especially the golden variety.

Between the lower beds and the upper section of the Southeast Garden is a series of eight raised beds on contour. These beds are planted as part of the garden crop rotations.

The Pond Garden

The Pond Garden slopes toward the southwest and borders the pond. This garden receives the force of winds from the west straight on, but it is partially protected from the prevailing northwest winds by the tree

The wildflower garden showcases native plants grown by Three Sisters Farm, and provides a quiet shady retreat.

line. The lower and eastern portion of the Pond Garden is subjected to the northwest wind due to a gap in the tree line where wet soil suppresses tree growth.

Native Plant Garden

The Pond Garden includes a perennial wildflower garden at the upper, northeast corner and a developing forest garden on the edge of the tree line to the north. The contours and shape of the Pond Garden create a triangular area in the northeast corner that is too small and oddly shaped to till often. This is planted with a native plant garden. It serves both as a public space for visitors looking at the wildflowers and a private space for intimate conversation.

The windbreak planting of old farm roses to the east and the fruitful tree line to the north create a sheltered and private setting with a view of the pond and Pond Garden. A large, flat rock serves as a bench beneath the shade of a staghorn sumac grove. Under the sumac is an understory of wild columbine, woodland sunflower and New England aster. A small 6-foot by 12-foot clay-lined earthen pool collects and holds rainwater. We often keep the pool filled during dry weather with water pumped with solar power from the pond. An overflow hose from an irrigation tank keeps it recharged as needed. This pool is surrounded by water-loving plants, including spearmint, cardinal flower, and swamp milkweed. Below the pool, a series of three beds curve from east to northwest. The paths between the beds are laid out to drain the beds toward the northwest edge. The east ends of the beds are planted with plants that prefer drier soil, including lupines and Virginia spiderwort. Plants preferring damp

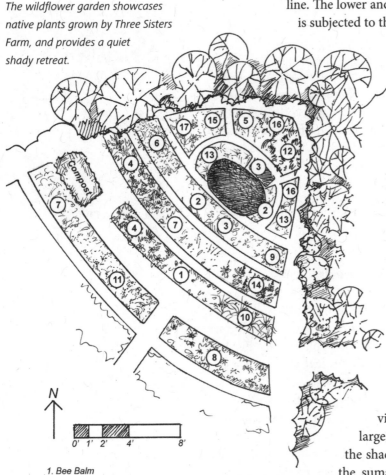

N

0' 1' 2'   4'   8'

1. Bee Balm
2. Cardinal Flower
3. Great Blue Lobelia
4. Green-Headed Coneflower
5. Jacob's Ladder
6. Joe Pye Weed
7. New England Aster
8. Purple Coneflower
9. Spearmint
10. Spiderwort
11. St. John's Wort
12. Staghorn Sumac
13. Swamp Milkweed
14. Wild Blue Lupine
15. Wild Columbine
16. Wood Geranium
17. Woodland Sunflower

soil, including bee balm and blue lobelia, are planted in the middle section of the beds. The wetter, northwest edge of the beds is watered by occasional overflow from the pool and rainwater falling on the paths. This section is planted with moisture-loving plants, including Joe Pye weed, green-headed coneflower and swamp milkweed.

A broad path curving around the south-most bed offers a clear view of the native plant garden. Several shorter beds, just beyond the path, are planted in sun-loving native plants, including bee balm, purple coneflower, New England aster, and the medicinal St. John's wort.

The native plant garden is weeded and then mulched with shredded bark each year. Seeds are collected for the nursery, and occasionally plants are potted for sale and replaced with new seedlings.

The main part of the Pond Garden is annual crop production beds. These beds, as with all the gardens, are laid out on contour. From the highest point at the native plant garden, to the lowest point at the southwest corner, this garden drops 5 feet in elevation.

We plant the Pond Garden in June, just after the crops in the South and East gardens are planted. Because the beds are long in this garden, they are often used for rotations of long-season crops, such as tomatoes, potatoes, squashes and sweet corn. The lower beds along the west edge of the garden are the wettest. They receive drainage from the upper beds and do not drain as well because the ground is more level.

This garden is mostly patrolled by redwing blackbirds that nest in the pond and swallows that prefer the more private birdhouses in the Zone 3 and 4 areas between the Pond Garden and the West Garden. Dragonflies and damselflies are also seen zooming about. A large patch of tansy along the lower edge of the pond and a patch of motherwort along the main central path promote beneficial insects. Leopard frogs, which breed in the pond, patrol the Pond Garden and the West Garden hunting for prey. Snapping turtles lay eggs in the beds and in the compost piles nearby.

Goldenrod dominates several thousand square feet of uncultivated land between the Pond Garden and the pond, feeding bees and other insects in late summer. As described in more detail in Chapter 9, the uncultivated space acts as a buffer zone, keeping excess nutrients from washing into the pond, and into Mill Creek. A similar space buffers the pond from the West Garden.

The West Garden

The West Garden slopes to the east and drops about 6 feet in elevation. Between the West Garden and the pond lay several thousand square feet of uncultivated land (currently filled with goldenrod) sloping toward the pond. It is somewhat protected from the wind by a mature tree line, a young orchard bordering its west side, and the slope of the land.

Forest Gardening

A forest garden is diverse, perennial, multi-species orchard. We have a long-term plan to develop the western portion of our farm as a sort of forest garden. Our plans are really a hybrid of a forest garden and alley cropping, which is an agricultural system that spaces tree crops wide enough to allow other crops to be grown between them. Our forest garden will group plantings in rows along contour beds.

We have been conducting trials of various fruits and berries to see which will work best in our forest garden. We have tried apple, pear, blueberries, raspberries, Juneberry, kiwi fruit, plum, currants and gooseberry. Based on these trials and customer demand, we will continue to proceed with development of the mixed species orchard, or forest garden.

We found an endorsement of our method of orchard groundcover management in the "neo-classic" on organic orchard management, *The Apple Grower*. The author, Michael Phillips, offers great advice on managing an orchard's groundcover to promote a healthy orchard. He says that individual sites and orchard management styles can and should vary. Some managers prefer to keep the ground mowed, others mulch the trees. In some of his orchards, Phillips plants patches of comfrey among his trees and allows goldenrod, selected weeds, legumes, and grasses to grow on the orchard floor. These help promote a diversity of natural predators and keep soil nutrients circulating. Mowing or scything is done close to harvest time. In our orchard, we scythe and mulch around the currants and other crops interplanted under some of the trees. Under the rest, we cut the groundcover as needed to ease the fruit harvest. This is, of course, a simplification of the discussion. We highly recommend Phillips's book as a detailed resource for orchard management.

David Jacke's two-volume *Edible Forest Gardens*, co-written with Eric Toensmeier, provides an extremely detailed study of perennials in the permaculture landscape. We developed our plan before *Edible Forest Gardens* was published, but we expect to draw heavily on both volumes as we implement and refine our plan.

The forest garden will increase farm profitability and spread the workload more evenly through the year. Fruit and other fresh products will be marketed to existing regional customers, both individual consumers and restaurants. The system will reduce the need for tillage and intensive cultivation of the soil. Perennials in permanent cover and mulch accumulate, rather than consume, carbon. Nutrient runoff and use of manure-based compost will be reduced. Instead, by-products of local sawmills — bark mulch, wood chips and saw dust — will be used for mulch and soil building.

As with the other gardens, the beds are laid out on contour to catch rain and irrigation from our pond. Our established pear and apple trees are interplanted with ☞

A gas pipeline was installed in this area before we started our farm. The path leading to and through the West Garden follows the course of the buried pipe so if the gas company ever needs to service the line, they can do so without disturbing plantings.

The West Garden is a Zone 3 and 4 area. We grow green manure there in the spring, then till it and plant the beds with crops in July. We

herbs, wild edible plants, and beneficial bird and insect habitat. Our more mature apple trees are interplanted with bush fruits, various species of currants and gooseberries. The grass is cut around the trees each June and used to mulch the bush fruit.

We are planting trees, bushes, trellised vines and other perennials in alternating rows. Semi-dwarf fruit trees, interplanted with bush fruit are alternated with shorter species, such as asparagus and raspberries.

Downslope is a bed of mulch crops, including purple vetch, red clover, alfalfa, and native grasses. Twice each year, in late May and in July, the mulch crops are cut with a scythe and used to mulch adjoining beds. The mulch crop beds are tilled and planted with annual crops as needed.

Newly planted and young trees and bushes are protected from gnawing rabbits and voles with tree guards and mulched with composted wood chips and bark, leaves, sawdust, and straw. We do this for several seasons. As the trees mature, permanent clover and grass groundcover will establish themselves. Nitrogen-fixing plants such as black locust, bayberry, bush clovers, alfalfa and red clovers will be interplanted or used as groundcovers.

Asparagus beds, to achieve maximum production, need to be heavily mulched with compost or leaves each year. Left unmulched, an asparagus patch can survive in grass for many years, but it will not be as productive. A semi-wild patch of asparagus should be mowed or scythed in the early spring and late in the fall to prevent goldenrod from overcrowding it.

As the system matures we will consider introducing ducks to the Forest Garden. Ducks can provide a valuable pest control service in the garden, and the direct application of their manure is beneficial. They also help to reduce pond weeds — and convert them to meat and eggs.

Several issues will need to be addressed when ducks are introduced. We will need to fence the other gardens and monitor and control the ducks' access to the Forest Garden when bush fruits are ripening. Predator-resistant housing will also be needed.

Research and Educational Value

Our mixed-species forest garden orchard will illustrate the value of integrating fruit production into market garden farms. Planting crops in the proper relationship to one another and to the landscape results in a healthy, profitable system. Diversifying farm crops with fruit and fruit-based, value-added products will increase the profitability of the small farm. The addition of ducks to the system adds a specialty meat product and makes use of the ducks' appetite for bugs, slugs and weeds.

Finally, the project furthers our intention to develop Three Sisters Farm as a model of diverse, small-scale intensive agriculture. The lessons we will learn about management of such a diverse system will expand the value of our farm as a learning center.

DARRELL FREY

Centennial apples, a small summer variety, are a favorite for fresh eating at Three Sisters Farm.

occasionally contract with a neighbor to till the whole garden with a tractor and then re-shape the beds with the rototiller. The green manure is a self-seeded vetch. The plant appeared on its own, either brought in by birds or left over from a previous farmer's seeding. Over the years, we have allowed it to self-seed and spread through most of the West Garden. The vetch and other self-seeded weeds, including amaranth, grasses and lambsquarters are mowed once or twice prior to tilling. Some years, we leave the West Garden fallow and cut it several times for mulch.

An irrigation tank sits on a level area at the high point of the garden. This space, between apple trees, is planted with ajuga and tansy for beneficial insect habitat and with asparagus for harvesting.

This garden is slowly transforming into a landscape of mixed perennials that will become our "forest garden" (explained shortly). We planted apple and pear trees at the upper elevations; along the western edge of the property we are planting new rows of fruit trees and shrubs every few years. (The plans are outlined in the accompanying sidebar.) We have also planted several *Ribes* species among the apple trees: red currants, black currants and gooseberries.

The West Garden is plagued by woodchucks and rabbits. We double our patrol for them when we begin to prepare the garden in late June.

Smaller Gardens

The Mailbox Garden is located in the far southwest corner of the farm (near the mail box). Measuring about 4,000 square feet, it is planted with annual crops in rotation. Left fallow much of the year, it serves as overflow location for main crops, potatoes, squash and sweet corn. Some years, it is fallow all year. Several apple trees border this garden on the west and south. The Mailbox Garden, like the neighboring West Garden, is subject to woodchuck and rabbit pests.

The Curve Garden is so named for a curve in the adjacent driveway and the curve of the beds on contour. This smaller garden is about 3,000 square feet. It is bordered on the south side by an uncultivated area of goldenrod, a self-planted spirea shrub and two apple trees along the driveway. A perennial patch of peppermint is encouraged in the wetter, lower bed.

Crops and Products

Crop selection for a small-scale farm needs to be diverse enough to keep customers happy and allow for good crop rotation. But it should not be so complicated as to be unmanageable. We were inspired by the writings of Booker T. Whatley in his *Handbook on How to Make $100,000 Farming 25 Acres*. In it, Whatley describes a strategy for farm design based on a series of crops that each bring in several thousand dollars or more of income. Planting, management, and harvest schedules are arranged so they do not overlap too much. Asparagus, followed by strawberries, followed by raspberries and blueberries, followed by summer and fall apples is one example. With this combination, the workload is distributed over the growing season, and the harvest times only overlap slightly. Combined with beekeeping and value-added processing, this fruit farm would be quite manageable.

Linda Frey tends edible flowers, lemon gem and tangerine gem marigolds.

A diverse mix of crops allows for security against crop failure, and diverse enterprises spaced through the year make a lot of sense. Each main crop — herbs, flowers, salad and mixed vegetables — all bring in good profits in their season. A succession of fruits adds to crop diversity.

Edible flowers, herbs and bouquets need sales of over $1,000 a year to justify tending. Some crops are not as profitable, but must be grown to please customers. For example, a few years back, as we were getting in full swing, our sales totaled $15,000 in salad, $12,000 in head lettuce, $1,500 in edible flowers, $1,200 in basil, and $1,500 in other herbs, but for potatoes and tomatoes we made only $300 each. But we continue to grow them because customers like them.

The problem with some crops (tomatoes, peppers and eggplant, for example) is that, in our climate, they have relatively small yields for the amount of space they take up and the time they take to plant.

CATHERINE PALADINO

A sampling of ingredients used in Salad of the Season.

Green Fennel

Anise Hyssop

Claytonia

Nasturtium

Chervil

Kale Blossoms

Lamb's Quarters

Chickweed

Oxeye Daisy

Pea Shoot

Mallow

Greens Include:

Baby Lettuces
Beet Leaves
Chard
Arugula
Tatsoi
Mizuna
Red Kale

After only six to eight weeks of harvest, production declines. In a warmer climate with a longer growing season, tomatoes, peppers etc. would be more profitable. Season extension (discussed in Chapter 4) can help increase yields, but only for a few weeks. With subscription sales however, these lower-yield crops are a part of the vegetable crop rotation, so altogether they can bring in several thousand dollars — or more — in sales.

Saving seeds is probably not economically worthwhile on a small farm. The time required to isolate plants, select prime fruit, prepare and save seeds is often only worthwhile for expensive heirloom varieties. However, there are certainly other reasons to save seeds. Most important is developing locally adapted selections. Crops can be grown specifically for their seeds that can themselves be sold. In some cases, when buying seed is expensive, it pays to gather them. For example, the high cost of buying organic seeds for micro-green production makes growing the crop unprofitable. Growing seeds in the summer for winter micro-green production, however, may prove profitable.

Salad of the Season — Food for a Permanent Culture

As we began to develop our market garden farm, we sought a "signature" crop that would be both healthful and profitable. We wanted a distinctive product that would provide a weekly harvest through most of the year. We wanted our product to be a food for a sustainable, permanent culture. Growing a mix of fresh salad greens and wild edibles fit the menu.

The name "Salad of the Season" describes the nature of the product. The mix changes with the seasons. Different varieties of lettuce, greens and herbs are planted to suit the growing conditions at different times of the year. Wild edibles and self-seeded crops choose their own timing.

After more than 20 years, I still take great pride in our salad mix. Fresh greens are full of life and vitality. Rich in vitamins and minerals, with a wide array of antioxidants and other micronutrients, salad greens fill a vital role in a health-promoting diet. It is a pleasure to offer our customers a diverse, seasonal blend of such good food. I feel also a sense of artistry and the satisfaction of a creative impulse. I find satisfaction in my work when the balance of ingredients looks good in the bowl and on the plate. A culinary experience can satisfy on many levels. With a good meal, other needs are nourished too.

The salad base is a mix of six to ten types of young lettuce, such as red sails, red salad bowl, green salad bowl, butter crunch, freckles romaine, speckled Bibb, Batavian endive, rouge d'hiver romaine, Paris cos romaine, and frisée endive.

Any edible leaf, bud, or flower can be added to this basic mix: cut baby greens, such as mustards, kales, mizuna, tatsoi; young pea shoots; cress; arugula; and other greens, such as nasturtium leaves and beet, spinach, and fennel tops. Inside the bioshelter, we plant and harvest the following crops weekly for the salad mix: arugula, kale, mizuna mustard, tatsoi, red mustard, pea shoots, Chinese mallow and shungiku chrysanthemum (an Asian chrysanthemum leaf).

As the seasons change, new ingredients appear. The spring salad may include violet greens and flowers, watercress, chickweed, ox-eye daisy greens, dandelion, spinach, broccoli rabe, kale florets, mache and Good King Henry.

The summer mix includes lambsquarters, amaranth, sheep sorrel, wood sorrel, purslane, anise hyssop, dock, orache, mallow, plantain, and pennycress.

By mid-summer, we can add the new dandelions that are sprouting. Late summer brings the first chickweed. Chickweed germinates in cooling soil, so it thrives in cool autumn weather. (The appearance of chickweed marks the time to plant spinach.) The sour acidity of wood sorrel, with its hint of lemon, is succeeded by similarly flavored sheep sorrel, which thrives in the cool months. Sturdy, bitter curled endive is traditionally balanced with feathery sweet chervil.

In the fall, red-veined, arrow-shaped Bordeaux spinach, smooth spinach and crinkled spinach are grown, along with chickweed, shungiku, pea shoots, various beet greens, red-ribbed dandelion, wild dandelion, red Russian kale, lacinato kale, chervil and sheep sorrel.

Fall salad may include kales, ornamental cabbage leaves, chervil, miner's lettuce, mizuna, tatsoi, spinach, chickweed and mache. Wild edible ox-eye daisy, sheep sorrel and dandelions continue to produce, so we include them as well.

The winter mix includes miner's lettuce, endive, beet greens, young chard leaves, shungiku, chickweed, chervil, salad burnet, sorrels, fennel greens, spinaches, nasturtium leaves and all the standard lettuces. We are always on the lookout for new salad greens. Many make their appearance for only a few weeks of the year.

Maximum Nutrition

The salad mix fulfils our wish to provide our customers and ourselves with maximum nutrition. Spring tonic plants like dandelion greens, violet leaves, chickweed and watercress are relatively unselected, if at all, by breeding for specific characteristics. Being in their more natural state, wild foods may contain a broader range of phyto-chemicals, antioxidants and other healthful compounds. Most provide calcium, iron and other minerals. Many wild edibles are rich in vitamin A and C. Young sprouts of Brassicas (the family that includes cabbage and broccoli), such as kale and mustard, contain a wealth of antioxidants and other phytochemicals. Purslane contains healthful omega-3 fatty acids. Watercress is rich in minerals and contains vitamin A, B_2, C, D, and E. Wood sorrel is rich in vitamin A and C and is a source of potassium and phosphorus. Chickweed contains potassium, phosphorus, manganese, as well as minor mineral nutrients. Dandelion leaves, rich in vitamin A and C, also contain calcium and substantial amounts of B_1, sodium, potassium, and some important trace elements. It seems logical to assume that we do not yet know all the healthful compounds in these foods.

A salad mix with many fresh and raw ingredients provides a broad range of nutrients while introducing "tonic foods" into the diet. Many wild plants are traditionally known as tonics, foods that rejuvenate the body and revitalize health after a long winter. Our own theory of a sustainable diet takes the concept of tonic foods and combines it with the concept of macrobiotics. A macrobiotic diet stresses eating fresh, local foods in their proper season. It is reasonable to believe that eating a variety of fresh, live foods, in their season provides us with the most direct access to nutrients at the time we need them most.

The use of wild edibles in the salad helps increase the profitability of the farm. Although wild edibles can be tedious to harvest, seed costs are non-existent. Young "weeds" growing between planted crops can be harvested for the salad, doubling and tripling yields per area. They are then tilled in as green manure or left as groundcover. Many wild plants act as accumulators, gathering nutrients from deep in the soil. This is good for the topsoil as well as the salad bowl. Some herbs self-seed, like kale, chervil and mache. Some are perennials, like bronze fennel, salad burnet, anise hyssop and Good King Henry. Many of these double as beneficial insect habitat.

Many salad plants naturalize and adapt to the site. Brassicas, for example, cross easily with one another. We watch for interesting hybrids. One spring, we found a red mustard/Russian kale cross in our polytunnel; it made a great addition to our mix. Mache has made a permanent home in all our gardens. For many years, miner's lettuce has self-seeded in the cold frames, providing a steady winter harvest.

We have observed a lot of variation in the leaf shape of many wild edibles: we have four distinct strains of lambsquarters (magenta, white, bronze and lemon) and several less identifiable crosses. An oakleaf lambsquarters appears in several garden beds; it's a cool-soil strain that grows only a foot tall and flowers faster than other strains. Sheep sorrel leaves range from 4-inch long, narrow arrows to 1-inch, round arrows. Dandelions have a broad range of variation, from rounded lobes with few teeth to very narrow leaves with lots of sharp, pointed teeth. These variations make salad growing more interesting and help keep the product in demand with top chefs and repeat customers.

Fall through spring, we reap a good harvest of wild edibles and self-seeded greens and herbs from the bioshelter, unheated cold frames, and polytunnels. The solar-heated greenhouse beds have their own seasons: chickweed, lambsquarters and miner's lettuce are prevalent in winter; orache, amaranth and purslane emerge in the spring as soil temperatures increase.

Precautions

The use of wild edibles requires proper identification. Please consult an experienced forager before eating any wild plant.

Some salad plants such as sorrels, beet greens and even spinach, while rich in nutrients, contain oxalic acid, which, in large concentrations, affects calcium absorption. Kidney failure has been associated with the excessive consumption of foods with oxalic acid and potassium oxalate. (Oxalic acid can occur as soluble oxalates of sodium or potassium or insoluble calcium oxalates.) Oxalic acid is probably not a problem unless low levels of calcium and B vitamins are combined with high intake of oxalic acids. Younger plants probably have lower levels of oxalic acid, and cooking them may help to nullify the effects. Mixing small amounts of these plants into a salad mix is a good way to enjoy and benefit from them in moderation.

Under certain conditions, such as low light and high nitrate levels in the soil, leafy greens can absorb more nitrates than they can process into plant tissue. When humans digest them, nitrates are converted to nitrites, which can be more toxic than nitrates. Nitrites replace oxygen in the hemoglobin and cause a disease called methemoglobinemia. Nitrates are also a potential source for the carcinogen nitrosamine and can cause gastroenteritis and anemia after a while. Studies have shown that nitrate accumulation can be reduced or avoided by keeping soil nitrogen low in winter gardens and harvesting near the end of a sunny day. In Pennsylvania, nitrate levels peak during cloudy weather from December through late January. Because of this, we have chosen this period as our down time. We rest, celebrate midwinter holidays and plan for the winter-spring garden. The small amount of salad we mix during this time is harvested only after periods of sunny weather. We also keep soil temperatures lower (under 50° F) during the winter solstice season to inhibit plant activity. Less active plants accumulate fewer nitrates. Growth resumes as the soil surface warms when the sun shines.

Sanitation is also a concern. We use only greenhouse-grown watercress in our mix. Thus we avoid potential health risks of parasites, waterborne contaminates and diseases. Manure must be fully composted before being used to grow raw foods such as salad greens. Customers are always advised to wash produce, whatever the source.

The Salad of the Season has evolved to be a perfect crop for our farm. With a diversity of potential ingredients, we always have a good mix, even if part of the crop fails. The seasonal variation in the salad keeps customers interested in the product. However, our determination to maintain a high-quality product limits our production capacity to about 40 pounds per week from May through November, and an average of 30 pounds per week from February through April. At the premium price we receive, our salad annually generates more than $15,000 in sales. Most important, use of what some consider a "weed" as a prime ingredient in our product demonstrates a major principle of permaculture, turning a problem into a resource.

Vegetables and Herbs

Producing a wide selection of vegetables is necessary for a successful CSA or subscription sales enterprise. Our vegetable crops are detailed in Chapter 8. From a permaculture design perspective, though, our vegetable production primarily takes place in intensively managed gardens. Incorporating smart rotations and successions are all standard practices with organic gardeners. Beyond that, permaculture design emphasizes the relative location of various gardens and crops and the creation of habitat for pollinators and other beneficial creatures. Other concerns include efficient use of resources, energy and time, and the development of successful nutrient cycles between the gardens and compost piles. We are always on the lookout for new crop varieties and new organic cultural techniques. We try to adopt and maintain several varieties of heirloom vegetables. Heirloom tomatoes are especially in demand. We generally purchase heirloom seeds from the Seed Saver's Exchange (visit them at seedsavers.org/) to help support their work in preserving rare varieties of garden plants.

Herb production is an important part of our restaurant and subscription sales. Rosemary, sage, mint, chives, thyme, oregano, parsley, dill, cilantro and several varieties of basil are all in high demand. We have a lesser demand for some herbs, including French sorrel, tarragon, and chervil. But we grow them because — as with vegetables — their placement in and among the bioshelter and gardens helps promote biodiversity and productivity.

The following herbs grow best in rich, well-drained soils and full sun. In the fall, it is best to mulch rosemary plants or bring them indoors so that they can overwinter:

• salad burnet	• mint	• sweet woodruff	• tarragon
• bronze fennel	• oregano	• tansy	
• chives	• rosemary	• thyme	
• lemon balm	• sage	• lavender	

DARRELL FREY

Red currants, black currants, raspberries and herbal bouquets.

Cut Flowers

Cut flowers are an excellent income-producing crop. The best reference we know on organic flower production is Linda Byczynski's *The Flower Farmer,* Her detailed information on growing and marketing cut flowers is highly recommended reading. Cut flower production is one of the many enterprises we currently do on a small-scale but hope to expand. Inside the bioshelter, cut flowers are mostly the herbal bouquets described below. We also grow freesia, snapdragons, African daisies and a few other cut flowers. In our gardens, we grow iris, gladiolus, roses, foxglove, lilac, black-eyed Susan, gloriosa daisy, and purple coneflower — to name but a few. Everlasting perennials, which are useful for various crafts, include yarrow, lamb's ear, artemisia and sages.

The Herbal Bouquet

Herbal bouquets add a high-value crop to the farm's offerings. An herbal bouquet is a bundle of assorted herb sprigs, edible flowers and ornamental vegetables. They are a total sensory experience, offering a mix of scents, colors, textures and tastes. There are many dozens of possible plants to include in these bouquets. Herbal bouquet production is another example of permaculture principles in action because the herbs and flowers serve multiple functions in the landscape. They promote biodiversity, which brings stability to the landscape. Seasonal succession and functional placement of the herbs and flowers are important planning elements.

Similarly, edible flowers fill a marketing, ecological and aesthetic niche on the farm. Chefs, caterers and individuals planning food events and dinners love to decorate their tables with fresh, seasonal culinary bouquets and garnish their dishes with edible leaves, flowers and stems of herbs.

A large diversity of herbs and edible garnish plants ensure a steady supply of cuttings for the bouquets. These, in turn, provide food and habitat for a wide range of beneficial insects, especially butterflies. Whether concentrated in an herb garden or used as companion plants to vegetables, these plants add beauty and interest to the garden.

Herbal bouquet favorites:

- African blue basil
- alpine strawberries
- anise hyssop
- borage
- bronze fennel
- calendula
- catmint
- chard leaves
- chive blossoms
- clary sage
- daylily
- flowering basil (cinnamon, purple, spicy globe, sweet, Thai, and others)
- flowering oregano
- garlic chives
- golden oregano
- Hopi red amaranth
- lavender
- lemon balm
- lemon verbena
- miniature roses
- nasturtium leaves and flowers
- pansy
- raspberry sprigs (with thorns removed)
- red clover
- rosemary
- sage (golden, culinary, pineapple, purple, and tri-color)
- scented geraniums sprigs and flowers
- sweet woodruff
- thyme (creeping, German, lemon and others)
- violets

A usable flower, leaf or stem is one that is edible, does not wilt quickly, and has a pleasing, but mild scent. Cilantro and dill are avoided because of their extra-distinctive scent. Chervil and fennel leaves can be used, but they will wilt if not kept in a vase of water.

Medicinal Plants in the Landscape

Production of medicinal plants almost occurs spontaneously on the permaculture farm. Yellow dock, dandelion, coltsfoot, St. John's wort and yarrow all came on their own and staked claims to their niche. Aspen, willow, alder, motherwort and boneset moved in around the pond.

Slippery elm and black cherry were present in the tree lines when we started. We continue to plant medicinal plants and allow them to naturalize on the edges of the gardens. Some medicinal plants we tend in the gardens.

Necessary Disclaimer and Some Good Advice

I am *not* prescribing or even recommending the use of particular plants as medicine. I do encourage readers to do what we do: consult several references — and perhaps your physician — to find the best remedy for what ails you.

I *do* seek to alert readers to medicinal plants' potential as a part-time enterprise on a market garden farm. Wise were the women and men who knew which plants to harvest and store for teas and tinctures for pregnancy and childbirth, for easing pain and calming the spirit. Locally sustained culture needs these wise people again.

A five-acre permaculture farm can easily provide an herbal entrepreneur with adequate plant material. I do not have the time to develop expertise as an herbalist or the investment capital to fund a start-up herbal products enterprise at our farm. I do, however, know that every neighborhood needs an herbalist — as well as a market gardener.

Plants as Medicines

The term *medicinal* covers a broad range of plants and applications, including tonic plants and more specific medicinals. Tonics provide a concentrated dose of minerals and vitamins. Dandelion and yellow dock fall into this category. When mild deficiency is the root of illness,

Calendula is a medicinal plant and provides edible flowers.

DARRELL FREY

such tonics can be excellent remedies. Historically, spring tonics rejuvenated our forbearers with valuable nutrients after lean winters of stored foods. Tonic plants are rich in phytochemicals. Many of these chemicals are used by the plants to fight microbial and fungal invasion. Our current understanding of vitamins (and our growing understanding of phytochemicals) only scratches the surface of the complex interaction of plant chemistry and our bodies.

Many common plants are more specific medicinal plants — those with particular concentrations of minerals and chemical compounds that our bodies can use to fight injury, infection, illness and maintain good health. Plants produce many more chemicals than we know about. Stephen Buhner presents a view of this complexity in *The Lost Language of Plants*. Buhner gives the example of yarrow, which contains over 125 chemical compounds. The isolation of salicylin, the chemical precurser of asprin, from willow bark led to the development of aspirin, which is a powerful medicine. But natural salicylin is but one component of the total complex chemistry of willow bark. It is likely to be the *synergistic effects* of these chemicals and nutrients that provide the best, most healing medicine.

At times, we want fast-acting synthetic medicine, but we may have to pay a physical price in side effects. At other times, natural plant-based medicine may give us a more complete and long-term healing.

Making herbal teas is the easiest way to use many tonic and medicinal plants. Many can be used either fresh in season or dried for later use. Others are best preserved as tinctures and extracts. The active constituents of medicinal plants may be soluble in water, alcohol or oil. A tincture is made by soaking a plant in grain alcohol or whiskey to extract the medicinal component for storage and easy use. Oil extracts are commonly used for medicinal ointments and body oils.

Damask rose.

We begin collecting medicinal plants for our own use in March and April, when we gather coltsfoot blossoms. Some we dry for tea and some we use to make a tincture. We use both for colds and bronchitis. All year long, we gather the roots of dandelion and yellow dock as we prepare garden beds. These are chopped and dried in the oven for teas or made into a tincture. We also collect and dry bee balm, peppermint, spearmint, lemon balm, chamomile, lemon basil, clover blossoms, alfalfa and stinging nettles. All are used for nourishing and tonic teas.

We make herbal body oils each year. Our cabbage rose hedge provides an abundance of fragrant buds and blossoms each spring. This rose is common on old farms in our area and is probably the damask rose 'Bella Donna.' Damask roses are especially well known as a source of rose oil for making perfume. For several weeks in June, we collect the freshly opened flowers and pack them in a gallon jar with apricot kernel oil and soak them for several days. Then the oil is strained off and we soak a second batch of flowers in it. This is strained again, bottled and then stored in a cool place until needed for use as a massage oil, bath oil or general body oil. Less expensive

Plants known to be medicinal at Three Sisters Farm

Annuals:
- calendula
- garlic
- onion

Perennials:
- angelica
- bergamot
- boneset
- burdock
- comfrey
- dandelion
- coltsfoot
- echinacea
- lilac
- mint
- motherwort
- nettles
- pennyroyal
- rose
- St. John's wort
- valerian
- violet
- yarrow
- yellow dock

Shrubs and trees:
- basswood
- willow
- alder
- black cherry
- slippery elm

oils, such as olive oil, may work just as well as a soaking solution, but we prefer apricot kernel oil because it has good texture, is resistant to becoming rancid and has little scent of its own.

We also make oil extract with lavender, the inner bark of the birch tree, rosemary, and thyme. The procedure is the same. Each of these oils can be used to make soaps and salves.

Crafts

We have dabbled in many crafts over the years. Wreaths, herbal vinegars, dried flower arrangements and herbal tea blends are all products easily made on a small farm. Combined in a gift basket with pressed flower cards, herbal oils and soaps, they have even more value.

Each season offers the opportunity for creating value-added craft products. The creative craft person can find many ways to use dried herbs and flowers for artistic creations.

Popular herbal vinegars include chive, nasturtium and tarragon. Herbal oils, for baths, massage and fragrance are made by soaking fresh herbs in oils, such as apricot kernel oil. After several days, the herbs are strained out and the oil filtered. Lavender, damask rose, rosemary and birch bark all make excellent body oils. Culinary herbal oils, made from soaking herbs in olive oil, can be made with thyme, rosemary, sage, and basils.

Herbal wreaths are made by wrapping fresh flower bouquets onto a wreath form and allowing them to dry. Any everlasting flower makes good wreaths. Some of our favorites include celosia, Queen Anne's lace, goldenrod, various sages, lavender, larkspur, annual clary sage, cinnamon basil, artemesias and sweet Annie. Wreaths can be made of single herbs or colorful combinations. Tansy flowers are good for wreaths and dried flower arrangements. Fragrant herbs and flowers can be also be dried in season for potpourris.

We keep a flower press in the bioshelter for easy access. Many flower and leaves are pressed in their season for later use in making greeting cards.

Plant Sales

When we first conceived of our farm, we wanted to develop a useful plant nursery. Once a permaculture farm is established it can be replicated, to an extent, by plant propagation. We have not developed this aspect of the farm to its full potential, but we do sell some plants each spring. Because many other area greenhouses sell vegetables and flower seedlings, we decided to focus on herbs, useful perennials, pollinator plants and native wildflowers. We also have developed a market for heirloom tomato seedlings. We sell about 400 tomato plants each year. We generally offer 12-20 varieties, which might include German Pink, Black Krim, Brandywine, Cherokee Purple, Aunt Ruby's German Green, Principe Borghese, Amish Paste, Green Zebra, Jaune Flamme, Austin Red Pear, Yellow Pear, Nyagous, Giant Syrian, and Hillbilly. Sungold are also good sellers. This selection of tomato varieties includes large slicers, small slicers, canning tomatoes, drying tomatoes, small salad tomatoes, green, and purple tomatoes.

We offer for sale many of the useful plants that grow in the bioshelter and herb gardens. Scented geraniums, with their edible flowers are popular. Herb plants, especially variegated varieties, are also popular. Stevia, lemon verbena, pineapple sage, and large rosemary plants are also big sellers. We also offer insectary plants: motherwort, bronze fennel, catnip, mints, tansy, and flowering oregano. Two of our most popular plants are the prolific self-seeders: sweet woodruff and forget-me-not.

Native Plants

Some wildflowers we start from seed and some we "rescue." By rescue, I mean we collect them from locations where they would otherwise be killed by road crews or utility line workers. Typically, we collect wildflowers within a few miles of our farm, along the roadsides in forested areas. The space between the road and the forest is regularly cut, and the ditches are cleaned out by the township. In these disturbed grounds, we have found many plants: purple-flowering raspberry, bergamot, Dutchman's breeches, wild bleeding hearts, red and white trilliums, wild ginger, blue and black cohosh, ramps, many types of violets, red elderberry, horsetail, jewelweed, wild phlox, wild geranium, asters, and goldenrod.

Spice bush, hazelnut, elderberry and a few other shrubs and wildflowers we gather on our own property from eroding stream banks and where they encroach on our driveway.

We also have several patches of ostrich fern and cinnamon fern established over 20 years ago. These spread quickly and can be thinned

Wood geranium.

each year for sale. The ostrich fern is a valuable wild edible, but selling the plants is more profitable.

Seed-grown wildflowers are started in December. Seeds are sown in flats of sterile soil mix. The flat is moistened, refrigerated for a month, and then placed under grow lights in mid-January. In a few weeks, most have germinated, and they are transplanted to 60-cell 6-packs. In March and April, they are potted up for sale in 4-inch pots. Seed-grown wildflowers that we sell include New England aster, columbine, Joe Pye weed, green coneflower, purple coneflower, black-eyed Susan, woodland sunflower, bergamot, great blue lobelia, showy tick-trefoil, Virginia spiderwort, and wild lupine.

We continue to plant more native plants all over the farm each year. These add diversity to the landscape and provide seeds and plant divisions for our plant sales enterprise.

Honey

Honey, beeswax and other bee products should be part of a small-scale intensive farm's product list. Our hive, however, lives untended. We have occasionally harvested some frames of honey, but we are too busy to tend them for high production. Many farms develop connections with area beekeepers, who maintain the farm's hives in trade for honey. We have found that honey production requires so much management time that it is incompatible with our intensive gardening. We simply do not have time. Bees and their role are discussed more thoroughly in Chapter 6.

Perennials

"There were so many things a tree could do: add color, provide shade, drop fruit, or become a children's playground, a whole sky universe to climb and hang from; an architecture of food and pleasure, that was a tree."

—Ray Bradbury, *The Martian Chronicles*

After the basic infrastructure for the farm was in place in 1989, we were ready to begin placement of perennials, gardens, and season extenders. We ran a hedgerow from the north tree line south, along the edge of the driveway. This provided space for berries and roses, and it would act as wildlife habitat. The hedgerow also served as a windbreak for the compost yard and bioshelter and as a snow fence for the driveway. Plantings, from north to south, include staghorn sumac, plum trees, thornless blackberry, fragrant damask roses, and serviceberry. After a gap that allows access to the Pond Garden, the hedge continues with highbush cranberry, another rose bush, Jerusalem artichoke, iris, royalty purple raspberries, August red raspberries and four varieties of blueberries.

On the north tree line we planted Heritage, Latham and Golden raspberries, North Star cherry, Reliance peach and Heartnut.

We extended an orchard begun by our neighbor along the western edge of the property and the driveway. We planted Seckel pears and apple varieties that included Northern Spy, Liberty, Freedom, Granny Smith, Red Delicious, and several wild seedlings. We spaced them so that the full-size trees would be about 30 feet apart. We later added a second row of trees along the west edge, at the top of the West Garden. These include apple varieties Golden Russet, Yellow Transparent, and Black Oxford and Summer Crisp pear.

Other apples are planted on the south property line bordering the public parking space. These delicious small apples include Chestnut,

Our Woodland Property

Our family home is located on a ten-acre lot near Carlton, Pennsylvania. The property is a land of much variety. Roughly a rectangle measuring 400 feet by 1,200 feet, the lot has two streams and a mix of bottomland, slopes and uplands. A seasonal stream crosses the southwest corner for about 500 feet and joins a year-round stream that also passes through the lot. The elevation rises 50 feet from the year-round stream to the upland building site. Fifty years ago, the property was part of an overworked and abandoned farm with degraded soil. The lot included two small fields and hilly pastures reverting to forest. The first trees to grow were quaking aspen on the wet slopes and wild crabapples and hawthorns in the drier areas. These trees provided shelter for many other trees to take root and grow under their protection. As the trees began to mature, native species of shrubs and understory plants suited to the site's varied aspects, soil quality and soil moisture moved in.

Today three forest systems are present and contain dozens of species of native shrubs, wildflowers, ferns, mosses and fungi, as well as many species of birds and wildlife. In the bottomland, there are riverine trees: black walnut, ash, butternut and willows; on the moist slopes grow aspen, ash, maple, tulip poplar, shagbark hickory and cucumber magnolia; on the dry upland there are cherry, maple and oaks. This forest is surprisingly diverse considering that only 50 years ago it was over-worked and abandoned pasture and field.

Centennial, and the tart Dolgo crab. (Another Dolgo crabapple is the centerpiece of the Spiral Garden.)

A row of mulberry trees is planted along the north edge of the East Garden. This acts as a windbreak for our bale-shelter polytunnel and provides shade for a compost area. Under the mulberries, common blue violet established itself. The violet flowers and leaves are harvested for our spring salad mix. After the spring harvest passes, we allow our chickens to forage the fallen mulberry fruit.

The Farm Woodlot

Every farm benefits from having access to a woodlot. Kindling and firewood and stakes, posts and saplings for trellising are all obtained in a well-managed forest. Three Sisters Farm has access to resources from our own tree line and our woodland home property, five miles from the farm. From the tree line, we gather fallen maple and dogwood branches to use for starting winter fires. But, because the tree line is relatively young, we do not thin out saplings. Instead, we get most woodland materials from our home property. Craft materials are gathered from both our gardens and the farm landscape. Grape vines and hardy kiwi vines are used for wreaths. Bamboo is used for trellising. Aspen and willow branches and other saplings pruned from the utility line are crafted into bentwood trellises and gates. Pruned wood from our fruit trees provides the base for our spring bonfire. Black locust and honey locust trees planted 20 years ago are now yielding posts for our garden fences and trellises.

We get our garden stakes for free from a local sawmill that specializes in making survey stakes. They give away the poor-quality stakes, which are just fine for our uses. We glean some firewood from our land, but also buy some and cut some (with permission from property owners) so that we can allow our own forest to develop naturally. Locust and cedar posts are purchased for most fencing needs.

Managing Plant Interactions

Plants affect other plants in many ways, some beneficial and some not. Understanding these interactions is important when designing a permaculture farm.

In a forest, spring wildflowers grow and begin to flower before the leaf canopy shades the forest floor. This brief period of growth is an

Wood and its uses on the farm:

- Bentwood gates: saplings of willow, hickory, birch, sassafras, aspen, ironwood, bamboo.
- Posts: cedar, black locust, post oak.
- Stakes and plant supports: hawthorn, oak, cherry, maple, bamboo.
- Trellises: willow, hickory, birch, sassafras, maple and ironwood.
- Shitake logs: white oak is best, but can also use other hardwoods that hold on to their bark for some years after inoculation.

important part of the forest ecology. These plants reduce erosion and capture nutrients that would otherwise leach away in heavy spring rains. They grow fast, set seed, and often die back by midsummer, returning some of the nutrients to the leaf litter. While growing, the spring understory provides early-season food for native pollinators and other insects. After the trees leaf out, their shade reduces competition from grasses and thins out tree saplings, maintaining the conditions preferred by the wildflowers. Over time, the complex mix of trees, shrubs, herbaceous perennials and associated fungi, insects and animals builds the soil's store of nutrients and organic matter. Burrowing animals, fallen trees and deep-rooted plants, over time, turn and loosen the soil, keeping minerals in cycle.

In our gardens, trees and shrubs create shaded microclimates for other plants and provide structure for climbing vines. Many plants delve deep and accumulate minerals contributed to the soil by dead plants.

A Note about Bamboo

Bamboo is an extremely useful resource on the farm. But it is a *highly invasive* perennial that is difficult to eradicate once established. Before planting bamboo, determine how it can be contained.

One method is to grow bamboo in heavy plastic planters, like a child's wading pool. We think a better approach is to establish the patch where it will meet a barrier: along a building's foundation, at the edge of a mowed area, beside a road, or at the edge of water (pond, stream, lake) or a dense forest — or any combination of these. Any of these will stop the spread of the rhizomes and act as a suitable containment strategy.

We have successfully contained bamboo between a stream and a dense forest, between a composting outhouse and a mowed path, and between the pond and a mowed path.

Dual-use bamboo is best, so pick a variety you will want to harvest for food as well as stakes. Early spring shoots encroaching into the yard can be harvested. Rhizomes spreading beyond the chosen boundaries can also be dug and cut to reduce spread. We grow yellow grove bamboo (*Phyllostachys aureosulcata*). The shoots are eaten both raw and cooked. Most winters, the wood dies back, but generally it re-grows well each year.

An enterprising farmer can get harvesting permission for an already rampant patch on a nearby property. Heavy harvest of mature canes can help limit the spread of a bamboo patch, keep a farm supplied, and make a property owner happier.

Bamboo may have unexpected uses. While harvesting a patch of bamboo near Pittsburgh, I noticed that the bamboo had overwhelmed and killed several large patches of the equally invasive, but much less useful, multiflora rose. The dense shade created by the bamboo was too much for it.

Dogwood trees perform this role in the Appalachian forest. Others, especially legumes, but also alders and *Elaeagnus* species, fix nitrogen and add it to the system. Joseph Cocannouer, in *Weeds: Guardians of the Soil,* discusses the action of deep-rooted weeds in loosening the subsoil that allows the roots of crop plants to follow them deeper to access water and minerals. Crabapple, hawthorn and aspen trees act as "nurse plants," reducing understory competition and providing the shade required for a succession of hardwood trees.

In describing the creation of complex multi-story communities of useful plants, Bill Mollison uses the concept of a *plant guild.* In this sense, a guild is a group of useful plants that benefit from close association. A guild may include a dominate tree and an understory of shrubs, vines, perennials, biennials and seeded annuals. The guild would require nitrogen-fixing legumes, accumulator plants and insectary plants for promoting pollination and pest control. A classic plant guild is the three sisters: corn, beans and squash.

Plants interact chemically in many ways with each other and with insects. In *The Lost Language of Plants,* Stephen Buhner documents the complex chemistry of plants. He says there are 10,000 alkaloids, 20,000 terpenes, and 8,000 polyphenols that are already known, and others are continually being identified. Buhner discusses how the "bioactivity" of these compounds increases when combined. David Jacke discusses the complexity of interactions between plant chemistry and the soil in *Edible Forest Gardens: Volume I.* As an emerging field, there remains much to learn about these complex interactions. As designers, we need to be conscious of the many ways plants interact — with each other, with insects and with animals — and keep alert for new discoveries.

Nitrogen Fixing

In addition to the application of compost made with animal manures, cycling nitrogen into the garden is accomplished by the inclusion of plants capable of establishing a symbiotic relationship with nitrogen-fixing bacteria. For the farm, there are two primary types of plant that do this: the legumes (bean relatives); and actinorhizal species, such as alder and *Elaeagnus.* When associated with these plants, bacteria can take nitrogen from air in the soil and convert it into a form plants can use. Nitrogen fixation by these microorganisms is the primary way nitrogen enters the soil. Lightning storms can also convert nitrogen to a form usable by plants, but the extent of that contribution is difficult to assess.

Before planting, beans and peas are inoculated with powdered legume bacteria. Using inoculated beans in crop rotations every few years allows beans to replenish nitrogen in the soil. Many other legumes have found niches on our farm. White clover grows well in permanent pathways. Red clover and alsike clover self-sow in our fallow garden beds. Yellow clover has found a niche in mowed areas between gardens and along pathways, as has bird's-foot trefoil. We have included tick-trefoils in our native plant gardens. Black locust trees around the gardens and alders along the pond's edge also add nitrogen to the landscape.

Siberian pea shrubs are a valuable multipurpose, perennial legume. Like the black locust, they provide good bee and chicken forage.

Elaeagnus species, including buffalo berry, Autumn olive, Russian olive and goumi, have associations with nitrogen-fixing bacteria. We have not established the potentially invasive Autumn olive at the farm and have yet to introduce any other *Elaeagnus*.

Mycorrhizal Associations

Virtually all plants form symbiotic associations with fungi. In these associations, the plants provide sugars and other products of plant chemistry to the fungus in exchange for minerals and other products of fungal chemistry. These fungi include both mushrooms and microscopic fungi. Some fungal associations are very species specific. Lady slipper orchids, for example, are difficult to propagate without their native soil. Other mycorrhizal plants can use a range of fungi. Application of commercial mycorrhizal inoculants is recommended for worn soils negatively affected by chemical farming. Well-established and organically managed soils are less likely to need inoculants. However, if in doubt, adding these valuable fungal allies is a good way to ensure a healthy landscape.

Accumulator Plants

Many plants are known for their ability to gather and accumulate specific minerals from the soil. Including these plants in green manures, fallow rotations, compost piles, orchards, forest gardens, hedgerows and poultry forage yards helps maintain a balance of available minerals in

Nitrogen-fixing plants at Three Sisters Farm:

Annuals:
- beans
- peas
- fava bean
- red clover
- white clover
- yellow clover
- alsike clover
- sweet clover
- bird's-foot trefoil
- tick-trefoils

Leguminous trees:
- black locust
- alder
- bayberry

the system. Robert Kourik's *Designing and Maintaining Your Edible Landscape Naturally,* and Jacke's *Edible Forest Gardens: Volume II* provide detailed lists of accumulator plants. *Phytoremediation,* the use of plants to clean soil of heavy metals, is another use of accumulator plants.

Allelopathy

Many plants produce chemicals called *allelochemicals,* that can inhibit the growth of other plants. Allelochemicals may be active for only a few weeks, or they can remain in soil for months or even years. These persistent chemicals are also known as *phytotoxins.* Straw mulch from oats, wheat and possibly other grains leach phytotoxins into the soil that may inhibit seed germination. This is good when using straw mulch for transplanted crops. But it may not be good if a direct-seeded crop is planted in the mulch or directly follows a mulched crop.

Goldenrod has been shown to inhibit germination. Some plants inhibit their own kind. We have observed that rosemary will not self-seed directly under or near a mature rosemary plant. Many weeds and crop plants are allelopathic, including include spotted knapweed, *ailanthus,* ragweed, and artemisias. Our observations suggest that the effect is probably more pronounced when plants grow in dense stands, as with goldenrod.

Black Walnut

Black walnuts are a good protein source and an excellent source of health-promoting oils. However, care must be taken when planting them on the farm. The roots of the black walnut, butternut, or any Persian walnut grafted onto black walnut root stock produce a allelopathic substance known as *juglone,* that is toxic to some plants. The toxic zone extends as far as the roots spread. This can be a radius of 50 to 60 feet around the trunk of the tree — or even more for larger trees. The toxic zone will increase each year as the tree grows. Juglone toxin occurs in the leaves, bark, and wood of the black walnut, but at a lower concentration than in the roots. Juglone is poorly soluble in water and does not move very far in the soil. Black walnut sawdust and chips from tree pruning can all contain juglone. Walnut and butternut leaves can be composted because the toxin breaks down when exposed to air, water and bacteria.

Some plants will not grow well or may be killed when planted within the root zone of black walnut trees. These include tomato, potato, blackberry, apple, crabapple and most azaleas and rhododendrons. But not all plants are sensitive to juglone. Some plants known to grow under or near the trees include eastern redbud, hollyhocks, rose of Sharon, black raspberry, calendula, pansy, peach, cherry, plum, sweet woodruff, and spiderwort. Vegetables such as squashes, melons, beans, carrots, and corn can be grown near black walnuts as well. Wildflowers that cohabitate with black walnut include violet, bergamot, Virginia creeper, trillium, bloodroot, Jack-in-the-pulpit and others. The late Tom Mansell, of Aliquippa, Pennsylvania, developed a productive tree crop guild that included black walnut interplanted with pawpaw, raspberry and asparagus. His forest garden landscape was managed from the late 1940s through the 1980s and included a hedge of filberts and hazelnuts enclosing the yard, Chinese chestnuts, 20 varieties of pawpaws and 80 apple varieties grafted onto 25 trees.

Invasive Plants

Great care should be taken before adding invasive exotic plants to the landscape. Thoughtlessly adding inappropriate plants to a system is contrary to the permaculture ethic of care of the earth. Each plant added to a system should be analyzed for its properties and ecological relationships and placed in a functional relationship to the rest of the landscape. Invasive plants should only be planted when and where they will be controlled.

Many of our most useful wild edibles and crops were imported from Europe and Asia. We do not want to eradicate most of them; at Three Sisters Farm we embrace them as good food for people and animals or as important parts of the farm ecosystem. But if you have done battle with Japanese knotweed or multiflora rose, you understand the need for careful consideration before adding invasive plants. Even plants we want to add to the landscape, including mints, comfrey, tansy and stinging nettles can be invasive and need to be planted where they can be controlled.

Many of our major pests and plant pathogens, including Chestnut blight, Dutch elm disease, zebra mussels, gypsy moths, honeybee mites and Japanese beetles have been imported from other, similar climates. Invasive non-native plants such as Japanese knotweed, garlic mustard, Japanese honeysuckle, multiflora rose, privet, spotted knapweed and Canada thistle are imported plants. If I were asked, I would suggest a moratorium on importation of biological materials from any other temperate climate. Enough of this continuous assault on nature! As climates change and habitats are fragmented, continued importation of biological materials will only add more diseases and pests and further stress bioregions struggling to adjust. As good stewards of the planet, we must preserve and protect as much wilderness, wildland and natural area as possible. At the same time, we must intensively cultivate and sustainably manage the land we do use. How to balance intensive cultivation with wilderness will be a continuous debate. We can only hope this debate is informed by an agreement on the need for good ecology and healthy bioregions.

Uncle Marlin's Garlic

Many times over a period of 15 years, my Uncle Marlin gave me bunches of his garlic in the fall. I would not quite get around to planting them. We always just ate it. Then, one year, I went to see Marlin after his house had flooded from heavy rainstorms. The flood had risen to within inches of the bunches of garlic hanging in the cellar stairway. "You better plant it this year," Marlin, age 73, recommended as he gave me my yearly supply. "I am not getting any younger you know." This time I paid more attention to the story of his red Italian hard-stem garlic.

Some of my earliest memories include eating the fresh tomatoes, onions and radishes from Marlin's garden. He has maintained the same plot in his backyard since 1964. Marlin retired from the Muncy Chief Hybrid Seed Company after 43 years of breeding field corn and feed grains. He always saved seeds from his prized crops in his home garden. The garlic was a gift from the grandfather of his lifelong companion, Judy.

Born in 1902, Judy's grandfather came from a small town near Naples, Italy to seek a new life in Williamsport, Pennsylvania in the early 1920s. Poppa Piccolo was a huckster, or neighborhood produce vendor, for much of the first half of the 20th century. The origin of the garlic is not clear, but when he gave Marlin a bunch of bulbs in 1985 he told Marlin he had been growing it for over 60 years. When he passed away at age 93 in 1995, Poppa Piccolo's garlic was in good hands. Now it is in mine.

This flavorful and relatively mild garlic has been a main ingredient in the oily hot peppers canned by ☞

Groundcovers and Lawns

Permaculture design seeks to mimic natural process, both to create fertility and to add diversity to a landscape. In nature, soil is protected with a covering of mulch, sod or other dense plant growth. Soil drainage, fertility, pH and amount of sunlight all affect the type of cover.

Under a canopy of trees, leaves and other organic materials accumulate in a mulch layer. The resulting leaf mold protects the soil from erosion, excess drying and temperature extremes. All the while, the mulch feeds the soil and the community it houses. In the home landscape, we mimic this process by planting grass and mulching plants with wood chips, bark mulch and leaf mold from composted yard waste.

On borders and banks and in foundations plantings, plants can replace grass and mulch. Perennial groundcovers for this purpose include useful plants like chamomile, sweet woodruff, oregano, thyme, and lemon balm, and ornamentals such as periwinkle, pachysandra and English ivy. On the north side of a building and under trees and shrubs, wild woodland plants do well. These include ferns, wild ginger,

Judy's mother for many years. The barter is simple. Marlin provides the hot "banana" peppers and garlic to Judy's mother. She sends back some jars of pickled peppers. Eating these peppers with Marlin is another vivid tradition from my youth.

I always counsel friends and students to save favorite seeds and to adopt heirloom vegetables. The list of seeds maintained at Three Sisters Farm includes purple pod soup peas, Seneca Indian pinto beans, favorite annual flowers and various tomato varieties. We budget our seed purchase dollars so we can support the efforts of The Seed Saver's Exchange and Seeds of Change. Each year we buy up to 20 varieties of heirloom tomatoes and other plants from these sources. As an advocate of varietal preservation by individual gardeners, I advise finding varieties that are suited to your local conditions and have special meaning to you.

I regret that it took so long to establish my own plot of Uncle Marlin's garlic. I was always so busy, and it was too tempting not to eat it. But the fact that Marlin's garden provided it every year made me take it for granted. After the most recent flood however, I was reminded — by the gardener himself — of the gardeners' mortality. The tradition of passing on prized crops to family and friends is an ancient part of being human. Uncle Marlin's health and vigor is excellent (no doubt due, in part, to good garlic), and he expects to continue gardening for years to come. Still, I think we both are a little happier knowing Poppa Piccolo's red Italian hard-stem garlic has found another gardener to carry on the tradition.

Doorstep Kitchen Garden

A doorstep garden can add ultimate freshness to a kitchen. Transforming the patio and adjoining yard into a living, breathing spice rack can transform a ho-hum cook into a world-class chef. Exploring new dimensions of flavor and aroma is a journey that can begin at your doorstep.

The doorstep kitchen garden adds life to a patio, enriching the landscape with scent and color. Contrasting greens of culinary herbs, mingled with multicolored chard and flaming red hot peppers beneath trellised orange Sungold cherry tomatoes, provide a savory setting to inspire culinary excellence. When designing a kitchen garden, placement is important. Keep planters away from the grill. At the same time, the closer the plants are to the patio, the more you will use them. By interplanting herbs and vegetables in pots and beds, a broad palette will be at hand and ready for the cutting board. Each plant can be grown in as little as 1 or 2 square feet of space. These plants like rich, balanced soil and a lot of compost. Most will thrive in a sheltered location with lots of sun. A raised bed system makes harvest easier. Perennials and annuals alike can be mulched with shredded bark if the fertility of the soil is high. Weekly application of liquid seaweed and fish emulsion will ensure productivity. A simple footpath through the doorstep garden is enough to provide access to the freshest ingredients. ☞

trillium, sweet woodruff, bloodroot, bleeding hearts, Virginia bluebells and others. Currants, gooseberries, and numerous other woodland shrubs do well in shaded borders. Wild plants should be purchased as nursery-grown plants or rescued from development projects, but not transplanted from the wild. Most prefer an annual mulch of leaf mold.

Lawns can be replaced with meadow plants. When selecting them, it is best to select a regional mixture. In the northeast, meadows include goldenrod, asters, arnica, American knapweed, moth mullein, ox-eye daisy and black-eyed Susan. Annual mowing helps keep out brambles and trees and maintains a space in the "meadow stage" of succession.

Where a lawn is desired, a more natural and native lawn can be developed. A lawn ecosystem includes a variety of plants. Some of my favorite lawns are a beautiful mix of mosses, grasses and wildflowers between mature trees. Another wonderful mix is violets, veronica, dogtooth lilies and wild orchids interspersed among merging patches of native moss and planted fescue or bluegrass.

Yarrow planted thickly and mowed makes a soft carpet, as does creeping thyme. American pennyroyal and peppermint will also persist through regular mowing near downspouts and other wet spots. White

The kitchen garden begins with the culinary herbs. Potted, in planters or in beds, a great variety of herbs can be grown around the patio and kitchen doorstep. Fresh sprigs of thyme — German, creeping or lemon — are easily stripped from the stem and sprinkled on roasting root vegetables. Branches of rosemary, with their evergreen freshness, complement any meat, red or white.

Fresh trout, bass, and other whole fish can be stuffed with chopped garlic chives and shredded French sorrel, wrapped in chard and mustard leaves and grilled to moist perfection without falling apart. Salmon steaks likewise can be topped with lemony sorrel and chives.

Chicken marinated in fresh tarragon and vinegar is a dish to relish. Chicken topped with a coarse pesto of parsley and basil or cilantro will have your guests inviting themselves back for more. Give them a fresh, cold beer and grilled flatbread pizza topped with fresh chopped basil, cilantro, parsley, oregano and tomatoes — and you will have them moving in.

Do not forget to include vegetables in the kitchen garden. Small plantings of a variety of vegetables in pots and beds near the patio can supply shish kabobs of baby squash, cherry tomatoes, eggplants, peppers, scallions, snow peas, beets and even new potatoes. The creative cook can add artistry to the presentation by gracing the plate with garnishes of herb sprigs and edible flowers.

Dutch clover provides nitrogen and, if you are lucky, the occasional four-leaf clover. Creeping yellow clover is an interesting addition if you can find it. Dandelions reach deep into the soil and bring nutrients to the surface for the other plants to use. They make great playthings for children as well. Violets, blue-eyed grasses and bluets and other small plants add interest and diversity. Of course, your choice of plants to use should be guided by your climate and soil type.

A permaculture landscaping plan considers the impact of the landscape on the surrounding community. By reducing lawn space, we save energy and reduce noise and air pollution. Diversifying the landscape with native shade trees, shrubs, meadows, and groundcovers builds the health of the local environment by providing bird and insect habitat. Natural mulches and diverse plantings enhance soil fertility. The resulting landscape is natural, organic, beautiful and functional.

Cultivation Guide for Native and Habitat Plants

Common Name	Botanical Name	Height	Color	Bloom Time	Moisture	Sun	Habitat
Black-eyed Susan	*Rudbeckia hirta*	1–3'	golden yellow	June–October	dry to moderate	full to partial	fields, prairies and open woods
Black cohosh	*Cimicifuga podocarpa*	3–8'	white	June–September	moderate to wet	partial to shade	woods with deep, loamy, moist soils
Blue cohosh	*Caulophyllum thalictroides*	1–3'	purple–brown, yellow green	April–June	moderate to wet	partial to shade	woods with deep, loamy, moist soils
Blue false indigo	*Baptisia australis*	2'–5'	blue	June–July	moderate	full to partial	thin woods and along streams
Common St. John's wort	*Hypericum perforatum*	1–2.5'	yellow	June–September	dry to moderate	full to partial	naturalized—found in fields, roadsides and waste places
Dutchman's breeches	*Dicentra cucullaria*	4–12"	white	April–May	moderate	partial to shade	rich woods — prefers to grow at the base of beech and maple trees
Dwarf ginseng	*Panax trifolius*	4–8"	dull white to pink	April–June	moderate to wet	partial to shade	moist woods and damp clearings
Evening primrose	*Oenothera biennis*	2–5'	yellow	June–September	dry to moderate	full to partial	fields and roadsides
Great blue lobelia	*Lobelia siphilitica*	3–4'	bright blue	August–September	moderate to wet	full to partial	rich lowlands, meadows and swamps
Green-headed cone flower	*Rudbeckia laciniata*	1–3'	yellow	June–October	moderate to wet	full to partial	fields, prairies and open woods
Hepatica	*Anemone americana*	4–6"	pink, lavender or white	March–June	moderate	full to partial	woods with sandy to loam soils
Jacob's ladder	*Polemonium van bruntiae*	1.5–3'	bluish–purple	May–July	moderate	partial to shade	swamps, bogs and rich woods
Jewelweed	*Impatiens capensis*	2–5'	orange and yellow	July–September	moderate to wet	partial to shade	shaded wetlands and woods

Cultivation Guide for Native and Habitat Plants cont.

Common Name	Botanical Name	Height	Color	Bloom Time	Moisture	Sun	Habitat
Joe Pye weed — plain	*Eupatorium fistulosa*	3–8′	pinkish lavender	July–September	moderate to wet	full to partial	damp meadows, shorelines
Joe Pye weed — spotted	*Eupatorium maculatum*	3–7′	pinkish lavender	July–September	moderate to wet	full to partial	damp meadows, shorelines
Miterwort	*Mitella diphylla*	8–18″	tiny white	April–June	moderate	partial to shade	rich woods
New England aster	*Aster novae-angliae*	3–7′	bright purple	August–October	moderate	full to partial	wet thickets, meadows, and swamps
Ostrich ferns	*Matteuccia struthiopteris*	Up to 5′	----------	----------	moderate	partial to shade	woods with moist rich soils
Ox-eye daisy	*Leucanthemum vulgare*	1–3′	white and yellow	June–August	dry to moderate	full to partial	waste places, meadows, pastures and roadsides. can spread rapidly.
Purple coneflower	*Echinacea purpurea*	1–5′	purple	June–October	dry to moderate	full to partial	dry open woods and prairies
Showy tick-trefoil	*Desmodium canadense*	2–6′	pink	July–August	moderate to wet	full to partial	moist open woods and edges of fields
Spring beauty	*Claytonia virginica*	6–12″	pink	March–May	moderate	partial	moist woods, thickets, clearing and lawns
Swamp milkweed	*Asclepias incarnata*	1–4′	pink	June–August	moderate to wet	full to partial	swamp, shorelines and thickets
Toothwort	*Cardamine concatenata*	8–16″	white or pink	April–May	wet	partial	moist low woodlands and damp thickets
Trillium — red	*Trillium erectum*	1′	burgundy	May–June	moderate	partial	woods with rich, well-drained soils
Trillium — white	*Trillium grandiflora*	6–12″	white	May–June	moderate	partial	woods with rich, well-drained soils
Violet	*Violaceae*	3–5″	lavender, white, cream	March–June	moderate to wet	partial to shade	damp woods, moist meadows, lawns and roadsides

Cultivation Guide for Native and Habitat Plants cont.

Common Name	Botanical Name	Height	Color	Bloom Time	Moisture	Sun	Habitat
Virginia bluebells	*Mertensia virginica*	8–24"	light blue	March–June	moderate	partial to shade	floodplains and moist woods
Virginia creeper	*Parthenocissus quinquefolia*	40'	light blue, white	Early Summer	moderate	full to shade	can spread rapidly; handling plants may cause skin irritation
Virginia spiderwort	*Tradescantia virginiana*	8–24"	blue–violet	April–July	dry to moderate	full to partial	woodland borders, thickets, meadows and roadsides
Wild bergamot	*Monarda fistulosa*	2–3'	lavender	June–September	dry to moderate	full to partial	dry fields, thickets and borders
Wild blue lupine	*Lupinus perennis*	1–2'	blue	April–July	dry	full sun	dry, open woods and fields
Wild blue phlox	*Phlox divaricata*	8–12"	blue	April–June	moderate	full to partial	rich woods and fields
Wild columbine	*Aquilegia canadensis*	1–2'	red w/ yellow	May–June	dry to moderate	full to partial	rocky, wooded or open slopes
Wild ginger	*Asarum canadense*	6–12"	red brown to green brown	April–May	moderate to wet	partial to shade	woods with rich, loamy, well drained soils
Wild oats	*Chasmanthium latifolium*	2–3'	enclosed in seed head	June–July	moderate	full sun to shade	woods with rich, loamy, well drained soils
Wood geranium	*Geranium maculatum*	1–2'	pink	May–June	moderate	full sun	woods thickets and meadows

CHAPTER 8

Seasons of the Garden

*Oh! The things which happened in that garden! If you have
never had a garden you cannot understand, and if you have had
a garden you will know that it would take a whole book
to describe all that came to pass there.*
— FRANCES HODGSON BURNETT, *THE SECRET GARDEN*

MANAGING AN INTEGRATED AGRICULTURAL LANDSCAPE, with productive orchards and gardens, windbreaks and wildlife habitat, farm animals and pastures, as well as homes and bioshelters, is a complex task. Planning for a manageable pattern of activities throughout the year provides the template for a successful farm. This chapter is a study of a seasonal cycle that allows for multiple farm enterprises to thrive. It also addresses in more detail the management of gardens and the crops grown at Three Sisters Farm.

Winter

When the wind blows hard and crystalline snow drifts over the cold frames and around the polytunnels, the farm seems dormant. Few things stir in the gray, frosty days and cold, clear nights. The tracks of rabbits and the hurried flights of small birds are the main signs of natural life. The tree line is populated by the usual crowd, black-capped chickadee, tufted titmouse, cardinal, blue jay and downy woodpecker. A Cooper's hawk makes regular assaults on the population of house finches and other small birds hiding among the dogwood twigs. Small flocks of dark-eyed junco rush about the farm, from tree line to fruit tree, foraging wild seeds. On still days, the red-tailed hawk patrols the skies above, and small groups of crows raise their call in distant fields. Beneath the snow and under the mulch and sod, moles and voles continue to tunnel. Mice tunnel under the snow and nibble seeds gathered in the fall. Woodchucks hibernate in their dens, dreaming of crisp lettuce and dog's teeth.

Extremely cold weather can freeze the tips of fruit trees and shrubs, but winter temperatures in our area mostly stay above 0°. However, in these times of changing climate, many cycles are being affected: warm winters disrupt the dormant cycles of many creatures; spring flowers sometimes begin to bloom in December; early thaws bring the frogs out early. Since just 1990, western Pennsylvania has jumped from hardiness zone 5 to zone 6. But do not go out and plant that commercial peach orchard just yet. There is not enough temperature stability in the spring and fall to count on good bloom set every year.

Winter Solstice

The outdoor gardens lie dormant under the winter's snow. The beds are planted in a fall-sown cover crop of winter rye or with a mix of self-seeded "weeds." At our farm, chickweed is an especially valuable self-seeding winter cover. The evergreen, branching stems quickly form large mats to cover and protect the beds while at the same time providing valuable winter forage for songbirds. Other beds are protected with the mulch from the previous summer.

We are happy to take fresh salad and a winter bouquet from the bioshelter to our hosts for a winter solstice dinner. The winter herb bouquet is a lovely and fragrant gift. It includes fresh sprigs of scented geraniums leaves, flowering rosemary, variegated sage, flowering pineapple sage, flowering Mexican sage, nasturtium stems and flowers, thyme sprigs, flowering African blue basil, lemon verbena, flowering borage and catmint.

The solstice salad includes lettuces (red oak, buttercrunch, green romaine and red leaf), frisée, endive, spinach, baby red mustard, kyona mizuna, tatsoi, arugula, mache, chickweed, red Russian kale leaves, five-colored chard and flat leaf parsley.

Culinary herbs to select from for holiday cooking include thyme, sage, rosemary, bay, lemon verbena, marjoram, and oregano. Fresh herbal tea is made with spearmint, anise hyssop and borage.

Deep winter is the time to review the past year and plan for the year ahead. Seed catalogues are consulted for new and interesting plants.

Three Sisters Farm Planting Schedule

Timeline (months): Jan. | Feb. | March | April | May | June | July | August | Sept. | Oct. | Nov. | Dec.

Season markers: —spring thaw— · frost danger ends · frost danger · frost certain · snowfall likely

- *start early Solanacea--*transplant as needed--*pot up--*plant out under cover
- *start main crop Solanacea--*pot up   plant out
- *start herb seeds------*transplant to pots and flats---*hardy herbs planted out
- *tender herbs planted out
- *pot up self seeded herbs and garden plants
- *make cuttings of hardy herbs------*pot up as needed
- *make cuttings of tender herbs in greenhouse------pot up cuttings
- *begin outdoor plantings hardy greens, alliums, peas
- *early brassicas started----transplant as needed----plant out early brassicas
- *start late season brassicas--transplant as needed--plant out
- *plant fall pea crop
- *direct seeded kales and other brassicas
- *replace winter cold frame crops with summer crops
- *begin planting fall/winter, bioshelter/tunnel/cold frame crops
- *start early Swiss chard--
- *direct seed Swiss chard
- *early melons and cucumbers in plug flats
- *direct seed and plant out melons and cucumbers
- *second planting summer squash and cucumbers
- *start ornamental and wildflower seeds--transplant as needed
- *start early edible flowers------transplant as needed
- *start late edible flowers------transplant as needed
- *direct seed succession of corn/beans/squash
- *bring potted herbs and edible flowers in before frost
- *direct seed bush beans------
- *last chance to plant garlic
- *harvest garlic when ready
- *plant main crop garlic
- *start scallion onion seeds bi-weekly
- *plant onions out bi-weekly
- *plant radish weekly
- *plant fall radishes
- *plant spinach and kale weekly
- *plant fall spinach and kale for spring crops
- *start early head lettuce--plant out in cold frames and tunnels
- *start head lettuces weekly--transplant weekly
- *begin planting out weekly succession of head lettuces
- *stop planting head lettuce out
- *head lettuce planted in cold frames and tunnels
- *direct seed cutting lettuce in tunnels and cold frames
- *direct seed cutting lettuce weekly
- *last outdoor cutting lettuce
- *direct seed carrots and beets weekly
- *direct seed dill and cilantro bi-weekly
- *plant potatoes in succession
- *plant spring turnips
- *plant fall turnips
- *plant summer green manure crops
- *gardens beds prepared weekly
- *plant winter cover crops

* indicates earliest possible time

-- indicates time spread over which task can be accomplished

Seed supplies are inventoried and sorted. Germination tests are done on seeds in storage. Sales and production records are reviewed and notes on successes and failures of specific crops and crop rotations are incorporated into the planning for the coming year. Applications for organic certification have to be made each year — just another annual farm task.

The farm plan documents the farmers' intentions for each year's garden management. A field history for each garden keeps track of soil input, including lime and other minerals, compost, green manures, and crops planted. Soil tests made the previous fall are used to plan which new amendments will be needed for the planned crops. For many reasons — weather, available labor, market variation, etc. — crops actually planted and work actually accomplished invariably differs from what was planned. Comparing the past year's plan to what was actually done is an important task.

Winter is the best time to take inventory of tools and supplies. Repairing and purchasing tools and supplies in the winter saves time and trouble later in the season. It is the best time to do equipment maintenance so everything will be ready to go when you need it. Most years, there is a need to repair and replace irrigation equipment. Hoses, sprinklers, watering wands and connectors are inspected and inventoried. Changing the oil in the tiller and lawn mower, sharpening the mower blades, and greasing moving parts are just a few of the tasks

Germination Testing

Count out 50 to 100 seeds and place them between layers of paper towel in a shallow dish. Moisten the towels and place in a location with lighting and temperature appropriate to the species. After germination, count the successful seeds to determine the percentage of viable seed. If germination rate is low, adjust the amount of seed you plant or purchase new seed.

Keeping up with our Customers

Customers are contacted early in the spring. A farm newsletter is sent to subscribers and other customers to remind them about our fine produce and get them thinking about the year ahead. As winter changes to spring, grocers and chefs are contacted to renew a dialogue. Usually they are looking forward to getting fresh, local produce and will have new requests and questions about our plans. Caterers and chefs plan weddings and other events many months in advance and may already have special requests. Personal visits (at a convenient time) help renew the farmer-customer relationship. This relationship benefits greatly when customers sense the farmer's enthusiasm. Most of our chefs and caterers are interested in hearing directly about what is new in the farm's crop plans, but individual produce subscribers generally do not want such a personal relationship with the farmer. We try to limit contact to the newsletter and the postcard or an email that alerts our CSA subscribers we will be back in the spring. By April 15, the year's delivery schedule and first billing is mailed to subscribers.

that should be done. Keeping a maintenance task list posted in the equipment shed helps organize these jobs.

Herb cuttings, seedlings and transplants are started on a schedule for spring plant sales and garden transplants, so potting room supplies need to be inventoried and replenished. As the spring rush of seeding and transplanting begins, it is best to be fully stocked. Plant labels, marker pens, flats, inserts and pots ordered well in advance will avoid delays later. Some material will need to be cleaned and sorted. For example, we use reuse flats, inserts and pots until they wear out. But they need to be cleaned after each use. We clean them by soaking and then rinsing them with water and hydrogen peroxide bleach; they are put outside in full sun to dry. Once dried, flats, inserts and pots are sorted by size and stored for later use.

Tending Trees and Shrubs

Fruit trees are pruned during the winter months. The common rule is that any month with an "R" is a good month to prune, so September through April are pruning months. Pruning in warmer months, especially in humid weather, can allow disease to infect open cuts. Fall through winter pruning allows the trees to adjust before fungal and bacterial diseases become active.

Pruning is a complex topic and is not dealt with in detail here. But, basically, we have found that the more you prune a tree, the more you will need to prune it. For example, the popular "open-vase" style of apple tree pruning requires intensive annual pruning to maintain that form. Espalier forms require even more annual maintenance. We prefer to allow our trees to develop a natural form. Once the tree is established, it is a simple task to remove a few branches from each tree to allow good light penetration and air circulation. Pruned branches are removed from beneath the trees in case they harbor pests or disease. Prunings are used for a spring celebration bonfire, cut for kindling, or used as the bottom layer of compost piles, to aid aeration of the piles.

First Outdoor Harvest: Late March

The first outdoor spring harvest of wild greens is a good time to walk the farm and look closely at the condition of the gardens. Beds are inspected for problem weeds, drainage, and any cleanup needed. Mental notes

Winter To-do List:

- Get some rest!
- Prune trees and shrubs
- Check and repair tools
- Stock up on supplies
- Order seeds
- Reconnect with customers
- Mulch paths and fence lines
- Plan tours and classes
- Apply for organic certification
- Prepare cold frames and seedling frames
- Turn compost

are made of repairs needed to fencing and gates. Locations of plants to pot up for sale are noted, as are perennials that should be divided. Birdhouses are examined and last year's nesting sites noted. Woodchuck holes are located and monitored for activity.

The harvest begins late on a Tuesday afternoon and extends into the early evening. The evening air is full of sound. Mourning doves coo, robins, phoebe, red-winged blackbirds, and the ever-present house finch all add their songs over the background rhythm of spring peeper frogs. The spring peepers and a few other frogs with deeper voices begin their evening song in the late afternoon, urgently calling for mates as clouds move in and distant thunder rolls through the heavy, humid breeze. Several male red-winged blackbirds swoop around the pond, staking territory where the cattail will emerge. Robins scurry about, collecting last year's dried grasses to build this year's nests. Crows fly overhead, chased by smaller birds. Soon they will begin to nest in the treetops along neighboring fields and chase away the red-tailed hawks that circle overhead. The sun sets and the birds grow calm in the twilight. A great blue heron squawks as it glides toward the pond, leaving its nest for its evening forage. The frogs grow louder as the daylight fades.

Fall-planted baby carrots are pale but sweet. Last year's spinach is still frost-tinged but will recover and produce nicely. Chives are just beginning to wake up and stretch from their winter sleep. Dandelion, ox-eye daisy leaves and mache all have sprung up quickly after a week of warm weather.

Cold frames that were watered and weeded three weeks before — in the first days of weather warm enough to melt the winter snow cover — yield a plentiful harvest of salad ingredients. These include self-seeded crops: miner's lettuce, mache, chickweed, red Russian kale, dandelions, ox-eye daisy greens and chervil. Fall-planted spinach, lettuce, lacinato kale, Swiss chard, and arugula have all wintered over. A few remaining weeds, grasses, red dead nettle (*Lamium purpureum*), yellow dock and thistle are removed. Young dock leaves are used in the salad, but they grow quickly and will shade and suppress other plants. Dead nettles are edible raw, but they are a bit coarse, so we avoid them in the salad mix. Red dead nettle is also quick to set seed and is an abundant hardy winter cover in the gardens. But, being a prolific seed producer, it can be invasive, so we keep it out of our tunnels and cold frames.

Top left: *Jenkins' garden with grape arbor.*
Bottom: *Jenkins' zone one garden.*

Top right: *Orb weaver,*
Araneus marmoreus.

Top left: *Bioshelter and grow frames viewed from west.*

Top right: *Ladybug with larvae.*

Bottom: *Bioshelter interior plantings.*

CATHERINE PALADINO

CATHERINE PALADINO

DARRELL FREY

Left: *Bioshelter viewed from the south east.*

Top right: *Bioshelter in the spring.*

Bottom right: *Lettuce and various wild edibles.*

TOM GETTINGS

CATHERINE PALADINO

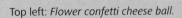

Top left: *Flower confetti cheese ball.*

Top right: *Wild columbine.*

Center right: *Bee balm edible flower, tea herb and a favorite of hummingbirds.*

Bottom: *Edible flower cake.*

Top: *Jujube Chinese date.*

Center left: *Wild crafted mushrooms, chanterelles are in high demand by chefs.*

Center right: *Fruiting shiitake mushroom logs.*

Bottom: *Sungold tomatoes in thyme.*

DARRELL FREY

DARRELL FREY

Top left: *Garter Snake.*

Right: *Spiral garden bench.*

Top left: *Ark 2010 south wall exterior.*

Center left: *Ark south wall.*

Bottom left: *Thermal mass pool.*

Top right: *Ark interior.*

LEO GLENN

CATHERINE PALADINO

DARRELL FREY

Top: *Butterfly weed, a native wildflower attracts beneficial insects.*

Bottom left: *Darrell Frey in the garden.*

Bottom right: *The three sisters, corn, beans and squash.*

Late Winter and Early Spring Chores

Walks around the gardens help reconnect the gardener to the land. Many tasks await the gardener in early spring. It is a good time to repair garden fences, so loose stakes, decaying posts and rusted fence are replaced. Mulching the fence lines with layered cardboard and sawdust early in the year discourages a tangle of weeds later in the season. Any row covers, garden stakes, flats or other items left in the gardens in the busy fall are gathered and put in their proper places.

Birdhouses should be cleaned of last year's nests. (When our children were young, we had a tradition of putting up several new bird-houses each Easter Sunday. Now we continue this tradition with our grandchildren.)

Perennials are tended as needed. Mulch is applied to small fruit trees and shrubs, asparagus, and other perennials. Dead canes are pruned from berries and roses. Plants we want to propagate are divided and replanted or potted. Pruning is an urgent task if not already completed.

Seedling frames and the seedling greenhouse are prepared for spring use. Flats of herbs and wildflowers that wintered over in the frames are inspected. Weeds are cleared out and the floors of the seedling frames are covered with a layer of cardboard and sawdust. Diatomaceous earth is dusted around the perimeter to discourage slugs, which love to hide under seedling flats. The seedling frames are checked for loose boards and glazing, and repaired as necessary. The seedling greenhouse shelves are cleaned and door hinges tightened.

Polytunnel and Bale-Shelter

The polytunnel in the South Garden and the bale-shelter in the East Garden are important tools for our spring salad production. Each spring, we clean dead plants out of the beds, add a little compost, till, and water. Red Russian kale and spinach that winters over is left in the beds. Head lettuces or bok choy seedlings are planted on 12-inch centers in the beds and the tunnel is watered heavily. Within a week, self-seeded salad crops begin to sprout between the planted seedlings. These include kales, mache, lambsquarters, chervil, chickweed and tatsoi. The heads of lettuces are cut weekly for salad, as are the self-seeded greens. Borage and cilantro also self-sow in the tunnel. After the spring salad is harvested, some plants are left to set seed for the following year.

The overwintered kale produces flower buds for most of the summer. These are harvested weekly for our salad mix.

The bale-shelter is planted each spring and fall with direct seeded salad crops, including mizuna, mache, tatsoi, red mustard, red mallow, and spinaches.

Bale-shelter

A bale-shelter is a low-cost, low embodied energy, micro-climate-enhanced polytunnel. Because it is insulated with straw bales fall through spring, it offers a more sheltered and more productive growing space than our other polytunnel. In our bale-shelter, lettuce, chervil, spinach, mache, kale and chickweed survive well in the winter months most years; our other tunnel tends to freeze each winter, killing all but the hardiest kale. When built in 2000, the cost for the lumber and posts was about $150, and the plastic about $75. It was built in two days by two people. It is a 700-square foot garden space that could easily keep a family in salad and other greens through all the winter months.

The bale-shelter design began as an intern project to build a low-cost greenhouse. The intern was a 50-something university professor gathering skills for his retirement. In addition to gardening, he was learning construction skills and wanted to help build a greenhouse. The project was something we would need to be able to build during his once-a-week workdays at the farm.

We needed a second polytunnel to expand our salad production in the spring and fall. We had a small budget and wanted to use natural and local materials for the frame. We were interested in experimenting with polytunnel design with the addition of an insulated north wall. Because straw is less expensive and more available in the fall than in the spring, we also needed a place to store up to a hundred bales of straw for our spring mulch.

The design is simple; it's a sort of pole building structure that is 6 feet wide and 45 feet long. Because the UV-resistant plastic comes in 100 feet rolls, a 50-foot by 24-foot length will cover two such structures, accounting for excess to fasten edges. Headers made of 2- by 8-foot lumber mounted on posts support the rafters. The endwalls have a door framed in that is big enough for a wheelbarrow and rototiller to pass through. Cedar, oak or black locust wood are used for the posts. All framing lumber is rough-cut red oak.

The north side, running east to west, has the header set on 6-foot posts, set 2 feet in the ground and spaced 6 feet apart. This makes the wall 4 feet high. The south side posts are 3 feet long, set 2 feet into the ground, with the header attached several inches above the ground; this reduces soil contact and decay. The center posts are 8 feet long, set 1 foot into the ground. The central header is made of 2- by 10-foot lumber.

Rafters made of 2- by 6-foot red oak are nailed to the headers. The south side rafters angle up from the south side header to the center header at a 45-degree angle. The north side rafters angle up from the north wall header about 60 degrees to the center header.

The rafters' edges and rough sides were hand planed so the plastic wouldn't catch and rip. The peak of the roof is cut flat and covered with small pieces of rubber roofing to protect the plastic further. Side walls are framed with 2- by 4-foot lumber for bracing. ☞

Polytunnels and cold frames are cleaned as needed of winter crop residue and fertilized with compost. Any needed repair from the winter's stress to the structures is done in early spring as well. Holes in the plastic are repaired with plastic repair tape.

All the framing and rafters are waterproofed with a mix of beeswax, turpentine and linseed oil. (The mix is discussed in more detail in Chapter 9).

Reinforced white plastic sheeting covers the end walls. The rest of the bale-shelter is covered with 6-ml UV-resistant greenhouse plastic. The plastic on the north side extends to the ground and can be rolled up in summer.

Each fall, 45 straw bales are stacked three high along the back wall. The space under the south header is stuffed with straw to reduce airflow.

A central path provides access through the bale-shelter from east to west. The north side bed is a 4 feet wide and 45 feet long. The beds on the south are arranged in a keyhole

pattern for easy access. (See top view illustration.) As of 2010, the headers and rafters are holding up fine. We need to replace the greenhouse plastic. When we do this we will replace several cedar posts, with locust posts, and we will recoat the frame with the water proofing mix.

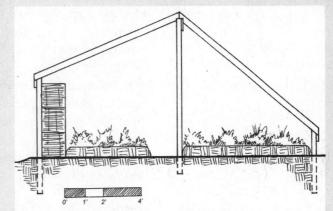

Bale shelter section and bale plan.

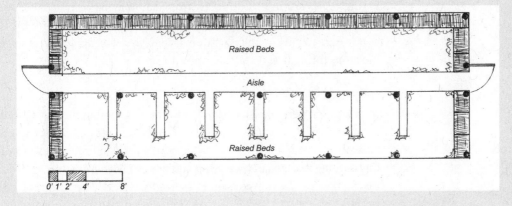

As winter breaks and new life stirs in the soil, a few hardy plants begin to grow. Mache, chickweed and chervil that survived under the snow are reliably producing by early March.

Spring garden preparation begins as soon as beds can be worked. In wet springs, this means preparing beds with the garden fork. After the soil drains and the beds are drier, they can be prepared with the two-wheel rototiller. Only after the beds are very well drained can a tractor be brought into the garden.

Early crops to go into the ground include alliums, such as onion, leeks, chives and garlic (if not planted in the fall). Peas, radishes, hardy cooking greens and direct seeded spring salad greens can be sown up to two months before the last frost date. Spinach, chervil, miner's lettuce, peas and mache are planted early for salad greens. Growth rate will depend on how soon spring warms the soil. Generally, we try to get an early start, willing to lose a few plantings to a late hard freeze in exchange for the chance of early spring harvests. Floating row covers help prevent loss from minor freezes.

Fall-planted cover crops should be tilled in before they get more than 6 inches high. Otherwise they need to be cut with the scythe before tilling. We often use the tractor-driven rototiller to till under large areas of cover crops. Spring-planted green manure, such as field peas, rye and other cold-season cover crops, can be planted in beds reserved for summer crops.

Spring

Spring cycles are in flux. The last frost date may be as early as May 1 now, but it can also come as late as the first week of June. More than ever, we are relying on the polytunnel, cold frames and row covers to get the crops through to certain summer.

Spring Propagation

Many garden perennials are divided for propagation in the spring. Our most intensively managed gardens are located within 150 feet of the bioshelter's potting room. This Zone 1 area contains stockpiles of potting soil ingredients and watering equipment. Pots are put in a garden cart so they can be moved around as needed. Recycled potting soil and compost is mixed in a wheelbarrow. Self-seeded plants are collected and potted up. Perennial herbs and flowers are dug, divided, and potted. Care is taken to prune back dead stalks and leaves and inspect plants for any pests. As we do for indoor cuttings, plants potted from the garden are fertilized with seaweed solution to promote vigorous root growth.

Most perennial herbs benefit from being divided and replanted with fresh compost or moved to new garden beds. In our area, April through early May is the best time to do this. In 20 years, we have relocated our main beds of perennial herbs at least five times.

Every year some of our bush fruit and brambles are divided and planted in new locations around the farm. Currants, raspberries, roses and Juneberries all produce new shoots readily. These are divided and replanted while still dormant to prevent transplant shock to young leaves.

New trees are mostly planted in the spring. As stated earlier, we usually limit new tree plantings to ten or fewer. With all the other work to be done, it is difficult to keep new trees watered, mulched and protected. Planting too many is a waste of money and effort.

When planting trees and shrubs, the soil should be well prepared in advance. As the saying goes, do not put a $10 tree into a $1 hole. With current apple tree prices approaching $30, the saying is even more instructive. The soil should be dug to at least twice the width of the root ball and the bottom and sides of the hole loosened with a digging fork. This gives them room to grow outward. In heavy soils, the hole will act as a pot and the roots will fill the hole before trying to spread.

Start by digging the hole wide and deep. Keep topsoil and subsoil separate. When the hole is large enough, loosen the bottom and sides with a digging fork or pick. Next, loosen the soil for several feet all the way around the hole with a U-bar subsoiling tool (described shortly). When planting the tree, replace subsoil first, then the topsoil. Mix compost, greensand, and rock phosphate and add it to the topsoil and in a 2-foot-wide band around the tree or shrub. Then mulch with shredded wood chips. Michael Phillips, writing in *The Apple Grower*, suggests that chipped saplings make the best mulch for fruit trees because they have a higher ratio of green wood. He also recommends a gravel mulch for the first few years, especially directly around the trunk. This helps reduce hiding places for the tree-gnawing meadow vole.

Late spring is a busy time. Plant sales, garden preparation, Earth Day events, garden tours, and garden workshops are all scheduled. Plant propagation, transplanting, harvest and marketing increase with the day length. More labor is required and longer hours are the norm.

Summer

When does summer begin? Memorial Day weekend is the official answer around here. But we know that full summer is near when robins sit on the nest and young swallows twist and turn overhead.

In recent years, we have had 80° days in early May and summers that were cool almost every night. Like the elusive springtime, summer weather, too, is changing. While hot weather is the norm, our mid-continental climate experiences both cool, wet summers and long, hot dry spells.

Many potted plants can be brought out of the bioshelter for summer. Rosemary and other herbs, ornamental sages, and hanging baskets are placed in Zone 1, in the arbor and Spiral Garden and around the pathways between the bioshelter and parking area. Here they benefit from direct sunlight, beautify the landscape, and can be watered daily. Larger planters act as hose divots around the seedling frames.

June brings intensive tilling, bed preparation and planting. The danger of frost is officially past on June 6. Long-season crops should be in their beds by the end of June. These include Solanaceaes, brassicas, basils, parsleys, sweet corn, squashes and cucumbers. Head lettuce seedlings and direct seeded salad beds are planted every week until September 21. Direct seeded summer crops that are planted every other week include beans, beets, carrots, cilantro, dill, bulb fennel and miscellaneous greens and roots.

The tunnels are either left fallow to self-seed, or planted with heat-loving crops. Peppers, basils and cherry tomatoes are planted in the tunnels as well to provide an extended fall harvest. In late spring, eggplant, cayenne peppers, Sungold tomatoes and occasionally sweet potatoes are interplanted with the remaining greens for a summer crop. Eggplants are far more productive in the tunnels and bioshelter. In the open garden, they are stunted by flea beetle damage. Flea beetles are not active under glass or plastic at our farm.

Plant each week in summer:
- head lettuce
- salad trays
- lettuce transplants

Plant every other week in summer:
- cilantro
- dill
- beans
- beets
- carrots
- fennel
- squash

Full-season crops:
- curled parsley
- flat parsley
- green basil
- red basil

- tomato
- pepper
- potato
- squash
- corn
- pole bean
- Swiss chard
- broccoli
- cabbage
- kale
- leeks
- onions

Partial-Season Crops:
- radicchio
- turnips
- spinach
- peas

The summer farm is a dynamic place. Young birds learn to fly, snakes slither among the long-season crops, cold frames and tunnels. Fish jump in the pond. Frogs and toads hop about. Butterflies and bees are busy from mid-morning until early evening.

We plan a work schedule that keeps the farm crew out of the mid-day sun. Morning and evening are the best times for most farm work, especially planting, harvesting and watering. Lunchtime lasts from around 1:00 until 2:30. After that, a rest or a swim, or indoor transplanting work is done. After 3:00, the garden work begins again. We keep sunscreen and natural insect repellant available for all workers to use.

Water

Almost daily in the summer, except after rains, we irrigate new transplants and seedlings with pond water. We water mature crops with well water. In dry spells, watering can be a full-time job for one person. But we have had several recent summers when it rained hard almost daily for weeks. These trying times limited our ability to prepare beds and tend crops. We were able to get more cuttings from our salad beds, but our planting schedules suffered.

Value-added

Through the summer a succession of craft items are harvested and processed. Flowers are pressed for cards and artwork. Herbs are harvested for making oils and vinegars and dried for teas and everlasting bouquets.

Excess crops are also processed weekly in the summer. Unsold basil and cilantro becomes pesto. Any not eaten right away is frozen for winter use. Tomatoes, peppers and salsa are canned. Fruit is picked and frozen. Applesauce is canned.

Weekly yard care goes on all summer. Weed trimming, and mowing are unavoidable tasks. We mow our garden's main pathways every week or two before harvest day. Other yard areas we try to mow as little as possible, but at least twice a month. Gardens are weeded and mulched again as needed in mid- and late summer.

Warm summer nights bring out the fireflies. Because bats and dragonflies keep the mosquitoes under control, we can enjoy the fireflies and summer stars unbitten. The return of the monarch butterfly in early July is a welcome occurrence. They add their orange and black wings to the dance of the tiger swallowtail, black swallowtail, hummingbird moths and other summer butterflies. We watch out for monarch caterpillars when weeding milkweed from garden beds so we can relocate them to a new home in a safe spot.

Fall

The first changing leaves in late August and the practice flights of Canada geese remind us to begin fall plantings. Tunnels and the bioshelter are prepared for fall plantings. The last plantings of cold-hardy crops are made by September 21. After that, cold frames, tunnels and the bioshelter are planted. Spinach, mache, and cilantro are seeded in early October for a harvest the following spring. Self-seeded chickweed, lambsquarters and chervil germinate in the cooling soil.

Frost can occur as early as August 31 or as late as October 15. Any plants needing to be brought back into the bioshelter are potted up and checked for slugs, snails and insects. These include semi-hardy herbs, calla lily, and hanging baskets.

Bioshelter Maintenance

Most maintenance jobs are done in October and November, when we have a little more time and the weather is still good.

Yearly Upkeep

Every building requires annual maintenance and general upkeep. Maintaining a bioshelter includes normal tasks and repairs plus a few special considerations.

Weatherizing

Proper weatherization is the key to successfully utilizing solar energy as part of a building heating system. Heat loss through air infiltration can significantly reduce building performance.

Each fall, there is a checklist of weatherizing tasks to be done. Windows, doors and vents are checked for proper closure. Loose or worn weather stripping is replaced. Latches should close tightly. Doorknobs and catches are inspected and replaced if worn. As winter sets in, exhaust fan vents are closed and sealed. We usually disconnect the large gable fan motor and place a plastic cover over the vent. In early

December, when cold weather arrives, we place clear plastic covers over all the roof vents (from the inside). These we leave on until March. We take down a few at a time as the days grow longer and the building warmer. These plastic vent covers are stored for reuse each winter.

Plumbing

Our bioshelter has a dozen plumbing fixtures; leaky faucets sometimes seem a constant. We keep extra washers and plumbing fixtures on hand. Broken and leaky faucets waste water and the energy used to pump it. Dripping water can also quickly cause wood to rot. We try to repair leaks as they occur. Exterior plumbing is drained before freezing weather can burst pipes.

Hoses also need to be repaired on occasion. Mostly they need new male ends because that end of the hose wears out from daily use.

Painting

Interior surfaces (except floors) in the bioshelter are waterproofed with a mixture of beeswax and turpentine. We put a fresh coat on critical surfaces (near planters and growing areas) each year. Other surfaces are re-coated every other year. Interior wood floors are painted with linseed oil as needed.

Exterior surfaces are stained with a linseed oil stain. Planters, arbors, seedling-frames, bale-shelter framing and other surfaces in contact with plants are painted with the beeswax mixture. The bioshelter exterior, cold frames and roof vent frames are repainted with linseed oil stain every few years. Doorframes are repainted as needed with water-based latex paint.

The Wood Stove

Many destructive fires begin with a dirty chimney. Chimneys need to be cleaned at the beginning of each heating season and again in late winter. Proper use of a wood stove prevents creosote buildup, but ash and soot need to be brushed out as needed. Chimney pipes should be inspected and replaced when they get worn. The single-wall steel pipes may last two to three years, but be sure to inspect them. The insulated stainless steel pipes that pass through the roof will last many years if properly cleaned each year.

Steel Barrels

Steel barrels used as thermal mass will eventually rust. To increase their life span, they should be cleaned and painted if necessary. A good design is to cap steel barrels with flashing to prevent soil and water from collecting on the top. Barrels should be set on gravel or other well-drained surfaces to prevent the bottoms from rusting.

Planters

We try to avoid wooden planters. They just do not hold up to rot. But plastic barrel planters sag and deform. Every year or two, we check bracing and support of the planter barrels to keep them secure.

Window Washing

Rooftop glazing and windows should be washed each fall and perhaps again in mid-winter. Dust and dirt buildup on windows reduces incoming light and can increase condensation on the glazing, which further reduces light. Roof glazing and glass windows can be washed with warm water alone or with a small amount of mild soap. Too much soap will leave a film that can attract more dirt. We use a cloth squeegee tool with a telescoping handle. We only wash the inside surface of the roof glazing. The outside of the glazing gets washed by the rain and by snow sliding off in winter.

Grow Lights

Florescent grow-light tubes need to be replaced after two years — at most. We date the tubes with a marker and replace them when they are too old for proper plant growth. Of course, you should replace burned-out tubes promptly.

Weekly and Daily Upkeep

Floor Sweeping

Floors need to be swept regularly. Moving flats and trays of plants inevitably leaves trails of soil mix and dirt. Working on planters and weeding also leaves dirt on floors. If not cleaned regularly, dirt collects and can absorb moisture that will rot wood surfaces.

The kitchen floor is swept daily, on harvest day or market days we sweep several times a day as needed. Countertops and sinks are cleaned and disinfected with liberal use of hydrogen peroxide bleach.

Floor Raking

The gravel floor and dirt floor sections of the bioshelter need to be raked regularly to remove plant leaves and other residue. Raking also redistributes the gravel, making a neater surface.

Dusting and Spider Webs

Spiders are welcome in the bioshelter. However, their old webs tend to collect dust and look messy. Periodically we clean the old spider webs from the rafters and corners of the building.

Pest Control

Bug and pest patrol is a constant activity. We want to promote beneficial insects and minimize pests through monitoring and control.

Mice and rats will try to invade any building. Both are most likely to invade during autumn and early winter as they seek shelter from the coming cold. Vigilant trapping and building upkeep are essential to reducing or eliminating these pests. We use a number of traps and occasionally resort to organic-approved vitamin D rat poison. The poison is placed in child- and pet-proof bait stations in out-of-the-way locations in our barn.

Firewood

Firewood is cut by midsummer each year to cure in time for fall. We try to cut or buy the following year's firewood in mid-winter. Wood is stacked on old pallets and kept covered until needed. Sticks and kindling are gathered as needed from the tree lines.

Managing the Permaculture Landscape
Tilling and Soil

Permaculture places a lot of emphasis on no-till methods such as mulch gardening, perennial food gardens, and the creation of *no dig gardens* built up with layers of cardboard, mulch and organic matter laid on the soil surface. Each of these reduces or eliminates tillage, therefore conserving time, energy and effort. These methods rely on earthworms and other soil fauna and decomposers to process compost and mulch and distribute nutrients through the soil. Fungi become the dominant component of nutrient processing and delivery, trading minerals and other compounds with plants for sugars and other things they need.

white sun warms the soil
seeds lie dormant in the earth
awaiting the rain

Tillage breaks up soil structure and adds air to the soil, which increases respiration of soil microorganisms. Aerobic bacteria break down organic matter, making more nutrients available to promote plant growth, and release carbon dioxide in the process. In nature, this happens when trees uproot. Plants attuned to disturbed soils, many of which are annuals and biennials, capture this rush of nutrients. When they die, they return their nutrients to the soil surface and begin the process of succession that leads to perennial cover. This process also occurs where burrowing creatures mound soil and subsoil, where herds migrate, and along naturally eroding watercourses. Many of our vegetable crops probably originated in such disturbed soil.

John Jeavons's writings (see the Resource List) have thoroughly documented the high productivity of intensive gardening. Also called *bio-intensive gardening,* this method relies heavily on deep tillage, crop rotation and the regular application of compost. Bio-intensive gardening methods have been used by cultures around the world for millennia. Garden crops have, in a very real sense, co-evolved to suit disturbed, or more precisely, worked soil. Thus, they prefer the balance of air, moisture, drainage and organic matter of prepared raised beds.

Intensive gardens, when well managed, improve with age. The annual addition of organic matter such as green manure, compost and mulch increases the soil's capacity to hold moisture and minerals while balancing the soil's pH. Soil tests are used to monitor mineral levels. Rock minerals such as greensand, rock lime and rock phosphate are slowly released to the plants by the complex interaction of the soil's biochemical processes. These processes include passage through the digestive system of an earthworm, distribution of nutrients by mycorrhizal fungi, and chemical interaction with other minerals. After a number of years of garden soil development, minerals become ever more available to the plants.

A Matter of Scale

My father used a spade to work the soil; my grandfather used a horse-drawn plow and disc; we use garden forks, rototillers and a tractor. Whatever the tool, the goal is the loose, well-drained and aerated soil that vegetables prefer. From an ecological point of view, though, tillage creates a disturbed environment. Digging, hoeing, turning, and discing all greatly disrupt soil structure and its flora and fauna — which can have serious consequences. For example, when immigrants brought traditional East German farming to the Great Plains prairie, they brought a farming method that relied heavily on plowing and disking. When drought and high winds came in the 1930s, millions of tons of disturbed topsoil were easily carried away in the dust storms.

When and where and how, then, do we till the soil? The answer is not straightforward. Use of tillage is a matter of scale and is site specific. Permaculture design encourages us to intensively cultivate only where we need to produce bountiful gardens and plant productive perennial landscapes; we should leave as much unmanaged wildland as we can. Our early attempts to rely on extensive mulching resulted in plagues of slugs, plentiful voles, and new weeds. Both straw and hay provide the perfect ecological niche for these pests and bring in weed seeds. Our current strategy is to rotate mulched crops with tilled crops. In any year, up to a third of our beds are mulched or fallow for part of the season. This allows for the continual addition of organic matter while also disrupting the life cycle of slugs and voles.

We limit our "involvement" when we can. Our perennial gardens are mulched only until the trees and shrubs are large enough to shade out competition on their own. When feasible, gardens are loosened and dug by hand with a garden fork to eliminate weeds — rather than using herbicides or power tools that would do more damage to soil structure.

Bed Making

Beginning in late March and continuing through spring and early summer, we prepare and plant an average of 2,400 square feet of raised beds each week. The unit size of beds is generally 4 feet wide and 50 feet long, so each bed is 200 square feet. Every week, we plant at least four to six beds. We plant adjacent beds so we can water with our overhead sprinkler. We plant a mix of crops each week. At the peak planting season in June, we may need to plant 12 to 18 beds each week to establish our long-season crops.

The first step is to rototill the block of beds to be planted. Tilling incorporates organic matter and helps loosen and aerate the soil to a depth of 6 to 8 inches. We till in any green manure or prior crop residue a week before planting. Tall green manure crops may need mowing before tilling.

Before tilling, it is wise to first check for lost plant labels, plastic flat inserts, or even lost hand tools. Sticks and stones used to hold down the floating row covers the previous year also have to be searched for and removed. Usually it requires two passes with our tiller to break the ground and incorporate the green manure. New ground may require several more passes or we may need to use the tractor-driven tiller to save time and wear and tear on the small tiller — and on our bodies.

The second step is to form the pathways with a hiller, or furrower, attachment on the rototiller. Initially the pathways were laid out 60 inches apart, on contour. This is a reasonable width for working by hand, tillable by two passes of the rototiller, and also the tire width of a small tractor. Several contour lines are measured with a water level and marked with stakes. (See Chapter 4 for more detail on how to do this). We then till along the contour line with the hiller attachment. The hiller is an adjustable triangular implement that pushes soil out on both sides — sort of a double-sided plow. It makes a furrow 6 inches deep and 12 inches wide. We then use a flat shovel to widen the furrow and flatten the bottom to establish the path. Soil removed from the path is added to the beds.

The final step is to add compost and prepare the bed for planting. After the green manure has begun to disappear thanks to the soil's various digesters and decomposers, usually in about a week (maybe two in cooler soils), compost is added, along with any necessary minerals and organic fertilizer. Beds are then tilled again.

After two passes with the tiller, the bed is raked to remove any crop residue or weed roots. Raking should not be a back-destroying chore. New gardeners rarely know how to use a rake properly. Raking raised beds is best performed standing erect and raking to one's side rather than stooping over and raking forward. To rake out roots and stones, one should begin a section with a deep pull and heavy pressure. The second and third pass over an area are progressively lighter, pulling the debris to the surface. The final shallow pass then rakes away the roots and stones, which are destined for the compost pile or the chicken yard.

If quackgrass has invaded the beds, we hand rake the roots out of the soil before adding compost. Starting with a three-pronged rake, we pull the grass roots up to the surface, then rake off the surface with a garden rake. This step is repeated as necessary to reduce the grass population.

Quackgrass is stubborn and hard to get rid of. Some hand pulling is usually required as well.

When establishing new beds, after tilling and hilling we also loosen the subsoil with a U-bar, or broad fork. This tool has a row of heavy rods on a bar and two long handles. Ours has two lengths of rods, 12 and 14 inches long. The rods are pushed into the soil by stepping on the bar. Then the handles are pushed forward and pulled back several times. The goal is to loosen the subsoil, not to turn it. Excessive strain is not needed. Most of the work is done by body weight, standing on the tines to push them in the soil, and leaning back and forth to break up the lower layer. Simply loosen a spot, and then move 6 to 8 inches forward and repeat until the bed is covered. This is not as strenuous — and not as thorough — as double digging, but it greatly improves soil aeration and drainage. This allows plants to feed from a larger soil area and will help increase the depth of the topsoil. Every three or four years, we repeat the subsoiling on most annual garden beds. The U-bar is also used in the bioshelter's deep beds to break up the hard pan that develops.

The method of bed preparation described above has many advantages. The beds are the right width to be worked from both sides and still maximize the use of space. As the path becomes compacted with foot traffic and wheelbarrows, they become small swales. The pathways catch any hard rain that runs off and slowly drains it under the beds. The extra soil from the paths increases soil depth in the beds, which stay loose because they are not walked on. Pathways are weeded with a stirrup hoe or hand weeded or mulched.

When an entire bed is replanted in succession, we often just till once and rake the plant residue.

Problems with Rototilling

One must be careful to not till up snakes and toads and the desirable plants. Often, we need to till around or move self-seeded plants. Excess tillage can make our soil *too* loose. The fine particles compact into a hard crust and prevent germinating seedlings from breaking the surface. The incorporation of several inches of compost in the final tilling before planting helps prevent this crusting. We manage the problem by keeping the soil moist with daily watering and covering direct seeded crops with a layer of compost and sand.

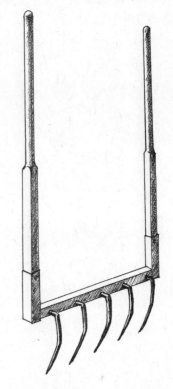

A broadfork, or U-Bar, is used to loosen subsoil, increasing soil drainage and root penetration.

Tending the Tiller

We have managed to keep a 1983 Horse Model 8HP Troy-Bilt rototiller running and in fair repair. This rugged machine, used for at least 100 hours each year for almost 20 years, has had the engine replaced twice, tines replaced every five or six years, and the gears rebuilt several times. The tiller came with a very useful, detailed manual that shows how to tear down and repair all key parts and keep the tiller working well.

In 2007, we purchased an 8HP BCS tiller with a sickle bar attachment. Learning to till with the new machine was a challenge. The Troy-Bilt tiller has variable speed gears that allow the operator to control how finely the soil is tilled. The BCS runs at only one speed and can make our soil too fine. We have at times used the BCS tiller to break ground and till in green manure, then used the older machine for the final tillage before planting. After adjusting to the new machine's design, we do find the BCS tiller a good tool.

We do not endorse any brand, not knowing if current models are built to the same standards. We do advise that the market gardener and small-scale farmer purchase rugged, well-built and easily serviced equipment. Make sure the manufacturer has repair parts available. It is advisable to keep spare drive belts, nuts, bolts and other small repair parts on hand. Regular maintenance of the engine, including oil changes and other lubrication, and maintenance of the tiller's controls, gears and levers should ensure a long life for a quality machine. We expect to eventually replace the tiller's engine with a diesel engine, which we could power with locally and sustainably produced biofuel.

Planting

A key aspect of intensive gardening is employing multiple strategies to make the most of space and bed preparation time. Below are some of the strategies we have found useful:

- Plant weekly, in blocks of four to six beds for watering and harvest efficiency.
- Use transplants to reduce the amount of time crops are in the garden.
- Use corn, squash, potatoes, and perennial beds to divide blocks of transplant crops and provide wind and pest barriers.

- Have two or more beds of perennial herbs in different gardens as insurance against pests and disease.
- Centrally locate the bioshelter, compost areas, and piles of sawdust, leaves and other garden supplies.
- Make central location on high ground, so seedlings and bulk supplies move out and downhill, and produce moves uphill, back to central packing and storage.
- Space plants closely.
- Use interplanting (companion planting).
- Do succession plantings.
- Over-seed beds.
- Space plants either in close rows or staggered to maximize plants per bed.

The three sisters, corn, beans and squash, with Hopi amaranth, at the Bioshelter.

Interplanting

Planting details for different crops are discussed below, in "Linda's Crops," but interplanting is a somewhat different subject.

There are many combinations of crops that can be usefully interplanted. We seed radishes, which mature in a few weeks, together with spring crops, such as peas, spinach, carrots and beets. We save on space and watering because the spring crops are still getting established when the radishes are mature. This form of multi-cropping allows for a succession of yields from one planting.

Many combinations are useful. Dill is often interplanted with cabbages and broccoli to attract braconid wasps. Carrots are planted in rows between maturing head lettuces, after the first harvest of wild edibles and a thorough hoeing. Carrots and basil are interplanted in the mulch between staked tomato plants. And rows of carrots, beets, or turnips can be alternated with spinach, dill, cilantro and other short-season crops. The root crops do not compete with the leaf crops, and they mature after the leaf crops are harvested, again allowing a succession of crops from one planting.

DARRELL FREY

Succession plantings without tilling are a common practice. Garlic is planted in the fall through the mulch after tomatoes are harvested. Spring crops are often planted in last year's mulch. Generally, these beds need a bit of weeding and additional mulch, but the beds are still soft and fertile. Succession planting makes the most of compost.

We often over-seed green manure into crops. White clover can be interplanted with unmulched brassicas as soon as the transplants are well established. Summer buckwheat and fall rye can be seeded as summer crops decline. Of course, we have plenty of self-seeded dynamic accumulators (a.k.a. weeds) acting as intercropped green manure as well.

Weeding

Weeding the crops has a very different meaning on a permaculture farm than on a commercial farm. Weeds are often either food for us or for the chickens. They are also green manure. Many accumulate minerals from the subsoil and keep them in the nutrient cycle. Many wild edible plants are allowed to grow for use in our salad mix. This complicates bed preparation, weeding, crop rotation and harvest. The term *weed management* at Three Sisters Farm often means encouraging the useful ones. One of the first things that interns learn at Three Sisters Farm is that a weed is not necessarily a weed. So, to clarify the discussion with the workforce, I use the term *weed* to mean any plant growing where it is not wanted, and wild edible to mean useful *weeds*. (Although, even inedible weeds are useful by acting as green manure and cover crops, collecting nutrients and hosting beneficial insects.) Interns are taught that the key to managing weeds is learning when they need to be hoed or pulled to allow a planted crop to grow — and when to leave them as the crop is nearing harvest. *However,* dense and fast-growing weeds

Radicchio and chickweed ready for harvest

CATHERINE PALADINO

can quickly outgrow seeded and transplanted crops. Perennial grasses, especially quackgrass, are aggressive and need to be eliminated as much as possible from garden beds.

Prevention is the first defense. Removal of weeds before they seed, as discussed earlier, is the most important control. Proper timing will help with some weeds. Quackgrass, for example, is easiest to remove in a dry spell, when the roots can be raked to the surface and killed by the hot sun. Heavy infestations can be killed by planting mulched crops for two or three years.

The second defense is mulching long-season crops. Long-season crops are those that are in the bed for most of the growing season. These include plants in the Solanaceae family (tomatoes, peppers, eggplants, and potatoes); the cucurbits (squashes, melons and cucumbers); the brassicas (cabbage, broccoli, collards, kales, brussels sprouts, and cauliflower); and Swiss chard, garlic, onions, parsley, basils, flowers and perennial herbs.

The third strategy in weed control is hoeing. Short-season crops are spaced in rows wide enough to accommodate the width of the stirrup hoes. Cilantro, dill, radish, carrots, cutting greens, and a few other smaller crops are planted in rows 6 inches apart and cultivated with a 4-inch stirrup hoe as weeds grow in the rows. Larger, direct seeded crops are spaced 8 to 10 inches apart for cultivation with a 6-inch stirrup hoe. These include turnips, beets, beans, larger carrots, peas, and sweet corn. Sweet corn is actually a long-season crop, but we do not mulch it. Corn needs to be hoed a couple of times when young and then will shade out weeds as it matures. Corn also provides a shaded understory for interplanted cool-season crops, such as chervil, in late summer, and of course, it is often interplanted with winter squash and pole beans.

Stirrup Hoe.

Cutting-lettuce beds may need to be hand weeded about two weeks after planting. Water beds before and after weeding to reduce stress on the seedlings.

Head lettuce is planted on 8-inch centers for baby varieties and 12-inch centers for full-size heads. The lettuce beds are also our best source of wild edibles for the salad crops. After setting the transplants in the prepared beds and watering daily for several days, the

Lettuces planted with intensive cropping spacing.

wild edibles begin to germinate — as well as weeds such as ragweed and galinsoga (a member of the daisy family). Weeds are pulled and fed to the chickens and the wild edibles are harvested for salad and cooking greens. Wild edibles are harvested young for salad ingredients for two or three weeks while the lettuce grows. After the lettuce is harvested, usually over a period of two weeks, the last of the wild edibles are harvested and the remaining weeds become a green manure crop that will be tilled in before they set seed. A few selected wild edibles, such as the magenta lambsquarter or colored amaranths, may be allowed to set seed.

A few weeds always get through the mulch. Hoed beds need to be hand weeded in the crop rows, and the pathways between beds also need weeding occasionally. The weeds are either left in pathways to return their nutrients to the soil or fed to the chickens. Each week, our chickens are given several wheelbarrows of garden weeds. This is an important part of their diet from spring through fall.

The main gardens have garden compost piles for large weeding jobs and fall cleanup. These piles accumulate organic materials for a few years and then are mulched heavily to kill any active weeds before the compost is added to the beds. This sort of compost is not technically considered compost because it is never heated by thermophilic bacteria. But, because no animal manure is involved, the difference is irrelevant.

Some weeds are extra problematic. Canada thistle needs to be mowed or dug out with a dandelion digger, a 12-inch-long sharpened rod with a handle that can pry out deep-rooted weeds (including dandelions, thistles, and dock roots). The thistles' sharp thorns make the use of heavy gloves necessary when removing them.

Linda's Crop Profiles

Much of the following text was written by Linda Frey. She draws on twenty years of experience as Garden Manager at Three Sisters Farm to share the details of cultivating fine produce.

Alliums

Plants in the genus *Allium,* known broadly as *onions* are nutritious and healthful. These members of the lily family provide vitamin C, potassium and an abundance of antioxidant phytochemicals including allicin, sulforaphane and quercetin. Garlic, a member of this group, is an especially health-promoting and medicinal food. I eat several cloves of garlic a day when fighting the common cold. Garlic, onions, and their relatives are all thought to fight chronic illness and certain cancers.

Of course onions, garlic, and their relatives are essential ingredients in cuisine. The great variety of alliums allows for a range of culinary options. Having some form of allium available all year is easy — with a little planning. Our main crop onions include scallions, slicing onions and winter storage varieties. We also grow wild ramps, Egyptian onions, garlic, garlic chives, leaf garlic, shallots (for both green use and storage), and leeks. Onion sprouts can also be grown for specialty markets.

Alliums like rich, moist soil and do well in cooler, mulched beds. It is best not to add too much compost to allium beds. We often plant them without adding compost because we rotate them with crops receiving heavy compost the year before.

Chives can be available all year if grown indoors. Potted chives need a rich soil and plenty of water. They can quickly become infested with onion thrips, so they need to be monitored. Soap spray will control thrips, but regular harvest should prevent infestation. Outdoor chives can be harvested from late March until November in Pennsylvania. Chive blossoms are in demand from chefs for their mildly flavored, pale lavender florets. We grow several large clumps to ensure a steady harvest. We also maintain full beds of chives, with clumps spaced about a foot apart. We weed our chive patches each spring, add some compost and mulch them heavily with leaves collected the previous fall. Some years, a midsummer weeding and second mulching is needed. We try to allow at least a few patches of chives in all our main gardens. It is a good idea to dig up and replant chives every few years to maintain plant vigor. Chives fit well in perennial polycultures as companions in orchards and forest gardens. They provide nectar for bees and some say they aid in repelling pests. I have found naturalized

Chives.

patches of chives that maintained themselves for decades in moist soil on the edge of a forest.

Scallions, also known as *green onions* and *bunching onions,* provide all the health benefits of the onion family, and they grow from late spring until late summer. While generally grown from purchased sets, scallions are easily grown from seeds. For the organic farmer, onion sets themselves must be organically grown. Onion sets tend to be the varieties best harvested and cured as winter storage onions. They will get large if not harvested green. Bunching onion varieties harvested in the green stage generally do not make good storage onions.

Many gardeners thin their storage onions for scallions. We prefer to use bunching onions for fresh table use. These include the Red Beard scallion and evergreen bunching onions.

We start scallion seeds in soil blocks or in plastic inserts. Inserts with 32 cells per flat are best. We plant six to ten seeds per cell and allow them to grow about 4 to 6 inches tall before planting out in the garden bed and mulching. Planting this way allows for whole bunches to be harvested in one pull.

Shallots are a form of *multiplier onion,* which reproduce by developing off-sets, similar to cloves of garlic. They can be harvested as fresh green onions or allowed to mature and the bulbs stored. Many shallot varieties are available. Because shallots are expensive, it is a good idea to propagate your own supply for several years and keep a patch for set production.

Summer onions are slicing onions; they do not store well, so are best used fresh. Large, sweet onions and early maturing varieties fall in this category. Summer onions come in many shapes and sizes. Seed Saver's Exchange members list hundreds of varieties.

Storage onions come in a wide range. They can be red, white or yellow, and range from mild to very hot.

Onions like fertile, well-drained soil with pH between 6.5 and 6.8. They can be seeded indoors in February or early March or direct seeded outdoors in late April to early May. Plant in rows about 4 inches apart, side-dressed and well mulched. For storage onions, break the flowering tops in mid-July to prevent flowering. Onions should be harvested shortly after the tops die back in the fall and cured in a partly shaded location to prevent rotting.

Basil

Basil is a sun-loving, tender annual. It is very sensitive to cold temperatures; it will be stunted if exposed to cold when young and will die at 35°. Basil needs a rich, well-drained soil, lots of organic matter, and a pH of 6.5 to 6.8. It requires high amounts of nitrogen for lush, leafy growth. We plant basil on 12-inch centers from transplants that are started in late April (we do succession plantings one or two times in summer for a continuous supply of fresh leaves).

Basil does well with a side-dressing of compost and thick mulch to suppress weeds. We put our transplants in the garden around June 1, unless it is a cold week. We harvest weekly by pinching or cutting a stem just above where fresh young leaves are ready to bush out and begin growing. Basil is an excellent example of a resource that increases with use. With weekly harvest, basil will produce until frost kills it. If left unharvested, basil will flower and then start to decline. Be careful not to bruise the leaves, and do not store them in a refrigerator that is too cold, or the basil will turn black.

Genovese is the variety of choice to sell by the pound to chefs. Thai basil is also popular. Another favorite is Red Rubin, a beautiful opal basil good for salads and vinegars. Cinnamon basil is good paired with fruit and in Thai cuisine. Lemon and lime varieties of basil are good with fish or salsa and are also excellent in herbal teas. African Blue is a nice variety to grow in solar greenhouses in the winter because it is not as sensitive to cold as other varieties are, and it is easily propagated from cuttings.

Beans

Bush beans are planted every other week, from mid-May until early August, for succession harvests. We plant several types, often a different variety each week to make the subscription produce bag interesting. We've planted yellow, purple, green and Romano beans. We do not often grow beans for restaurants because chefs prefer smaller beans, which are more costly to harvest. However, if you have a customer willing to pay a premium for the time and effort, it may be worthwhile. We mostly harvest more mature beans for subscribers.

Pole beans are planted once or twice a year, usually as companions to corn. The need for trellising is a deterrent to large-scale market production, but we highly recommend them for the kitchen garden. Pole beans will yield over a longer period if they are regularly harvested. Runner beans, both scarlet and painted lady are grown for their showy, edible flowers.

Broccoli

Start seeds indoors or in cold frames in March and April for an early crop and late June through mid-July for a late crop. For early planting, start broccoli seed in a row flat and transplant into a 60-, 48-, or 36-count tray of six-packs, or directly seed into the six-packs.

Broccoli requires a very rich soil with a pH around 6.8. It can be transplanted outside by May 1 and either side-dressed with compost and then mulched, or over-seeded with a living mulch of inoculated, nitrogen-fixing white Dutch clover sown around it. Broccoli takes approximately two to two and a half months from transplant to harvest. Plant a second, midsummer crop for a succession of harvestable heads and side shoots. The first planting will lose its vigor, but a second planting will last into late fall.

We like the varieties Arcadia and Emperor for their side shoots after the main crop. We also like the variety Minaret as a late-season variety. It is very sweet, especially after a frost.

Some pest control is needed for broccoli crops. Cabbageworms (larvae of cabbage butterfly) and flea beetles like broccoli. If an outbreak is not too severe, we control cabbageworms by hand picking them (the chickens consider them a delicacy). We usually spray *Bacillus thuringiensis* (Bt) when we spot the cabbage butterflies laying their eggs. Row cover can help young seedlings resist flea beetles.

Cabbage

Like broccoli, cabbage is a member of the Brassica family. Because they like a good amount of nitrogen, brassicas do well when following legumes or with compost dug into beds.

Cabbage likes a soil pH between 6.6 and 6.8. We start our early cabbage plants indoors or in a cold frame in March or April; late cabbages get started in April or May. The seedlings can be put outside in April and May (six weeks from seed, generally). Protect cabbage from late frosts and spring winds with row cover, which also keeps flea beetles and egg-laying cabbage butterflies out.

Depending on the variety of cabbage, we plant 18 to 24 inches apart, using a diamond pattern that allows maximum heads per bed. We side-dress cabbage with 1 to 3 inches of compost and mulch with straw. As with broccoli, we plant a living mulch of white Dutch clover between plants and rows as a low-growing, leguminous cover crop. If the bed is rich enough in nutrients, scallions can also be interplanted.

Depending on the variety, cabbage plants mature in 68 to 120 days. Sometimes we seed a mid-season variety in June for a September/October harvest. As with broccoli, Bt can be used to control cabbageworms.

Chinese cabbages and bok choy can be grown in much the same way. Depending on expected mature size, they are spaced 12 to 16 inches apart, but only 12 inches apart for baby heads. Early or late planting will avoid excessive heat and flea beetles. Direct seeding and thinning is best, as these plants have deep taproots.

Carrots

Carrots prefer a well-prepared, sandy loam with a pH between 6.5 and 6.8. They like a soil high in organic matter that is well drained. Carrots cannot have manure added that is not completely broken down because excess nutrients cause them to develop excess root hairs. After the bed is prepared and raked, we hoe furrows about 2 inches deep and sow the seed as thinly as possible, then cover with about a half inch of fine soil (the furrows catch and retain water).

Planted seeds take about two weeks to germinate and must be kept moist. If conditions are hot and dry, cover the moist bed with a layer of cardboard to aid germination. Start checking for germination at seven days and remove the cardboard as soon as germination is underway. We often plant radishes in with the rows of carrots to mark the rows and get a quick harvest. Radishes also need plenty of water to grow quickly.

Carrots can be planted outside in our region as early as mid-April and as late as August for baby size and/or storage carrots. Carrots planted before June 1 are susceptible to carrot fly maggot damage. A side-dressing of wood ash may help deter the carrot fly. After one month in the ground, carrots should be thinned to about 1 or 2

Carrots.

inches apart. I like to mulch between the rows because the retained moisture seems to help them size up. For maximum yield, succession plant carrots every two to four weeks all summer long.

We often sow carrots in diagonal rows between maturing lettuce plants. When the lettuce has been in the bed about two weeks, we hoe between the plants and scatter and rake in the carrot seeds by hand. The lettuce provides shade to aid carrot germination. After the lettuce is harvested, the spaces between the carrot rows are again hoed. Another successful strategy is to plant carrots for the fall crop between tomato plants. After the tomatoes are staked and mulched, narrow rows are made in the mulch and thinly seeded with carrots. These are usually the large fall varieties that can be harvested after frost finishes off the tomato crop.

Sweet Corn

Corn likes a pH around 6.8 and needs to be grown in soil high in humus and plenty of nitrogen. This can be achieved by planting in a soil that has had a good stand of alfalfa or red clover tilled in one month prior planting. You could also till in inoculated field peas or vetch. The bed could get an application of compost, blood meal, or organic alfalfa meal tilled in at planting time.

Beginning about May 1, we plant seeds 8 inches apart and in rows that are spaced 1 foot apart and then we cover. Short-term varieties germinate in one to two weeks. For an extended crop, we do succession planting every two weeks until mid-June or later. Sometimes we plant corn using the traditional Native American method. This is done by planting seed in 2- to 3-foot-diameter hills that are six to eight inches high. Once the corn plants pop up, we interplant beans and squash in the "Three Sisters" style, and then plant a nearby stand of corn rows for insurance of good pollination. Corn planted in rows is a good interplanting candidate for summer lettuce or chervil because corn protects both crops from excessive heat or sun. The cool, shaded microclimate of the maturing corn provides a good environment for these cool-season crops to germinate and grow with less heat stress.

Eggplant

Eggplant likes a rich, well-drained soil. Eggplant is a challenge to grow outside because of the heavy damage flea beetles inflict on them. We usually grow them indoors or in a polytunnel, where flea beetles do not bother them. One trick is to get them off to a vigorous start indoors by potting them up to 4-inch pots in the spring and allowing them to grow over a foot tall before setting out in the garden or polytunnel. When they go into the beds (after all danger of frost is past), about 18 inches apart, they should be covered with row cover for several weeks and kept well watered to promote fast growth. As long as the plants can outgrow the flea beetle damage, they will do fine. Our favorite market varieties include Rosa Bianca and the white-and-purple striped Calliope.

Greens

Greens is an informal name giving to plants used primarily for their leaves, rather than their fruits or flowers. These include Swiss chard, spinach, lettuce, endive, kale, turnips, dandelion, and collards. As a general rule, plants that are grown for their green leaves like a sandy loam, high humus content, and a pH between 6.5 and 6.8 with plenty of nitrogen. We try to add 6 inches of compost each spring to take care of all of that, but we also add rock phosphate, potash, or lime as needed.

Endive is a type of chicory, similar to lettuce. The seedlings are planted 14 inches apart, allowing 2 inches of airspace around the intended full-size 12-inch heads. Our favorite varieties are très fines curly endive (frisée) and Batavian broad leaf endive (a.k.a. escarole). Endive can be planted from mid-April to mid-May with good results, but after that it is best to skip planting till about mid-July so as to miss the "bolting" period. Late-sown endive is best harvested at baby size. Summer-grown endives tend to be extra bitter and are subject to tip burn. In the cool fall, endives provide weekly cuttings for inclusion in our salad mix.

Dandelion is an overlooked, tasty and healthful green. We usually direct seed Red Rib dandelion for our salad mix. Harvest four to six weeks after planting and chop remains into the soil. Note that Italian dandelions are usually classified as a type of chicory.

Fennel is grown for the bulb, which forms above the ground. It likes high fertility, a pH between 6.5 to 6.8, and good drainage. Fennel is best direct seeded, but can be transplanted if you are gentle with it. Full-size fennel should be spaced 12 inches apart.

Our customers ask us to grow all sizes of bulb — from the 1-inch baby size, to 2–4 inches, to full-size 6-inch-wide bulbs. The smaller sizes yield a tender bulb with a more delicate anise flavor. We usually cut the leaves from the bulbs and add them to the salad mix.

Baby fennel takes about 4 weeks; allow 6 weeks for medium fennel; and expect about 10 to 12 weeks for full-sized fennel. If you allow the fennel plants to flower and set seed, you will get edible flowers and fennel seed, which is used as a seasoning.

Kale thrives on the same culture as other brassicas, such as broccoli and cabbage. One variety, Red Russian, self-seeds rampantly. Kale has the best flavor early in the season and after frost. We've decided that the best varieties are the blue kales, Lacinato, Red Russian, White Russian and Winterbor. Kale plants tend to withstand cold and are very hardy. As with other brassicas, Bt works well for controlling caterpillars. For the best plants, side-dress or use compost tea midseason. When planted in mid-June, kale comes into its prime in September and will provide good yields until heavy freeze stops its growth. Kale is a great crop for cold frames and winter polytunnel crops. Over-wintered kale provides delicious flower buds well into summer and can be picked weekly. We add these to our salad mix.

Kohlrabi is another member of the Brassica family; it is grown for its edible stems. We plant it in both the spring and fall, either directly sown or from transplants. When direct seeded, we plant in rows about a foot apart so we can hoe or mulch between rows. Transplants can be spaced about 8 inches apart in a hexagonal pattern.

Collard and turnip greens require the same planting conditions as kale.

Rutabagas require the same care as other brassicas. Space 12 inches apart and mulch.

Swiss chard plants are spaced 1 foot apart; however, the baby-sized plants that are used in our mesclun mix are an exception. These get spaced more closely. Swiss chard is an excellent food producer. Plants will produce year-round in cold frames and also in the greenhouse. Occasionally they winter over in the

Kale.

garden. Favorite varieties include Five Color Silverbeet from Seed Savers Exchange (a good, open-pollinated, full-color variety), Bright Lights (nice in mesclun), and our favorite for indoor seed saving, Charlotte (very beautiful with deep red stems and leaf veins). We allow chard to live for several years in our cold frames and bioshelter beds. These perennial chard plants will continue to produce bountiful leaf harvests if they are cut back when they flower in midsummer. We allow a few plants to go to seed and to self-seed.

Side-dress and mulch outdoor Swiss chard crops.

This hardy crop is resistant to most pests. Slugs are one of the few pests able to bother this sturdy plant, but they can easily be trapped with beer. Another way to prevent slug damage is to place boards in pathways between beds; throughout the day lift the boards and check the undersides for attached slugs. Gather and feed them to the chickens. Diatomaceous earth spread around plants is another option for repelling slugs. This can be cost and time prohibitive on a market farm, but it is a good strategy in the home garden.

Leeks

Sow leeks in February or March through June. Grow in rich, deep soil about 6 inches apart. Plant in small furrows, leaving only a few inches of the seedling plant uncovered above the soil.

Lettuce

Lettuce is an excellent market garden cash crop. Find the market you will be selling to before sowing because lettuce needs a precise harvest schedule. Direct sow for salad, but use transplants for heads, bunches, or for winter harvest under cover. There are many varieties. When planning salad harvest, choose a mix of colors and textures: light, bright, and dark green; different red shades; and smooth, semi-smooth, and frilly textured. We like a mix of Red & Green Oak, Tango, Buttercrunch, Speckled Bibb, Lollo Rossa, Majestic Red or Valentine, and Rouge d'hiver and Green Ice for mesclun. My favorite varieties for heads include Lollo Rossa, Red Oak, Nancy Speckled Bibb, Blushed Butter (only in spring), Tom Thumb, Diamond Gem, and Pirat.

Succession planting every week provides a regular supply for market every two or three weeks from the home garden. Lettuce is in the ground from four to eight weeks, depending on whether it is harvested as leaves, baby lettuce, full-sized heads or romaine-type heads. It likes a pH between 6.5 and 6.8, as well as lots of organic matter with plenty of nitrogen and a medium amount of phosphate and potassium.

Head lettuce is not usually mulched because it grows quickly (it is in and out of the soil fast), and mulching attracts slugs (lettuce is very susceptible to slug damage). We have successfully interplanted carrots, onions, leeks, and fennel with lettuce after it has been weeded, a couple of weeks prior to harvest. But when harvesting lettuce heads, be very careful not to trample on any young, interplanted crops. Side-dress and mulch the new crops after the lettuce has been harvested.

Cutting Lettuce

At Three Sisters Farm, we plant cutting lettuce weekly from April until late September. We use six or seven small handfuls of seed in a 200-square foot bed. These three to four ounces of seed yield about 20 to 24 pounds of 2-inch to 3-inch leaves in four to five weeks, depending on the season and the variety of lettuce. A dense lettuce planting will suppress weeds. In the spring and fall, however, too-dense plantings can get a mold disease, so then we sow thinner and plant more beds. Usually, we get two weekly cuttings of 20 pounds each week. On rare occasions, we can get a third cutting, but the quality may drop.

Baby lettuce leaves can be from 1 inch to 5 inches long. Planting densities, germination rates, temperature, soil moisture and fertility all affect the yield. We recommend a salad mix that includes both the baby leaves and the larger 4-6 inch leaves — and finding customers who do not mind variation in the size of their lettuces. This allows the grower greater flexibility in harvest schedules. It also allows for the inclusion of unsold head lettuces, especially smaller varieties such as Lollo Rossa and Tom Thumb.

The price of lettuce seeds varies considerably. We buy them by the pound and half-pound. Some more expensive seeds — such as dark Lollo Rossa and speckled Bibb — can be planted in smaller amounts to accent larger plantings of less expensive seeds, such as Red Salad Bowl and red romaines. Other crops planted in the cutting beds include the shungiku (an edible chrysanthemum), peas, bull's blood beets, kales, red dandelion, Chinese mallow, and mache.

Peas

We generally grow snap peas because they have always been a favorite with our customers. We have come to like the dwarf varieties because they do not usually need staking or trellising. Sugar Lode and Sugar Ann are my two favorite varieties. We also grow several varieties of snow peas and a soup pea. The Capucijner pea is an old heirloom soup pea that produces a highly ornamental flower and purple pod. Edible peas also have edible shoots and flowers. Dwarf gray peas, which we plant thickly in beds and in trays for cutting greens, also provide a colorful edible flower.

Peas like a pH between 6.2 to 6.8 and produce well in average fertility. I seem to have better luck with germination if I pre-sprout them by soaking the seeds for four to six hours, then rinsing two to three times daily until they sprout, which is usually in about three days. They can then be planted directly into outdoor beds. I furrow out three rows down the length of the bed and plant the sprouted seeds, inoculated with symbiotic nitrogen-fixing bacteria, about 8 inches apart, then cover (generally it is good to bury seed as deep as it is big, but no more than twice as deep). Keep moist until above the sprouts appear at the surface.

To control weeds, we cultivate our pea plants shallowly with a stirrup hoe. Sometimes, if there is room, I plant rows on the outer edges of beds too. You can mulch, if desired. Peas are cold hardy and can be planted early April if daytime temperatures are above 50°. Succession planting is done every other week until mid-June or early July.

Snap Peas.

Parsley

As with most leafy greens, parsley likes a rich, well-drained soil with plenty of organic matter. We rotate each year, avoiding beds where other crops in the same family have been planted in the previous three years (such as carrots, fennel or dill, which are related to the parsley family). I plant parsley on 10-inch centers, in a hexagonal pattern for space efficiency. Parsley plants can be side-dressed a few weeks after planting if there wasn't already enough compost in the bed, then mulched with good hay. Weed the plants one or two times over the summer.

When harvesting, select the outer green leaves and discard discolored ones. Given good weather (plenty of sun and water), parsley can be harvested weekly all summer long. To maximize yield, go through your parsley patch at least monthly and clean off broken or yellow branches and weeds. We often plant a midsummer succession of parsley because some will bolt in a hot summer.

Peppers

Peppers like well-drained, average garden soil without an excess of nitrogen; they need moderate levels of phosphorus and potassium for good fruiting and disease resistance. Smaller, hot varieties can be planted 12 to 16 inches apart, or 20 inches apart for the larger bells peppers, such as Ace and King of the North. Mature plants set fruit better if their leaves are touching neighboring plants. Hot varieties are planted several hundred feet from sweet varieties to prevent cross-pollination. Peppers can be mulched with shredded leaves or hay/straw mulch. Sometimes we interplant head lettuce or scallions when planting peppers if the soil is rich enough for a quick crop; then we mulch. We sometimes interplant the peppers in beds of maturing lettuce.

Favorite hot varieties include jalapeño, Joe's long red cayenne, garden salsa, cayenne and habernero (for wreaths), and Hungarian banana (good for canning and freezing). Favorite sweet varieties include Ace (small, but productive), Keystone (nice red), Karlo (nice yellow), and Park Whopper (I got these plants from my neighbor and they are very large and productive). Last, but not least, Anaheim is a good variety to sell for chili rellenos.

Potatoes

Potatoes are an important food crop. While they are not a big income producer for us, customers look forward to them each year. Some of our favorite varieties include French fingerlings, Russian banana, Yukon gold, Peruvian purple, rose gold, and the old standbys Kennebec and Red Pontiac. Potatoes grow best when planted shallowly in loose beds that had compost the previous year. Rock phosphate and greensand are added when preparing the beds, but not lime, which will cause potato scab. After the potatoes send up their leaves, we mulch them heavily. As the plants mature, extra mulch may be needed. Potatoes should be harvested as soon as possible after the tops die to prevent rot in wet weather and damage by voles.

Radicchio

An Italian chicory, this bitter green is best seeded in row flats and then transplanted one to two weeks later into six-packs. After four to five weeks, plant the transplants on 12- to 14-inch centers in a well-prepared, fertile bed with a pH between 6.5 and 6.8. Radicchio can be mulched.

Radicchio is in the ground outdoors for 10 to 12 weeks. Many varieties are not uniform, so mature plant size can vary. I start putting plants outside in early May and put out a crop once a month, stopping in August. It is important to plant the right variety because many of the older, more traditional varieties can only be planted in the fall. Cold weather intensifies the color. The varieties I have success with all season long include Indigo, Radicchio di Treviso, and Red Preco. Generally, you can expect to weed one or two times per crop. Harvest when heads are firm, before they start to bolt.

Asian Greens

This group includes mizuna, tatsoi, and red mustard, but I'm also including arugula here; although it is not considered Asian, its culture is almost identical with that of the Asian greens.

This group of greens, while somewhat cold hardy, all have tender leaves when young and are irresistible to the flea beetle. There are several ways to remedy the flea beetle problem. Asian greens can be grown as an indoor crop, in frames or grow tunnels, or under row cover as a spring or fall crop when the fleas beetles are less active. There is a fairly effective biological control on the market — a flea beetle-consuming nematode. These can be applied every few years as needed. (We have not used this predatory nematode, not wanting to introduce a new species into our gardens.)

Asian greens like a rich soil with a fair amount of nitrogen. They are members of the Brassica family, so rotate accordingly (see section on broccoli and cabbage). Asian greens are sown in the greenhouse all winter. We like to grow these plants as cutting greens in plastic trays with drainage holes. The potting soil for these cutting greens is amended with an extra one third-part of compost. We harvest at four to six weeks as sprouts or baby greens. Often, we get a second cutting, after which the leftover roots are fed to the chickens or composted, layered into fresh manure.

Our next favorite choice for planting is to put four-week-old seedlings or fresh seed into cold frames or polytunnels about mid-September until mid-October. Generally, these plants will produce until sometime in December, depending on when the deep freeze hits. They will start producing again by mid-March and during mid-winter thaws. Asian greens will flower in mid-April through early June, providing excellent early bee forage, as well as edible and cut flowers.

Spinach

Spinach should be direct seeded in beds that have been well prepared with abundant compost and have a pH from 6.5 to 6.8. We grow several varieties of spinach — a few each of smooth and crinkled types. We especially like the new, red-veined Bordeaux. Beginning in mid-March (soil can be pre-warmed with clear plastic mulch for two weeks), sow seed in 3- or 4-foot-wide beds with rows 10 to 12 inches apart and seeds about 1 inch apart. Cover with one quarter inch of fine soil and water; cover with row cover to keep soil moist and warmer. Hoe or mulch to control weeds. Side-dress or use compost tea, if needed.

We plant spinach every three weeks or so until mid-May. Day length will make plants bolt in mid- to late June, but can be re-sown starting in mid-July (if protected from heat) and sown every three weeks until mid-September. September-sown spinach winters over beautifully and resumes growing in March. It is a good crop for cold frames, tunnels, and indoors.

Shungiku

Shungiku is a member of the chrysanthemum family that has edible flowers. It is direct seeded in rows weekly to be cut young for salad. As the plants grow larger (they can grow to almost 3 feet high), they can be cooked in stir-fry dishes. We plant shungiku thickly in the bioshelter beds for cutting greens. In the winter, we get several weeks or months of harvest from one planting.

Summer Squash, Cucumbers and Winter Squash

These members of the cucurbit family are planted in well-composted and mulched beds or hills spaced 2 to 3 feet apart. Because the inevitable cucumber beetle and squash bugs bring a wilt disease, we plant several successions of these.

Summer squash varieties, both green and yellow zucchini, patty pan, crookneck and others are planted in two or three successions, one each in May, June and early July.

Winter squash is planted in May and June. We grow delicata, acorn, blue hubbard, kuri and butternut squashes — and a few pumpkins.

Cucumbers are available in wilt-resistant varieties. We grow both slicing and pickling varieties.

Wild Edible Greens

Amaranth comes in many varieties that are good for salad (Hopi Red Dye, Joseph's Coat, and many assorted-hued, wild varieties). Self-sowers can be rampant, but they are good in salad or for cooking. Amaranth has iron, vitamin C, and protein.

Dandelions have a variety of uses: the leaves are used in salads and cooked; the iron-rich roots are good for tonic teas; and the edible flowers can be used as the base for a great wine. Dandelions tend to be invasive, of course.

Orach can be deep red or green. This plant is ornamental and will self-sow. We use it in our salad mix.

Lambsquarters.

Ox-eye daisy is good for mesclun, common in the wild, and ornamental, but it is also invasive.

Purslane is found wild locally and there is also an upright golden variety, but all varieties are very high in iron, omega-3 fatty acids and vitamin C. It can be invasive, but it is a good companion to corn because it is shallow-rooted and prevents soil compaction.

Sheep and wood sorrel are very ascorbic (high in vitamin C) and delightful in salad.

Violets are a wonderful spring tonic green. The delicious edible flowers are known to be high in rutin, which strengthens capillaries, and vitamin C.

Lambsquarters, like purslane, is known as a beneficial companion crop with corn, if not too rampant. Multi-hued plants include greens, golds, reds, and hot pinks and are good in salad or as cooking green (lambsquarters and amaranth are both fine substitutes for spinach).

Watercress is an excellent salad ingredient in spring and is nice for cream soups. Care must be taken, when harvesting watercress from a natural spring, that the water is free of contaminates and pathogens. A good rule is not to harvest watercress from water you would not drink. We avoid this problem by growing watercress in soil inside the bioshelter. It likes plenty of water, but produces well in soil.

Tomatoes

Tomatoes are a complex, varied crop with differences depending upon varieties and location. They all require a rich, well-drained soil, and, above all, a sunny location. We start our tomato plants indoors in late March or early April. They can also be started in row flats or seeded directly into six-packs or four-inch pots if you are sure that the seed germination is high. Don't give in to the temptation to start earlier, unless you have a patio variety or other very small, determinate variety. The latter can be planted 16 to 18 inches apart, but the huge, sprawling indeterminate varieties should not be closer than 2 to 2.5 feet apart. Also, trellising should be in place when you begin planting. Trellising can range from staking and tying to boxing or trellising.

Tomatoes lend themselves well to companion planting (as long as the soil is well drained and fortified with well-rotted compost). We have had good success interplanting basil, beets and carrots with tomatoes. When the root crops come out, side-dress with compost and mulch heavily with hay or straw.

Favorite tomato varieties include Celebrity and Oregon Spring for early tomatoes, and heirlooms, such as Brandywine, Pineapple, Hillbilly, and Grandmas Green Gold, for excellent flavor. Cherokee Purple and Black Krim are grown for their excellent flavor and dark color. Jaune Flame, Red Zebra and Green Zebra three small tomatoes that look and taste great together in a tomato salad. Amish Paste or Oxhart are grown for our home canning. Sungold orange cherry

Sungold tomatoes in thyme.

tomatoes are a high-yielding mainstay in our gardens. We grow them year-round in the bioshelter.

Tomatillos

Basically they have the same culture as tomatoes, but can be direct seeded. I like to simultaneously plant one crop from transplants and one crop from direct seed. They are spaced about a foot apart.

Turnips

This member of the Brassica family is very cold hardy and can be direct seeded in March or April for a spring crop and in July or August for a fall crop. Chefs like fall crops best, it seems, for their fall menu, and often they prefer turnips to be picked at the 2- to 4- inch stage for the sweetest crop. We often grow three varieties of turnips: white, red and golden. If mulched heavily as freezing weather arrives, turnips can be harvested well into winter.

Rosemary

Rosemary is a sun- and sand-loving Mediterranean native and handles heat and dry weather well. Rosemary does appreciate 3 to 4 inches of good compost worked into the bed before planting, along with a little kelp meal. The preferred pH is between 6.5 and 6.8. When growing in pots, add extra sand, perlite, and well-rotted compost. Potted rosemary also needs good ventilation, plenty of sunshine, and probably extra moisture if overwintering in a dry house. Rosemary is difficult to over-winter north of zone 6. We have kept some over-wintered with a heavy straw mulch applied before a heavy freeze in early November and removed in mid-March. A more successful strategy is to keep them potted and move them indoors. A potted rosemary can be sunk in a garden bed for summer and then dug up for winter.

Thyme

Thyme comes in many varieties and most are hardy perennials. New plants are best seeded in February, such as German, English, French, or creeping thymes. Some varieties are propagated from cuttings (cuttings may be taken any time there is growth, but for commercial spring sales, it is best to take the cuttings no later than January or early February.)

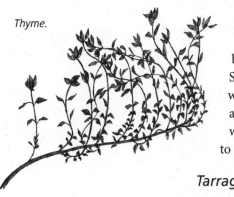

Thyme.

Varieties generally propagated by cutting include Golden, Lemon, Creeping Lemon, Silver, and Orange Balsam.

Thyme grows best in a sandy, well-drained soil with a pH between 6.5 and 6.8. Plant about 18 inches apart to allow room. Side-dress with 2 to 4 inches of well-rotted compost and mulch with weed-free straw or hay. Compost and mulch should be added each spring to maintain productivity. Thyme winters best when trimmed back and mulched between November and March to protect the plants from hard freeze.

Tarragon

Tarragon is a hardy perennial. It can be difficult to establish, but is worth the effort. Dividing the plants every three to five years propagates French tarragon. I have never tried to propagate by taking cuttings, but I do not see why it wouldn't work. Tarragon likes the same pH that most herbs enjoy (between 6.5 and 6.8) and likes the soil to be well drained. Side-dress with compost and mulch.

Tarragon plants can be spaced a good 2 feet apart because they like to spread out and can reach from 18 inches to 2 feet in height. Tarragon dies back by mid-fall and re-sprouts in April or May.

Other Herbs

Cilantro is best direct seeded starting in April, weather permitting, in rows 8 to 10 inches apart. The seed should be sown every inch or two, sometimes thicker. This allows for cutting right down the strip. Usually, I get two bunches per row this way. We sow every two weeks from May until early September. Cilantro needs to be sown frequently because it flowers and goes to seed quickly, especially in the heat. Before it goes to seed, we harvest for two to three weeks (four to six weeks from planting). If you are growing for the "coriander seed," let the seeds mature. I like to plant cilantro in a bed that has compost or fertilizer added for good leaf growth. If allowed to flower, it is a good plant for beneficial wasps and other beneficial insects. Plus, the flowers are pretty in old-fashioned bouquets and are also edible. We occasionally make a special sauce of the whole plant for Thai recipes. The roots, leaves and seeds of cilantro are all ground together into a paste and brushed onto meats or tofu. Cilantro will self-seed and come up early the following spring.

Dill is similar in culture to cilantro. Dill flowers are used as an edible garnish and salad ingredient. The seeds, of course, are used for pickles. Dill is self-sowing and good for beneficial insects. Dill and cilantro are therefore both great for scattering around in and on the edges of your garden.

Oregano can be grown from seed, usually started in February or March, or grown from dividing plants (best done in spring or from cuttings taken any time during the growing season). I like the Greek oregano best for cooking because it has the full flavor loved in pizza, spaghetti, and other ethnic dishes.

I have a variety of golden oregano that was labeled "golden marjoram," when I bought it, but it is definitely an oregano leaf. However, these two plants (marjoram and oregano) are related, hence the frequent mix-up. Oregano is a hardy, sprawling mint family plant that will spread each year. Marjoram is a tender perennial similar to oregano but with a lighter fragrance. We only grow marjoram indoors and in pots that are brought back indoors each fall.

Bronze fennel has a similar culture as bulb fennel, but bronze fennel is highly ornamental and lends itself nicely to use as both garnish and salad ingredient. Be warned, however, bronze fennel can be somewhat invasive if allowed to set seed. Bronze fennel is considered a biennial, but it can be grown as a perennial if it is cut back before setting seed.

The Edible Flower

Edible flowers have been a main crop in our bioshelter and market gardens since we began our business. While not every chef uses edible flowers, those that do often find them indispensable in their culinary art. Whether giving grace to our sons' wedding cakes, or adding a splash of color to a green salad, edible flowers have found a regular place in our kitchen as well.

Edible flowers can be a significant part of the farm's annual sales. It is difficult to know exactly how to price edible flowers. We keep our prices slightly lower than the national average because they are so perishable. Our edible flower sales have been as high as several thousand dollars each year.

Nasturtium.

A Few Words of Caution

This book describes only the flowers we eat and sell. Other good reference books list additional edible flowers, but because of conflicting information on some flowers, it is prudent to speak only from our own experience. I recommend that readers seek out several sources and decide for themselves before eating any flower. Be aware that while nectar, pollen, and perhaps the pigments and phytochemicals of flowers can be healthful, some people have allergies that restrict their use of edible flowers. In addition, *extreme caution* is necessary to avoid accidental ingestion of poisonous flowers. For example, we never grow sweet pea flowers on the farm for fear that someone will confuse these toxic flowers and stems with edible pea flowers, pods, or shoots. Also, we feel it is best not to give edible flowers to children too young to understand the dangers of eating the wrong flowers.

All that being said, plants with edible flowers have many roles in the sustainable landscape. Many have additional culinary value as herbs and vegetables; most are food and habitat for beneficial insects, butterflies and hummingbirds. Edible flowers specifically planted for harvest are maintained in Zones 1 and 2, (the bioshelter and the immediate surroundings) and in the home kitchen garden. They belong in high-use areas because most varieties need regular deadheading to maintain production. After weekly harvest, old and excess blossoms are cut off and added to the buckets destined for the chicken yard. Another reason we plant edible flowers and other culinary herbs in Zone 1 is because they are pretty and add color and interest to the areas seen by visitors and customers.

Anise hyssop.

Culinary Beauty

Anise hyssop is on the list of top beneficial insect habitat plants, as discussed in Chapter 6. Anise hyssop's culinary uses are as a dessert garnish and tea flavoring. Anise hyssop's mild licorice flavor makes its leaves and flowers a delicious base for medicinal teas, or a nice addition to mint tea or black tea. As a garnish, anise hyssop flowers are best when small. Their light purple tops can be chopped and sprinkled on cakes and cream pies. When they get larger, the furry florets may not be as enjoyable.

Arugula is grown all year in the bioshelter. We allow small patches of arugula plants to flower in the bioshelter in the spring and summer. These creamy white petals have the distinctive arugula flavor. Stalks of the four-petaled flowers are harvested for use on salads and pasta dishes. They are also a lovely garnish.

Borage produces clusters of bright blue, six-pointed flowers that are a sharp contrast to the coarse and hairy leaves. The flowers fade to pink as they age. A white-flowered variety is now available. Borage blossoms have a sweet cucumber flavor. Most of the flavor is in the calyx, which is the star of green sepals that support the blue petals. Borage flowers are a good nibble anytime. They are used in salads and teas. Because they do not keep more than a few days in the refrigerator, borage flowers are not a big seller to chefs except for special events. They are best picked early on delivery day (or the evening before) and kept moist and cool.

Beans have delightful flowers. Scarlet runner and painted lady are two large-flowered runner beans. These should be planted as edible ornamentals, on prominently placed trellises and along mulched fences. Any *Phaseolus* bean flower is edible, with a delicious, sweet bean flavor. These sturdy flowers go well in salads, pasta salads and as a garnish for cooked vegetables.

Wild bergamot flowers, or bee balm, burst forth in a blaze of scarlet to celebrate midsummer. The red variety, *Monarda didyma,* is commonly named bee balm or Oswego tea. When in their prime, they are cut and dried for a tea that is quite similar to Earl Grey tea, which is flavored with the bergamot orange, *Citrus aurantium.* The top half of the bergamot plant is cut, bundled and hung to dry (about a week), then the leaves are stripped and stored for later use. Fresh leaves and flowers are added to tea bouquets and sprinkled on desserts. Bee balm flowers are well known for attracting butterflies and hummingbirds, so grow enough to welcome them to the garden. A few weeks after *M. didyma* peaks, *M. fistulosa's* light purple flowers continue the task of feeding butterflies and hummingbirds. Commonly known as wild bergamot, *M. fistulosa* is perhaps too strong for good tea, but its flowers make a beautiful and fragrant garnish. *Monarda didyma* will naturalize along stream banks, ditches and near wet soil, especially along the edge of a wooded area where competition with grass is minimized. *Monarda fistulosa* does better in dry soils and more sun and where competition with grass is minimized.

Calendula's bright orange and yellow petals enliven any salad, pasta or vegetable dish. Petals are plucked from the composite flower and sprinkled generously over the plate and bowl. Calendula is demanding on the gardener, preferring a rich soil and requiring weekly deadheading. Planted in mass, however, they reward the diligent gardener with a steady harvest until heavy frost. They are used in bouquets and in herbal healing salves. Calendula grows well in the winter bioshelter. We start fresh plants in September and pot them up several times. They grow best in a large, fertile pot and begin blooming in February.

Chives were discussed earlier in detail. Their spherical clusters of flowers are highly prized by any cook who has used them. Light purple, sturdy and flavored like a sweet onion, chive blossoms are also a favorite of honeybees, metallic bees and other bees.

Cilantro's delicate white flowers belie their sweet pungency. Any dish improved by cilantro will be complemented with its flowers. Small and delicate, they are best used sprinkled on top of cool dishes, like pasta salad. They lend a mild cilantro flavor to dressings and cooked dishes. Cilantro seed is the culinary herb coriander. It readily self-seeds and will come up early the following year.

Chamomile is not commonly used as a raw edible flower, though it can be. More commonly, chamomile is gathered and dried for tea. If you appreciate chamomile tea, you will want to grow many plants. The tedium of harvest can be eliminated with a chamomile rake. This hand tool is a gathering basket with teeth; it is like a dust pan with long, narrow tines to strip the flowers from the stem and catch them in the pan.

Many chrysanthemum flowers are edible. Pinks and dianthus are the ones most commonly used by chefs. The five-petaled dianthus comes in many forms. Colors range from white to dark pink. Many are frilly or ruffled and come in numerous color patterns and combinations. Like calendula, dianthus needs weekly deadheading to extend production. Dianthuses are usually perennial and should be fertilized and mulched annually.

Clover blossoms, primarily red clover, are gathered and used fresh or dried and used in tonic teas. While they are edible, clover is not often used fresh, although the florets can be pulled from the flower heads and sprinkled on salads or other dishes. Red clover self-seeds in most of our gardens. We harvest and dry enough flowers to fill a gallon jar for the year's tea supply. When grown for a larger crop, they are planted in Zone 3 gardens, with other main crop varieties.

Dandelion flowers are probably used less than they should be. The yellow wispy petals can be trimmed from the cluster and tossed into many dishes. More commonly they are used for wine or cooked in batter.

Daylily flowers come in many shades of orange, yellow and red. All are edible. Most often, it is only the buds that are used, because the flower wilts quickly. The buds can be stir-fried and added to grain and pasta, or served as a side dish. Daylily flowers can be stuffed with cheese or grains and baked like squash blossoms.

Dill flowers are highly recommended for adding a sweet dill flavor to dressings, salads and pasta salads. The small bright yellow flowers are cut from the main stem for a delicate garnish. Dill is highly prized as beneficial insect habitat. Of course, the seeds are also a marketable crop.

Fennel, both green and bronze, has flowers similar to dill flowers — a flat cluster of smaller flowers. These bright yellow flowers, with contrasting brown stems, are used in tea and desserts. Their mild licorice flavor makes them a good garden nibble. Just be sure not to bite into a predatory wasp, which loves to forage fennel.

English daisy blossoms are small and usually pink or white, although some varieties come in other colors. These perennials will naturalize in lawns and gardens. Petals can be cut from the flowers and sprinkled on food.

Garlic chive's starry white flower clusters are eagerly sought by our chef customers. The mild garlic flavor and the fine shape make up for the lack of color. They are used to flavor dressings, garnish salads, grain, pasta and roasted chicken.

Hollyhock and its woody relative *Rose of Sharon* have large, papery flowers perfect for stuffing and roasting. Colors range from white to a deep purple that is almost black, plus many shades of yellow, pink and red. The closely related *Malva* (mallow) is also useful. The zebrina variety is half hardy in cold frames and can yield lovely purple blooms from mid-spring through late fall. These are used in salad and as a garnish.

Kale buds and flowers are a wonderful addition to salads. The clusters of yellow flowers appear the second year. Red Russian kale will overwinter in the polytunnel and cold frame. Beginning in the spring, we can pick the flowers weekly for months. Heavy harvest will prevent kale from going to seed. The plants get bushier as they grow through spring and summer. Kale buds are also used in many other dishes.

Marigolds are edible, but many are not palatable due to their strong, distinctive flavor. Lemon and Tangerine Gem are the most commonly used marigolds. They are small and have a milder flavor. These bush plants need weekly harvesting or deadheading to remain productive. Plants are started in early spring for June flowering. They can produce hundreds of flowers for most of the summer, but it is wise to plant a late spring succession to ensure a supply until frost.

Mint flowers taste like mint. There are dozens of varieties of mint, and each has a different flowering style. Spearmint tends to have flowering spikes at the tip of the stem that make them especially useful on desserts and in tea. Other mints have flowers spaced along the stem and may not be as ornamental, but make a good garnish.

Nasturtium flowers are a delight to the eye and a spicy surprise to the palate. Growing as a vine or a bush, and in a wide range of colors, a succession of nasturtiums in the greenhouse and garden will keep customers supplied most of the year. Preferring cool weather, nasturtiums may not be as abundant in the heat of July and August, but will be revived in the fall until killed by frost. Given too much fertilizer, nasturtiums will produce more leaves than flowers. This is fine if you want the edible leaves for the salad bowl, but not if you want to market the flowers. Nasturtium flowers have a "hot" taste, and add color to salad dishes. They can be stuffed, with a mix of cream cheese and herbs, for a lovely and colorful appetizer. Nasturtiums flowers also make a pretty and spicy vinegar. All parts of the nasturtium plant are edible. Fresh seeds can be ground to a paste that is very similar to the sushi condiment wasabi.

Pansy, viola and violets all add cheer to the salad bowl. Pansy flowers are high in the phytochemical rutin, which helps strengthen our blood vessel walls. Researcher James Duke has reported that they can help prevent varicose veins. The many colors and sizes of pansy and viola make them versatile. They add a festive flair to food. Johnny jump-ups are a popular and heavy-producing variety. Violets also come in various shades of blue, purple, yellow and white. Some violets are locally rare and should only be harvested when abundant.

Pansy.

Pea blossoms taste like peas with a touch of sweetness. But there is a difference between peas and sweet peas. I repeat: *Do not eat sweet pea seeds or flowers.* Sweet pea (*Lathyrus odoratus*) seeds and flowers both contain toxic elements. But the flowers of snap peas, snow peas, shelling peas and soup peas are all edible, and some are quite striking, in shades of red and purple and patterned with white.

Pineapple sage is a favorite with chefs for its bright red color and mild pineapple flavor. This sage is especially popular around Christmas, when it is in full production in the greenhouse. Children also love to taste the tubular flowers. Pineapple sage is a tender perennial shrub, growing to a height of 4 feet. In cold areas, it is grown in pots and brought inside for winter. It usually begins to flower in the fall, and can continue to bloom all winter. Dwarf pineapple sage grows to 2 feet tall and will bloom in late summer.

Raab is a brassica grown for its buds and flowers. Similar to kale flowers, but more delicate, raab is great in salads, but is also an excellent side dish; it can be substituted for broccoli in any dish.

Roses come in countless varieties, and they are all edible. However, unless you want the rose scent on the dish, unscented roses are the best choice for culinary use. Miniature roses bloom spring through fall, and right through the winter in the cold frame and greenhouse. Scented roses are used for cakes, desserts, teas and other foods that are complemented by the floral perfume.

Rosemary blooms all winter in the greenhouse. The flower colors range from lavender to blue. Tasting like sweet rosemary, they are used as a garnish to meats and other dishes that rosemary complements.

Scented geraniums come in dozens of varieties, each providing flowers of unique color, size and shape. They can be red with black stripes, pale lilac, apricot, white, or combinations of these colors. Many are quite fancy, reminiscent of orchids. These natives of South Africa also each have uniquely scented leaves. Children especially like chocolate mint

DARRELL FREY

Snap peas provide edible shoots, flowers and pods.

geraniums and the citrus-scented varieties. Easily grown from cuttings, scented geraniums are a good plant to use to teach plant propagation.

Most bloom for a month or two in spring, but a few varieties will bloom through the summer. Scented geraniums are damaged by frost.

Flowers are used for many dishes, but are especially lovely on desserts. They also candy well. Scented geranium leaves can be layered in sugar for several days to flavor the sugar for icings. While it is possible the names may change depending on the source, we recommend Scarlet Unique and Apricot because they bloom all summer. Other popular varieties include Chocolate Mint, Mabel Grey, Rose Scented, Nutmeg, and several citrus-scented varieties.

Squash blossoms are popular with chefs. Both the male flowers and the female flowers attached to tiny fruits are used. These are commonly stuffed with grain and cheese and roasted. Some varieties of summer squash are grown for their profuse flower production. Native Americans used squash blossoms. In *Buffalo Bird Woman's Garden,* author Gilbert Wilson reports that Plains Indians harvested only the male flowers, allowing the females to fruit. These were dried for winter and then added to soups and stews.

Sweet woodruff flowers are the prime ingredient in May wine. The starry white umbels of delicate flowers, along with the leaf whorls, are soaked overnight in white wine. Sliced strawberries can also be added. The smoky-flavored phytochemical coumarin gives May wine a distinctive taste and adds to the sedative effect of wine. Because coumarin can be harmful in high doses, moderation is the rule. A handful of flowers and leaves per quart of wine is considered a safe dose.

Thyme flowers are also in demand among chefs. Growing in clusters along the tips of the stems, thyme flowers taste and smell of thyme. They are either stripped from the stem and sprinkled on cooked dishes, or added as a garnish while still on the stem. Thyme flowers vary among the varieties. Creeping thyme and lemon thyme both have distinctive and useful flowers.

Yucca flowers have sturdy, pale white petals that make a good addition to southwest dishes. They can be cooked or eaten raw. Yucca plants grow well in hardiness zone 6 and flower early each summer.

We use a number of other flowers as well. *Watercress* flowers are a common ingredient in our salad mix. *Chickweed's* tiny white flowers on a green leafy stem are common in the salad from fall through spring. *Miner's lettuce* is harvested from the time it begins to sprout in January through the flowering stage in April and May.

Edible flowers

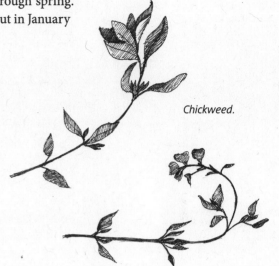

Chickweed.

- anise hyssop, *Agastache foeniculum*
- arugula, *Eruca vesicaria sativa*
- borage, *Borago officinalis*
- sweet basil, *Ocimum basilicum*
- beans, scarlet runner, *Phaseolus coccineus*
- bergamot, *Monarda didyma, M. fistulosa*
- calendula, *Calendula officinalis*
- chive, *Allium schoenoprasum*
- cilantro, *Coriandrum sativum*
- German chamomile, *Matricaria recutita*
- dill, *Anethum graveolens*
- dandelion, *Taraxacum officinale*
- daylily, *Hemerocallis fulva*
- fennel bronze fennel, *Foeniculum vulgare dulce* 'Rubrum'
- sweet fennel, *Foeniculum vulgare dulce*
- marigold, *Tagetes tenuifolia* 'Lemon Gem' and 'Tangerine Gem'
- mints, peppermint, *Mentha x piperita*
- spearmint, *Mentha spicata*
- pineapple mint, *Mentha suaveolens* 'Variegata'
- dianthus, *Dianthus*
- English daisy, *Bellis perennis*
- garlic chive, *Allium tuberosum*
- hibicus, *Hibiscus syriacus*
- hollyhock, *Alcea rosea*
- kale, *Brassica napus*
- raab, *Brassica rapa*
- nasturtium, *Tropaeolum majus*
- pansy, *Viola tricolor* hortensis
- Pineapple sage, *Salvia elegans (S. rutilans)*

- roses, *Rosa* spp.
- rosemary, *Rosmarinus officinalis*
- scented geranium, *Pelargonium* spp.
- squash, *Cucurbita moschata, Cucurbita maxima*
- sweet woodruff, *Galium odoratum (Asperula odorata)*
- thymes, *Thymus vulgaris*
- violets and viola, *Viola cornuta* and other *Viola* species
- yucca petals, *Yucca glauca*

Fungi

Wild edible mushrooms are occasionally found on the farm. Those we find on or near the farm and on our woodland property include meadow mushroom, giant puffball, brick top, honey mushroom, chanterelle, chicken-of-the-woods and various boletes.

Most of these we harvest for our own use. At times, we do harvest chicken-of-the-woods and chanterelle for our chef customers. *Extreme caution* should be used when identifying and using any wild mushroom.

Wild crafted mushrooms, chanterelles are in high demand by chefs.

DARRELL FREY

Shiitake on the Farm Woodlot

Shiitake mushrooms are a highly valuable crop for the farm woodlot. Native to Japan, these fungi are highly nutritious and have an unsurpassed flavor. Shiitake naturally grow on dead hardwood under the shade of a full forest canopy. Since their commercial introduction to the US from Asia in the 1970s, the popularity of shiitake has continued to grow. They can be used in any mushroom recipe, added to soups, scrambled eggs, pizza or lasagna. Generally, the market price for mushrooms is set by the large producers, who mainly grow them in climate-controlled facilities. But once chefs and other customers taste fresh-harvested, naturally grown shiitake, many are willing to pay a premium price for the higher quality.

Shiitake are approximately 18 percent protein and 80 percent carbohydrates and 2–3 percent fat. They contain B vitamins and significant amounts of calcium, phosphorus, potassium and magnesium. They are rich in iron, selenium and vitamin C. Shiitake have been shown to strengthen the immune system, especially aiding those with chronic illnesses.

Shiitake are excellent fresh and can be stored several weeks in the refrigerator. For longer storage they can be dried or frozen.

Shiitake grower Dan Wilcox has developed a small-scale production system at his home, Round Oak, near Carlton, Pennsylvania. His 70 acres of forested property produces hundreds of oak seedlings each year. These are thinned for shiitake logs as needed.

Shiitake cultivation begins in the late winter. New logs are cut and stacked when dormant (which is the same time that the air contains few spores of other decomposing fungi). Generally, the logs are 4 to 10 inches in diameter and are cut to 4-foot lengths. Once gathered, holes drilled into the logs the logs are packed with shiitake mycelium spawn. Mycelium spawn is sold as inoculated dowel rods or as inoculated sawdust. ☞

*Fruiting shiitake
mushroom logs.*

DARRELL FREY

Dan prefers to insert sawdust spawn into the holes using a plugging tool. The holes are spaced about 6 inches apart around the log and are 1 inch deep. Once inoculated, the plugged holes and log ends are coated with wax to seal moisture in and other spores out.

The logs are stacked in a forested propagation yard for a year. They are kept covered and moist while the fungus becomes established in the log. The shiitake mycelium grows into the wood and fruits from the cambium layer, which is the layer between the bark and the older wood. Developing under cover in the propagation yard helps keep undesirable fungus from getting established.

After a year, the logs are moved to a fruiting yard, where they will produce heavy crops for three to five years and then smaller crops for several more years. Under the cover of mixed hardwoods, Dan's logs are stacked in the traditional Asian style. An inoculated log is held against a tree trunk several feet above, and parallel to the ground. Two more logs are laid against this log to hold it against the tree. Then the stack is continued, two narrower logs horizontal: one near the top, one a foot above the ground; and two heavier logs leaned against it. The stacks, supported by the trees, are 10 to 12 feet long, in even rows, with approximately 24 logs each.

Shiitake come in a number of varieties. Different types fruit at different times of the year, being sensitive to seasonal temperature, light levels and humidity. Dan mostly grows summer varieties because they give the most reliable yields in his region. A small crop of early spring *donko shiitake* are relished by enthusiasts. These have thicker flesh and attractive tortoise shell caps due to slow development in cool weather. Whether donko is truly superior in flavor, or just seems that way because it is the season's first fresh fruit, is up to the connoisseur to decide.

For many years, Dan carried stacks of logs to a small pool in a nearby stream. Here they soaked for several days and then were restacked. Because of the intensive labor required, and the difficulty of stacking the logs, he sought a better way. Now, when he wants to induce fruiting in a stack, he sets up a small sprinkler. Pond water, flowing by gravity, drops several feet in elevation down to the fruiting yard, and is gently spread by a slowly spinning sprinkler. When the logs are thoroughly soaked, Dan hits each log with a small tree branch. This vibration somehow encourages the swelling mycelium to produce fruiting bodies. One theory is that the impact may open tiny cracks in the log. Other possibilities are that the mycelium has evolved to respond to the shock of the tree falling on the forest floor, or perhaps the fungus is responding to the heavy clap of thunder, which heralds a flush of lightning-generated nitrogen. A heavy lightning storm will, in fact, induce a mass fruiting.

When they begin to fruit, clear plastic sheeting is suspended over the logs, to protect them from rain. Rain at the wrong time can reduce the quality of the harvest. If the mushrooms are too moist they do not store as well. Top quality shiitake are picked soon after the cap breaks free of the stem and while still curled. By the time the caps open wide and flat, flavor is reduced and fungus gnats may have invaded the mushroom. Once picked, they are layered in shallow cardboard boxes and kept refrigerated.

Dan seems to have a special relationship to the mushroom spirits. In addition to shiitake, he annually harvests chanterelle, chicken-of-the-woods, sheep's head, morels and several other wild edible mushrooms from private wild patches. Ever-observant, and a good steward of his forest, he is rewarded with an endless supply of his favorite food.

CHAPTER 9

Bioshelter Defined and Designed

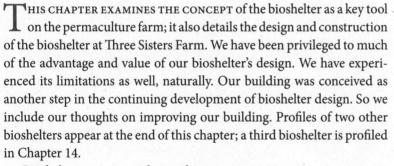

T HIS CHAPTER EXAMINES THE CONCEPT of the bioshelter as a key tool on the permaculture farm; it also details the design and construction of the bioshelter at Three Sisters Farm. We have been privileged to much of the advantage and value of our bioshelter's design. We have experienced its limitations as well, naturally. Our building was conceived as another step in the continuing development of bioshelter design. So we include our thoughts on improving our building. Profiles of two other bioshelters appear at the end of this chapter; a third bioshelter is profiled in Chapter 14.

Bioshelters can enrich modern communities in many ways. Bioshelters are wholesome, biologically rich corridors between the wild and the cultivated, linking natural and man-made environments. They also captivate the imagination and the spirit. A design that collects, stores, and releases the sun's energy, and returns composting heat, moisture and CO_2, to the soil's cycle is a realization of the goals of good permaculture design.

The Bioshelter Defined
Living Shelter

A bioshelter is a solar greenhouse managed as an indoor ecosystem. The word *bioshelter* was coined at the New Alchemy Institute by consulting solar designer consultants Sean Wellesley-Miller and Day Chahroudi.

EARLE BARNHART

Ark 2010 South Wall Exterior

The term was created to distinguish bioshelters from old-style petrochemical-fueled monoculture greenhouses. Bioshelters are a major departure from traditional greenhouse design and operation; they represent a synthesis of energy-efficient architecture and ecological design.

New Alchemy's pioneering work in ecological design is documented in their published journals and reports, and in several books published by John Todd and Nancy Jack Todd. In 1976, the Alchemists built the Cape Cod Ark bioshelter and her sister, The Prince Edward Island Ark. For the next 15 years, the New Alchemy Institute studied and reported on the use of these prototypes for food producing ecosystems.

The term bioshelter (life-shelter) derives from two points of view. The first is that of *ecological architecture.* A bioshelter is designed to be similar to the design of a living cell. The structure and glazing act as the cell membrane, controlling gas exchanges and energy absorption. The building "takes in" nutrients, gases, and energy from the surrounding environment, and it produces food and wastes. Thermal mass used for energy storage relates to glucose molecules stored in a cell. The plants and animals within the building act as energy processors and nutrient recyclers. From this perspective, the built environment becomes a living, breathing component of the surrounding landscape.

The second point of view is that of the *ecological gardener.* Earle Barnhart of the New Alchemy Institute aptly compared a bioshelter to a miniature *ecosystem:* the bioshelter's systems control air temperature and air movement, the flow of water, evaporation, humidity, and other environmental factors. Solar heat is absorbed and stored for later use. Beneficial insects are provided habitat. Gases are exchanged between the animals, insects, microorganisms, soil and plants. Within the bioshelter there are a variety of microclimates. The south wall bed and central beds receive the most direct sunlight. The north wall receives full sun from the autumnal equinox until the spring equinox and is partially shaded through the late spring and summer. The northeast and northwest corners are shaded for a portion of the day, morning and evening, respectively. In some cases, nutrient cycles are developed between fish and plants, and/or poultry and plants. Ecological relationships between pests and their predators are encouraged. The system is controlled by good design and human intelligence to be a shelter for a diverse community of life-forms.

Both views — the bioshelter as architecture inspired by biology, and the bioshelter as a managed ecosystem — intermesh with permaculture concepts. Bill Mollison and David Holmgren developed permaculture design as a system built on integrated components that would function as ecosystems. In permaculture, nature and natural systems inform design. Modeling architecture on the living cell and managing the greenhouse as an ecosystem in miniature is a realization of these concepts.

Why Build A Bioshelter?

The transition from a fossil fuel economy to a solar economy is inevitable. The longer we delay the change, the more urgently we need it. The society of the 21st century will, of ecological necessity, find a way to a permanent, sustainable culture. Once humanity moves beyond the present high-energy addictions of western culture, every temperate climate community will benefit from having at least one of these "life shelters" and a bioshelter market garden.

We see bioshelter market gardens as a key to sustainable living in our climate. There are many applications of bioshelter technologies. Bioshelters can serve as seed and plant material banks for testing and propagating marginal species and for greatly extending the harvest season. From garden cafes to health spas, living spaces, educational facilities and neighborhood resource centers, the possibilities for bioshelter gardens are many.

Energy Conservation

As discussed in previous chapters, the energy costs of our modern food system are high. Energy costs for producing food include energy invested in manufacturing farm machinery, energy invested in fertilizers, energy to heat farm buildings and greenhouses, and fuel for farm equipment. There are also processing and packaging costs. And transporting food across continents and oceans is costly in terms both of money and in energy.

Most of this energy is currently derived from non-renewable fossil fuels. Every step in the process — from tilling soil to stocking the supermarket shelf — adds CO_2 to the atmosphere and increases pressure on oil reserves in marginal ecosystems. Finding ways to produce our food with reduced energy input is a vital necessity in planning for sustainable communities and a stable global climate.

The bioshelter can reduce energy input in food production. Solar heat is free. Biothermal energy is also free heat — when that heat is a byproduct of composting. (And making compost from animal manures also reduces the amount of methane that livestock add to the atmosphere.) Firewood as a backup heat source is locally inexpensive and does not add to the long-term carbon burden in the atmosphere.

Local food production has the potential to reduce energy costs in a several ways. Storage time and transportation costs are reduced considerably. Packaging and processing is reduced when food is bought local and used fresh.

Fresh — Local — Organic

The superior quality and nutritional value of fresh, locally produced food cannot be denied. Whether professional chefs or homemakers, our customers consistently make this point. A diversity of seasonally fresh foods is a key component of a healthy diet. Organic agriculture that nurtures and sustains the local environment is a key component to sustaining healthy bioregions and a healthy planet.

Bioshelter Daydreams

We were inspired to dream of building a bioshelter as part of our permaculture farm by the research being done on bioshelters at the New Alchemy Institute in Massachusetts. Their work in developing composting greenhouses also caught our attention. The microbial activity of composting organic matter generates heat and releases large amounts of CO_2. The New Alchemists found a way to capture what would otherwise be lost by building a greenhouse that incorporated compost chambers. Heat and CO_2 from the compost was blown through ductwork to the garden beds in their greenhouse. Excess ammonia, another byproduct of compost, was captured in a peat filter and recycled back to the compost.

Study of the composting greenhouse and the New Alchemy's other project, the Cape Cod Ark bioshelter, led us to realize the value of an agricultural bioshelter to a market garden farm. Equally, we felt that a bioshelter could only reach its full potential within a garden landscape because the building would need sufficient outdoor garden space to make full use of its production capabilities. We decided that a bioshelter should be at the heart of our permaculture farm.

As we were developing plans for our bioshelter, we studied the Solviva Winter Garden bioshelter built by Anna Edey in the early 1980s on Martha's Vineyard. Edey had integrated poultry into the Solviva bioshelter; this enhanced the bioshelter environment by closing a gap in the nutrient cycle and adding body heat and CO_2 to the solar-heated structure. The reported success of Solviva as a commercial bioshelter was a strong encouragement for us to pursue our dreams.

Bioshelter Design

Design begins with a study of the site and a survey of available resources. These are then compared with the farmer's goals, plans and budget. Regional, sustainable design techniques are assessed for applicability. The design that emerges should provide for the needs of the farm and the farmer within the framework of available resources and local conditions.

Site Considerations

A permaculture farm and bioshelter require a site with the following characteristics or capacities:

- Good drainage for the bioshelter foundation.
- Access to sufficient water.
- The ability to process graywater.
- A source of electricity.
- Full solar exposure and good air circulation.
- Protection from winds or the space to put in windbreaks.
- Several acres of garden space surrounding the bioshelter.
- Zoning regulations that allow agricultural enterprises.

If a bioshelter is to serve also as a community space, learning center, or a site for other enterprises, additional site requirements may include a location close to the community served, public transportation, controlled access, and visitor parking.

Climatic Factors

A bioshelter must be designed for the local climate. Data on average solar income in winter sets design criteria for insulation, thermal mass and backup heating systems. Summer solar gain sets design criteria for shading and ventilation requirements to prevent overheating of the building. Glazing angle is partly determined by latitude. The lower the sun's angle in winter, the steeper the glazing angle should be to allow maximum light penetration. Snow load and wind load must also be figured into roof and glazing angles and the positioning of windbreaks and berms.

The building's productivity relates to seasonal sunlight variation and outdoor growing season. Greenhouse crop selection is determined by the local climate. Seedling production is timed to fill garden and plant sales needs.

Existing Structures

A bioshelter can be added to existing buildings or built to stand alone. Barns or other farm buildings can be integrated into a bioshelter design. Integrating farm buildings into one structure can save on building materials — and can result in many beneficial interactions between the various parts. Workflow patterns should be planned into the design. Layout of building components and related spaces should allow for efficient movement of materials and workers between them.

Brainstorming: Other Possibilities for Bioshelters

Winery bioshelters could be designed so that the fermentation vats provided both thermal mass and CO_2 enrichment. The vats would be filled with the wine in the fall and ferment through the winter, allowing a long, cool winter maturation after a long, warm fall fermentation. In the late winter and spring, solar heat would speed up the process and finish the wine.

Miso fermentation could also be integrated into a bioshelter. Again the fermenting vats of miso would benefit from the long, cool aging while providing CO_2 and thermal mass.

The bioshelter as a health spa has great potential because it has a stabilized heating and ventilation system to keep temperatures moderate. Combining garden spaces with massage tables, a sauna, a hot tub, and tai chi or yoga space would create a very healing environment.

When the combination of climate and production goals of a bioshelter requires wood-fueled backup systems, other functions of the heater could be exploited. Wood stoves can be designed to heat water, cook food and co-generate electricity as well as heat space. When the days are too cloudy and cold for good plant growth, ☞

Building Materials

Selection of building materials for a bioshelter depends on a number of factors. Budget, available resources and the skill and expertise of the work force are primary considerations. At the time we built our bioshelter, part of our grant-funded mandate was to use construction materials and techniques commonly used in our area. Were we to do it all again, we would build a post-and-beam framework of local lumber. We would try insulating some of the walls with straw bales or clay-coated straw and using recycled slate for the roof. Each bioregion has its most appropriate building materials. We highly suggested the reader investigate natural building and green design practices appropriate to their region when designing a bioshelter.

Goals and Plans

A small-scale intensive farm requires a number of buildings, structures and other components. These may include animal housing, tool storage, crop processing areas, potting sheds, seedling frames, office space and greenhouses or other season-extending structures. Starting our farm on an open field, we had the opportunity to plan for maximum integration of our farm components.

Different sites and situations will, of course, have different needs. Educational facilities need classroom and group working space. Markets co-generated electricity could power grow lights for seedling and cutting propagation.

Pottery kilns might integrate well with a bioshelter. In cold, cloudy weather, the kiln would be fired to provide supplemental heating. The pottery could be sold alongside the plants come spring. Similarly, combining a bakery's massive masonry oven with a bioshelter would create heat for the bioshelter as well as the daily bread.

Innovation and creative design are badly needed in the 21st century. While the general public and governments waste time debating how many trees to plant to offset driving a car, much more serious changes must be made in our total consumption. Integrating productive solar design into all our buildings is one of the most important steps we can make. Integrating productive plantscapes into these solar designs can bring buildings to life in many ways.

need storefront space. Community projects need gathering space. Restaurants need kitchens and dining areas. Composting facilities are needed anywhere livestock is raised. Some situations will combine several or all of these functions. All these areas can be integrated into a bioshelter's floor plan.

A Bioshelter for Three Sisters Farm

Following the principles of permaculture, we designed our bioshelter to serve multiple functions and have redundant systems. The building is energy efficient, uses biological resources for heat and nutrient processing, and is a small-scale, intensive system.

The bioshelter at Three Sisters Farm is a combination solar greenhouse, barn, and composting facility. The bioshelter serves as the heart of the farm, allowing year-round production of vegetables, herbs, flowers, eggs, meat, and high-quality compost.

The building is designed to capture and store solar energy to reduce the need for external energy input. Much of the growing space has deep soil beds, designed for mature crop production (rather than the standard shallow greenhouse benches). Poultry and compost bins within contained areas provide gas exchange, heat production and storage, food and compost. A storage area and processing kitchen along the north side of the building helps buffer the greenhouse area from winter winds and provides cooling ventilation in the summer.

The 4,000-square foot building contains, under one roof, a two-story solar greenhouse, classroom space, a potting room, poultry housing,

Bioshelter floor plan.

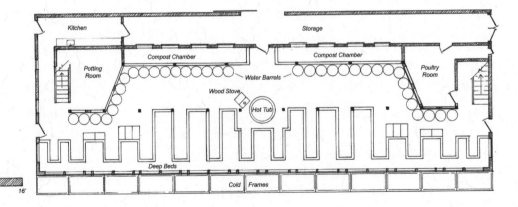

office space, kitchen, storage barn, compost chambers and cold frames. Integration of farm components allows us to maximize beneficial interactions and pathways between them because the various parts of the building are all connected to one another.

Design Criteria

We had two primary goals: (1) to design a low-maintenance structure that would provide maximum agricultural productivity with a minimum of external energy input; and (2) to have a positive impact on the quality of our soil and the surrounding environment. We wanted to use the latest information available on solar design, build with the most appropriate materials, and follow standard building construction practices. All aspects of the final design are the result of extensive research into the various options. Twenty years later, we have had some repairs to perform. These are noted below. But, for the most part our bioshelter has performed very well and is in good condition.

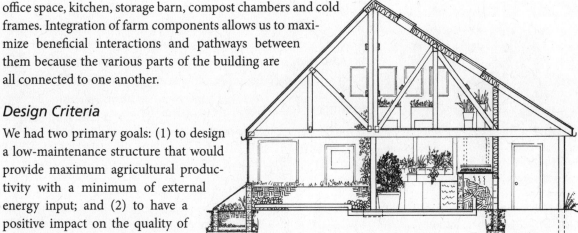

Bioshelter cross section.

Building Systems

Heating systems include passive and active solar, biothermal heat (from poultry and composting manure and hay) and backup wood heat from both a wood-heated water tank and a cast iron stove.

Ventilation systems include windows in the east, south and west walls; 11 small roof vents; 11 large roof vents; and a large exhaust fan in the east wall peak. Depending on the weather and time of year, different ventilation strategies are used. The bioshelter is designed to be half shaded in summer to reduce cooling costs. It receives full sun from the fall equinox until the spring equinox.

Biothermal Heating and CO_2 Enrichment

The poultry room is in the northwest corner of the greenhouse area. Two compost chambers are located along the north wall. We ventilate the poultry room *through* the east compost chamber. The result

DARRELL FREY

The ventilation pipes in the deep beds delivery warm air to the soil at Three Sisters Bioshelter.

Bioshelter foundation and wall.

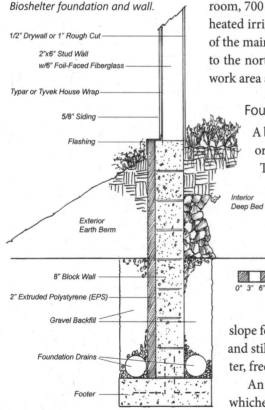

1/2" Drywall or 1" Rough Cut

2"x6" Stud Wall
w/6" Foil-Faced Fiberglass

Typar or Tyvek House Wrap

5/8" Siding

Flashing

Interior
Deep Bed

Exterior
Earth Berm

0" 3" 6" 9" 12"

8" Block Wall

2" Extruded Polystyrene (EPS)

Gravel Backfill

Foundation Drains

Footer

is biologically heated air that is rich in CO_2. This air is then pulled from the compost chamber through a tray of peat moss (to absorb ammonia) and blown through ductworks into rock storage beneath the soil. The rocks absorb the heat and release it slowly to heat the growing beds. The tops of the compost chambers also give bottom heat for seedlings. This system is explained in more detail in Chapter 11.

Design Details and Construction Process

The bioshelter at Three Sisters Farm is 105 feet long and 40 feet wide. The 28-foot by 105-foot main section contains 3,000 square feet of growing space on two floors, a 175-square foot poultry area, a 175-square foot potting room, 700 cubic feet of composting chambers, and a 500-gallon wood-heated irrigation tank. Thirteen cold frames attached to the south side of the main section provide 420 square feet of growing space. Attached to the north wall of the main section is an 8-foot by 20-foot kitchen/work area and an 8-foot by 85-foot pole-framed storage area.

Foundation

A bioshelter's foundation should be cement block or concrete — or otherwise substantial enough to support the walls and roof. The foundation must extend below the frost line and have a drainage system. Some municipalities allow the use of frost-protected shallow foundations. These are packed stone and gravel trenches laid over a drainpipe. Foundations of a greenhouse, as in any building, should be drained with foundation drain pipes (French drains) inside and outside. As is best with any foundation, the drain of a frost-protected shallow foundation should lead to "daylight." This mean the site must have sufficient slope for the foundation drain to lay at a moderate downward angle and still emerge at the ground's surface. Otherwise, in a severe winter, freezing water and soil can damage the foundation.

An architect or a construction professional should approve whichever foundation system you decide on. (An important consideration for the bioshelter: the building foundation needs to be rodent proof.)

Three Sisters' bioshelter has a concrete block foundation set on concrete footers. The main section and work area have an 8-inch block foundation on an 8-inch concrete footing. The east, west and south block foundation walls are 64 inches high — 32 inches above grade and 32 inches below. Many of the block cores are filled with concrete to strengthen the walls. The north block foundation wall is 48 inches above grade and has eight 4-foot openings into the compost chambers and three openings for doorways into the greenhouse section.

Raised deep beds inside the bioshelter are constructed of 6-inch concrete blocks on poured concrete footers.

A 6-inch perforated flexible-pipe drains the foundation, both inside and outside the wall. Exterior block walls are insulated with 2-inch extruded polystyrene insulation board (EPS) to a depth of 2 feet below grade. Aluminum flashing extends from the sill plate to below grade to protect the foam insulation. The exterior foundation along the kitchen and on the east and the west and are earth bermed to further reduce heat loss.

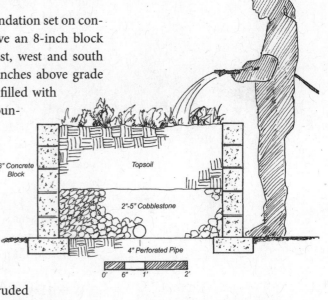

The cold frame foundations run the length of the bioshelter's south wall. Cold frame foundation walls rise a few inches above the surface and extend 32 inches below grade. They are constructed of 6-inch block on a concrete pad. The foundation is divided into twelve 4-foot by 8-foot growing areas by walls made of 4-inch blocks. The cold frames drain into the foundation drain.

Framing and Siding

A bioshelter can be framed with any number of building techniques, provided walls are weather tight and rodent proof. Stud framing, post-and-beam, adobe or concrete blocks all have their place. Concrete, brick, or stone walls, when insulated on the exterior, provide valuable thermal mass. Each bioregion dictates the best materials and practices.

The block foundation of our bioshelter is topped with a compressed sill seal and a bolted-down sill plate of treated wood. This sill is sealed in

Deep bed planters combine thermal mass, and growing space in a well drained easy to work raised bed.

polyurethane and covered with metal flashing to prevent copper chromium arsenate from leaching from the treated wood and contaminating the soil. All potential air leaks are caulked or filled with spray foam insulation. Doors are tight fitting and weather stripped, as are windows and vents.

The east, west and north walls are constructed with standard stud framing, using 2 x 6's on 24-inch centers. Six inches of fiberglass in the walls, with housewrap and textured plywood siding (T-111) sheeting outside, give an insulation value of R-30. The 4- by 8-foot T-111 plywood siding provides a strong wall with minimal air infiltration. The housewrap (Typar) further reduces infiltration and heat loss. The north roof is insulated with 10 inches of fiberglass (foil-faced for reflectivity and to act as a vapor barrier), giving an insulation value of R-30. The interior seams of the roof and wall insulation were taped with foil-backed insulation tape and covered with nylon-reinforced, UV-resistant white plastic sheeting. This reflective sheeting has not degraded in 20 years.

Which insulation you use can vary as long as the walls are, again, weather tight. We used fiberglass batting and ridged foam in our design. If we had it to do over again, we would try straw bale or other natural insulation materials. Of special concern with any insulation is the need for a secure vapor barrier inside the building. With the high humidity of a bioshelter, interior air leaking into the walls and ceiling can quickly condense against cooler outer walls, eventually leading to rot and structural failure. (This is especially important with straw-bale walls, because they would absorb this moisture and rot fairly quickly without a sealed vapor barrier.) Creating a wind-tight seal is also very important for preventing heat loss. Air loss and wind infiltration must be controlled to maintain desired temperature and control moisture.

Siding materials and exterior finish is, again, a regional choice. Our bioshelter was built with a siding of T-111 panels to save time and money. The low cost and ease of installation of the sheeting panels make them a common material for agricultural buildings. The building is painted with a semi-clear oil-based stain.

As we plan upgrades and remodeling of the bioshelter, we expect to eventually cover the T-111 with a siding made of local hemlock boards and battens. They will provide a more durable and aesthetic siding.

Glazing Systems and Roof Design

Special consideration is needed when designing the roof. For a successful bioshelter, informed choices need to be made not only on style and materials, but also on the roof angles. As discussed in Chapter 5, the desired solar gain must be balanced with the need for insulation and summer shading. The floor plan and building dimensions set the stage, and the glazing and roof style set the lighting. The more area of the walls and roof that are glazed, the more supplemental heat will be required, because glazings are poor insulators. Similarly, a fully glazed structure has a higher need for ventilation in the spring, summer and fall. Roof design should also allow for as much open space inside as possible, while allowing for structural integrity under stress from wind and snow loads. Trusses and other roof supports should minimize interior shading of thermal mass and growing areas. Partial summer shading of the interior can help keep the building cooler, providing more comfortable work and gathering spaces.

Choice of glazing is dependent on budget, roof style, and the quality and quantity of light desired. Choices include double-paned tempered safety glass, acrylic panels, and polycarbonate panels. Polyethylene plastic sheeting is not substantial enough for a bioshelter. Our building has a glazed south roof. The north roof has some skylights but is mostly covered and insulated. This layout allows for full light penetration of low-angled winter sunlight from mid-September until late March. During the hot months of June, July and early August, the building is partially shaded from the high summer sun angles by the north roof.

The roof of Three Sisters' bioshelter is largely framed with prefabricated trusses. Trusses are spaced on 48-inch centers to accommodate the glazing system. The south trusses extend from a center beam (running the length of the main section) to the south wall and angle up 45 degrees to the roof peak. The north roof has two rafter systems. The 900 square feet of both the east and west ends are roofed with 2-inch by 12-inch rafters on 24-inch centers. These extend from the north wall of the storage barn and connect at the peak to the glazing support trusses. The middle section of the building is roofed with trusses, which extend from the rear storage area/barn and angle up 30 degrees to the roof peak. The roof is sheathed with 1-inch boards and covered with felt paper and asphalt shingles.

The north roof trusses in the middle section have been substantially modified since construction. The original truss system, which spaced trusses on three-foot centers, was difficult to work in and gave too much shade. We have added a second post-and-beam to support the bottom chord of the trusses and a header beam under the roof rafter section of the truss. This allowed us to cut out many of the interior truss members. We left a truss intact every 12 feet, building up the vertical truss members into 6-inch by 6-inch posts. In the process we created a series of "bays" through the second floor. It would have been better to build an open floor plan at the beginning.

Passive Roof Vents

The north roof has two rows of manually operated skylight/vents: eleven 22-inch x 32-inch vents and eleven 45-inch x 65-inch vents. These vents are glazed with flat fiberglass, which is sealed against 2–4-inch frames with a sticky glazing tape. They are accessed from inside by a walkway through the second floor areas. During the winter months the roof vents are sealed with clear plastic sheeting to reduce heat loss. These vents are sufficient to cool the building nine months of the year, and are supplemented with a whole building exhaust fan on hot summer afternoons.

After 20 years of operation some of these roof vents need to be repaired or replaced. Their frames and glazing have aged. We expect to replace them when it is time to re-shingle the roof. We plan to install two continuous roof vents near the peak of the roof. This will allow more passive ventilation and reduce or eliminate our use of the exhaust fan.

Glazing

The south roof is glazed with 26 Exolite acrylic panels, each 4 feet wide and 18 feet long. This is a total of 1,872 square feet of roof glazing. Acrylic panels are installed with a polyvinyl chloride (PVC) glazing system. We originally had a glazed roof of double-paned glass units. These failed due to different rates of expansion between roof trusses, glazing system and glass panels.

Cold frames are glazed with a single layer of flat fiberglass reinforced plastic (FRP). The removable cold frame covers measure 4 feet by 8 feet and are built from 2-inch by 4-inch lumber. The lids are hinged to the bioshelter's south wall and mounted on wood framed at a 45-degree angle.

The south wall is glazed with double-paned insulated glass units: 18 fixed units, measuring 50 inches by 50 inches and 9 framed and hinged windows, measuring 18 inches by 52 inches. These windows are rough framed with double 2 by 6's and continuous headers, and have finished frames of 1-inch pine with 1-inch cedar sills; they are held in place with 1-inch by 2-inch molding strips.

Special care must be taken to design windowsills to be rot resistant. During winter, water will condense on the inside of glass and run down to the sill. When watering, it is inevitable that the sill will get wet. Sills should slope away from glass and be kept coated with waterproofing stain or paint.

Interior Design

Interior spaces are laid out to best use the various zones in the building. The bioshelter contains 3,000 square feet of growing space (including deep beds, second floor areas and benches, plus 240 square feet on top

Wax Treatment

All exposed framing is painted with a mix of beeswax and turpentine to resist decay. Twenty years after construction, we have seen no signs of decay. Outdoor structures, such as arbors, gates, seedling frames and the bale-shelter are also painted with the mixture.

Our wax treatment recipe:

Place ½ pound of beeswax in a 5-pound coffee can. Add enough turpentine to cover the wax, and put a lid on it. Place it in a very warm place (we put it on the second floor of the bioshelter where it can melt in the 90° heat.) *CAUTION: This mixture is flammable! Do not heat the mix of turpentine and wax on a flame or stove burner.* The initial mix will form a paste. Stir it every few days. The paste is then mixed with more turpentine to make it the right consistency for application. Or, it can be mixed half-and-half with boiled linseed oil.

Use this mixture to protect wood surfaces. The turpentine in the mixture helps to carry the wax into the wood's pores. Added linseed oil will help seal in the wax. The volatile parts of the turpentine will evaporate, leaving a waterproof finish. Too much wax will leave a slippery finish and should be wiped off or rubbed in.

More caution:

Turpentine is distilled from the resin and sap of pine trees. Mineral spirits, a common substitute for turpentine, is derived from crude oil. Both are volatile organic compounds and can cause a number of health problems when inhaled or ingested. Always use turpentine in a well-ventilated space, and take all proper precautions to prevent skin contact and accidental ingestion.

of the compost chambers and 400 square feet in the cold frames attached to the south wall). Planters and structures are water resistant and massive for heat storage. The deep growing beds are all connected and laid out for ease of use. They are framed with 6-inch concrete blocks to a height of 32 inches above the floor. The beds are mostly 5 feet wide and vary in length. (See floor plan.) The holes in the block walls are planted with perennial herbs and flowers. A variety of planters are used inside the bioshelter. Used 50-gallon plastic juice barrels are cut lengthwise to make planters measuring 2 feet by 3 feet each. Drainage holes are drilled in the bottom and the planters are set up on wood frames covered with flashing. Various-sized pots hold plants for sale and harvest.

The shaded northwest corner of the bioshelter has a 175-square foot potting room. Here we store potting soil supplies, pots and flats, small tools, catalogs, and reference books. Above the potting room on the second floor is a 200-square foot growing-bench area.

In the shaded northeast corner of the bioshelter is the 175-square foot poultry area, housing 30 to 50 chickens, with access to an outside forage area. Above the poultry housing is a 200-square foot second floor growing-bench area.

The second floor is 14 feet wide by 105 feet long. A walkway through the second floor connects a series of planting benches and allows access to the roof vents. Additional water storage, planters, seedling benches and hanging plants are located along this walkway. The second floor area is accessed by 3-foot-wide stairways on the east and west ends of the building.

Compost Chambers

The two sets of compost chambers are located along the north wall of the bioshelter and are accessed from the attached barn through four 40-inch-square doors on each side of the center rear door. The chambers are built of six concrete blocks, 30 feet long, 4 feet wide and 4 feet high and topped with lids made of plywood and 2 by 4 lumber (waterproofed with our beeswax mixture). These lids have begun to deteriorate after 20 years of use. We have been collecting used and discounted floor tiles and will replace the wood covering with these tiles and concrete board. Management of the compost chambers is detailed in Chapter 11.

Compost Chamber.

DARRELL FREY

Heating

We strive to keep the bioshelter's inside temperature above 40° F on even the coldest nights and in the 60s and 70s during the day. This is accomplished through solar heat management and use of supplemental wood heat. The chickens and compost chambers also contribute heat to the building seasonally. (More on this in Chapter 11.)

Solar

The vast majority of heat our bioshelter needs is provided by sunlight. The building is designed for both passive solar and active solar heating. Passive solar means that sunlight enters the structure and is absorbed by the building's mass as heat. Active solar heating is the use of electric-powered fans or pumps to move heated air or fluid to storage. With both passive and active solar heating, when the surrounding air is cooler than the storage medium, the heat radiates from the storage to the air. Because solar design is a complex topic, the details are discussed in the chapter 5 and in the appendix.

Thermal Mass

Thermal mass in the bioshelter includes insulated block walls, block framing of deep growing beds and compost chambers, fifty 50-gallon water barrels, all the interior wood framing, 3,000 cubic feet of soil, 30 tons of rocks in the deep growing beds, the 500-gallon irrigation tank, and 50 planter barrels. Altogether, the thermal mass in the building has a storage capacity of over three million Btus.

Thirty tons of stones provide thermal mass in Three Sisters Bioshelter.

DARRELL FREY

Plastic drum planters.

Most of the building's mass acts as passive solar heat storage. A total of fifty 50-gallon water-filled drums are located on the first floor to provide thermal mass and support plant benches. The second floor planters include ninety 25-gallon planters, which provide additional thermal mass.

Woodstoves

Two woodstoves stoves use about three cords of wood each year to supplement the solar and biothermal heat systems. A 60,000 Btu woodstove sits in the middle of the bioshelter. The stove is used to heat the air in the bioshelter at night when temperatures outside are below freezing, and on cold, cloudy days. By maintaining a minimum building temperature, the stove helps keep the temperature of thermal mass from falling too rapidly. The stove is adjacent to a large mass of concrete and barrels of water, which absorb some of the heat and radiate it back to the building when the stove cools down.

A 600-gallon reinforced concrete tank is also in the center of the bioshelter. In it, we installed a *Snorkel Stove* — a submersible,

Solar Aquaculture as Thermal Mass

Our initial plans called for four solar algae tanks in the bioshelter. With these, we had planned to raise 160 pounds of catfish and 1,500 lettuce plants each year. We designed floor space for eight tanks 5 feet high and 5 feet in diameter that would hold approximately 600 gallons of water each. This would be a significant contribution to the heat-storing thermal mass of the building. Solar algae tanks were developed at the New Alchemy Institute, where extensive research on their use and performance was conducted. Their tanks are constructed of .040 mil Kalwall fiberglass reinforced panels. Sunlight enters the translucent walls of the tank allowing algae to grow. Either catfish or tilapia raised in the tanks are fed fish food, supplemented with the algae, worms and insects. Aquarium aerators are used to aerate the tanks. Each

week, 20 percent of the water in each tank is changed to maintain the water quality. The water removed is used to fertilize and irrigate crops. Lettuce grown hydroponically in floating trays in the tanks can reduce or eliminate the need to change the water.

We decided to put off the installation of the solar tanks. We have yet to find affordable organic fish food — or an alternative. Also, we have reservations about the potential of chemicals from the fiberglass and its adhesives leaching into our bioshelter's soils. But the bioshelter's thermal performance would benefit from the additional thermal mass of the fish tanks. We do hope to someday resolve the hurdles and develop an organic fish production system.

wood-burning stove. In cold weather, the heated water and tank provide supplemental thermal mass storage for the bioshelter. The submerged stove can produce up to 120,000 Btus per hour. Once the fire is burning hot, and if fuel is added hourly, this stove can heat the water ten degrees per hour. Because the heat from the fire transfers directly through the stove's aluminum walls to the water, it is extremely efficient. The stove is also designed to burn a very hot fire. The strong draft and design allow for a complete combustion of the wood and wood gases, greatly reducing smoke and soot emissions.

The warmed water is used to irrigate crops via by porous irrigation hoses, buried 6 to 8 inches below the soil surface of the deep growing beds.

We have plans to eventually upgrade the bioshelter's wood heat system by adding a massive masonry stove. The masonry will add significantly to the building's thermal mass and increase the efficiency of the wood heat. Because a massive masonry stove takes a few hours to heat up, the freestanding woodstove would still be used at times for quickly warming the air.

Composting Chambers

The addition of composting chambers to the Three Sisters' bioshelter is based on the composting greenhouse at the New Alchemy Institute. Research there indicates that inclusion of these chambers in a biosphere is justified if the farm "has strong incentives for composting and using greenhouse space."

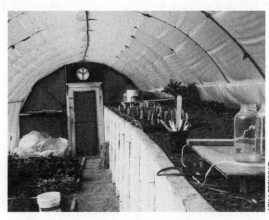

Compost Chamber at New Alchemy.

There were several reasons for including these chambers into our bioshelter design. Primarily, we needed large amounts of compost to use as organic fertilizer in the bioshelter and the outdoor gardens and nurseries. The chambers potentially allow us to produce over 100,000 pounds of compost (prepared from poultry manure, bedding and shredded hay) each spring. Our actual production in most years is about 32,000 pounds of compost.

In addition to the compost, we wanted to capture the by-products of the composting process. As previously mentioned, the tops of the compost chamber lids are used to stage seedlings and potted plants that thrive on the bottom heat provided by the compost. The compost also provides heat and CO_2 to the deep growing beds. A ¼ HP Dayton blower is ducted to draw air from above each compost chamber through a network of 6-inch flexible, perforated pipe to 30 tons of rock storage beneath the growing beds. Air is drawn into each of the compost chambers through one of three manual vents: one vent allows air to be drawn from the growing area into the 8-inch air space below the compost and up through the compost, aerating it; the other two vents draw air from the adjoining animal areas: one into the air space below the compost and one directly across the top of the compost.

A biological filter is necessary to remove excess ammonia from the chamber air. The filter contains a mixture of peat soil, potting soil ingredients and compost. The filter mixture and the microorganisms in the filter mixture trap the ammonia while allowing CO_2 and heat to pass through. We monitor the "biofilter" simply by smelling the filter mixture. For more technically minded folks, numerous ammonia-measuring devices are available. When strong ammonia can be smelled, part of the filter mixture is replaced and the saturated peat is put into the compost chambers or is placed in the outdoor finishing compost piles. Chapter 11 examines the compost chambers and their use more closely.

Ventilation

Ventilation is critical in the summer. As the temperature inside the building rises through the day, the wooden trusses and the glazing expand. It is important to keep the temperatures on the second floor below 100° to prevent over-expansion of the glazing. The 18-foot-long acrylic panels will expand more than 1 inch from 0° to 110°. This temperature range can easily occur on a sunny winter day. Over-expansion can lead to cracked panels.

The bioshelter has a series of passive roof vents as described above. These are located across the length of the north roof, near the peak; they open when temperatures reach 90° at the vent level. A second set of 11 manually operated vents are located just below the peak roof vents. These are opened as needed to further ventilate the roof.

At the peak of the east wall gable, a 36-inch exhaust fan draws warm air out of the bioshelter when the two sets of roof vents are insufficient.

Each gable end wall has four 2-foot by 3-foot hinged windows which can further increase ventilation.

Side wall windows were custom made in our neighbor's wood shop. These hinged windows are difficult to secure in strong winds. Double hung windows or horizontally sliding windows would allow for ventilation and hold up much better. We plan to soon replace the second floor windows on our west wall with a custom-built sliding glass door. This will allow more controlled and secure ventilation and provide visitors a better view of the farm.

On the ground floor, at each of the end walls, we have a door, a 2-foot by 3-foot hinged window and a 50-inch by 50-inch removable hinged window to provide cross-ventilation through the bioshelter, east to west.

Along the south wall, the nine 18-inch by 52-inch hinged window vents allow a chimney effect to take place when the top vents are open.

Planned Improvements

We plan to upgrade the bioshelter's ventilation system soon. We want to replace the existing roof vents with two continuous vents, each 4 feet wide by 30 feet long, near the roof peak. These will be opened as needed to passively cool the building and will eliminate the need for the exhaust fan altogether.

Interior Air Movement

Several small fans provide air circulation inside the bioshelter. These are located near the wood stove during the winter and run at night to help distribute heat through the building. In the spring, they are relocated to help ventilate seedlings. Interior ventilation is not necessary in the summer when windows and vents are open.

Cold Frames

The cold frames are ventilated by opening them manually. In the frost-free months, they are removed and stacked until fall. In the winter on sunny days, they may need to be ventilated. We plan to install a series of 6-inch vent pipes from the cold frames, through the south wall and into the bioshelter. These will allow us to capture excess cold frame heat and vent it into the building instead of losing it to the outdoors.

Integrated Animal Areas

The poultry room is located in the northeast corner of the bioshelter. The daily care required by poultry fits in well with the attention required by a greenhouse. Both are Zone 1 activities. Produce surplus and wastes supplement the poultry's diet. An outdoor forage area allows the birds to leave the indoor area during warm weather and naturally prevent overheating the bioshelter.

The key to having this integrated system work efficiently is in the monitoring and control of the gas exchange between the animal housing and the growing areas. Excess ammonia gas from the animal wastes can harm plants, chickens and people. The gas exchange is controlled in several ways. The animal areas are separated from the growing areas by vapor barrier walls. Chicken droppings are covered with a layer of hay each morning and evening to prevent the loss of ammonia. The animal area is isolated from the growing areas by vapor barriers and weather-stripped doors. Controlled inlets allow air to enter the animal housing from either outside or from the bioshelter growing areas. Air from the animal areas is pulled through the compost chambers and blown into the rock storage under the growing beds. In the process, the blower ventilates the poultry room. During the winter, the chickens can provide a significant amount of heat and CO_2 to the indoor environment, enhancing productivity. Chapter 12 closely examines the role of poultry on the farm.

Processing and Storage

The kitchen work area, located in the northwest corner of the bioshelter, contains a stainless steel counter and sink, storage space for packing supplies, counter space and refrigeration. Here we prepare and pack harvested produce for delivery. A scale is used to weigh produce as we pack bags of herbs and salads. A large mixing container is used to mix the greens for the salad mix.

The rear barn is designed for storage of supplies, equipment, and tools. The space, 8 feet wide and 80 feet long, provides sheltered access to the compost chambers, poultry area, kitchen, and greenhouse sections of the bioshelter. This space is also an important part of the building's climate control, acting as a windbreak in the winter and a source of cool air in the summer. During the coldest days of the year, this wind-sheltered space serves as the main entrance to the building.

Plumbing and Electricity

The bioshelter is plumbed with nine hose faucets located at key points in the building, two outdoor hose faucets and the kitchen sink faucet. Two faucets service the second floor, one on each end. The rest are located across the first floor. Each faucet has a dedicated hose and watering wand. Well water is used for all greenhouse and seedling watering and for drinking water.

The bioshelter is connected to the utility grid and is fully wired for lighting, fans, blowers and other uses. Ground fault protected circuits lead to outlets on both floors, in the barn, and outside, on each end of the building. These outlets allow a lot of flexibility for locating fans to circulate air and power tools if needed. Dedicated circuits provide power to blowers and fans, water pump, cooler and grow lights. Separate circuits power lights and receptacles.

Farm Design

Inherent in the design of the bioshelter is its placement within the larger farm and nursery of the Three Sisters' operation. The building is oriented to face true south and is set on the high point of the field to allow for drainage away from the building. Windbreaks of multi-use plants and trees reduce heat-robbing winds, and provide market crops as well as food and shelter for beneficial songbirds and insects. Compost produced in the bioshelter compost chambers and outdoor windrows is used both within the bioshelter and in the surrounding market gardens.

Three Sisters' bioshelter is a pioneering effort to develop innovative agricultural technology. We have created a synthesis of several existing bioshelters and greenhouses, a synthesis that is designed to fulfill our needs for greenhouse space, animal shelter and composting facilities and to function in our local climate as the heart of a productive, year-round, low-input agricultural system.

Three Sisters' bioshelter is but one example of bioshelter design. By following the premise

Bioshelters can provide a transition between the home and the animal housing. Imagine doing the chores in the winter and never going outside.

Prevailing Winds

House

Compost Chamber

Water Barrels

Bioshelter

Chicken Coop Barn

Deep Beds

Forage Yard

Cold Frames

Garden

N

0' 2' 4' 8' 16'

of designing and maintaining agricultural systems based on ecological principles, many variations can be created: from the home-scale chicken coop/greenhouse to larger urban eco-structures combining growing spaces with living and working areas.

PROFILE

A Bioshelter Permaculture Farm and Homestead for the High Desert and Mountains of the North American Southwest

Mt Elden rises to an altitude of 9,299 feet, towering 2,300 feet above Flagstaff, Arizona. Ancient volcanic activity created porous soils, leaving the area with no perennial streams. For nearly 2,000 years, Native Americans journeyed each

A "Living" Room

How to explain the experience of a bioshelter in winter? The smell of rich earth and moist air accented with the fragrance of rosemary and evening scented flowers. The vision of a green oasis in the midst of blowing snow . . . . the sound of a rooster crowing as the winter wind rushes overhead . . . the crackle of the fireplace and laughter of friends sipping wine around the fire...the buzz of the carpenter bee pollinating the tomatoes . . . the taste of mint and dill, of bay leaf and lemon grass, pungent arugula and hot mustard . . . the vibrant orange and yellow of pansy and nasturtium, bright blue borage and the many colors of scented geraniums . . . the feel of soft leaves and warm earth and the sun on your face ... One must experience for oneself a bioshelter in winter to truly appreciate the beauty and bounty of the winter garden.

Imagine a bioshelter attached to your home. Your "living" room measures 24 feet long and 16 feet wide. You enter the space through a door from your kitchen. Inside the bioshelter, you sip fresh mint tea before a fireplace, warming yourself and the plants. Perhaps the hot tub is heating, to act as a radiator through the coming cold night. An aquarium is bubbling in the corner; catfish lounge in the cool water. ☞

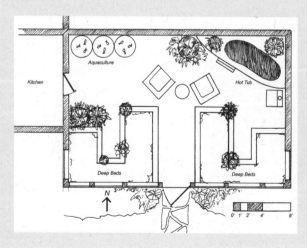

spring from their winter homes at lower elevations to forage and tend their gardens in the rich alluvial soils at the mountain's base. When it fell in sufficient volume, rain and winter snowmelt draining off the mountains provided the necessary moisture for a good crop. The summers were cooler than at lower elevations and the surrounding land also offered abundant resources for foraging, game for hunting and materials for crafts and construction.

Located where the high desert meets the San Francisco Mountains, Ponderosa pine, Pinyon pine, juniper, Gambel oak, prickly pear cactus and yucca dot the landscape along with native and introduced grasses and wild flowers. This diverse landscape supports many native animals, including tree swallows and western bluebirds, horned lizards, mule deer and bobcat, as well as many bee species and other insects.

Today, a new local culture is germinating among the potsherds of the Anasazi. A modern experiment in permaculture and urban subsistence farming

Your plant collection is an expression of your personal tastes and interests. Perhaps pots of scented geraniums line the shelves. Culinary herbs are set about, bay leaf and basil, dill and cilantro, thyme and oregano, curled parsley and flat parsley. Cooking greens, such as multi-colored Swiss chard, spinach, assorted kales and mustards are planted in the deep beds. Salad plants, too, are planted in the beds. Many lettuces, red romaine, crisp buttercrunch Bibb, speckled Bibb, bittersweet endive, frilly Lollo Rossa and red oak leaf are interplanted among other greens with lovely Latin names, like claytonia, stellaria, and portulaca. Wild edibles growing among the cultivars include lambsquarters, amaranth and sheep sorrel.

Vines, passionflower, beans, and brilliant nasturtiums climb the posts. Ripening tomatoes hang down from planters. A bumblebee flies by on its way to pollinate the new tomato flowers. You reach up and pick a soft, ripe, Sungold cherry tomato from the vine. The flavor is sweet and incredibly delicious.

You also have a collection of interesting and useful plants — aloe vera, with its stalk of subtle and graceful orange flowers, sits beside the flamboyant clivia, with its flaming orange trumpets. The amaryllis replies with its own scarlet blooms. Your jade tree is now 3 feet tall. The camellia (tea) plant and citrus are showing new leaves.

The sunshine warms your skin as you sip your tea. You contemplate the evening meal. Perhaps it will be roasted potatoes with thyme and rosemary chicken, or perhaps it will be rosemary-marinated tofu. Maybe you will make pesto pizza, the crust fresh baked in the fireplace oven. Of course you will have mixed salad greens, maybe arugula and baby chard with dill tonight. Certainly you will sample the wine fermenting under the seedling bench. Outside, the clouds of a late winter storm move in. You put a little more wood on the fire and then sit back and leaf through the new seed catalogues, planning the summer garden.

is beginning to flourish at Mountain Meadow Farm. The homestead occupies two and a half acres along the southern base of Mt. Elden. A series of swales containing garden beds fortified with abundant compost catch runoff and occasionally fill with moisture from the mountain's watershed. A low-tech irrigation system covers the property. Irrigation is essential for the survival of young transplants and other plants during periods of drought. In 2009, only 10 inches of precipitation fell on the site.

The landscape, by trial and error, is developing into a bountiful oasis of food and useful plants. Several dozen varieties of fruit trees and shrubs and other useful perennial plants, and many varieties of herbs, flowers and garden crops are grown in the gardens and in the greenhouse portion of the farm's 2,600-square-foot bioshelter. The swales are planted with many annuals, including garlic, onions, corn, pumpkins and potatoes. There are also perennials in the swales, including asparagus, Jerusalem artichokes, day lily, raspberry and rhubarb. Inside the bioshelter are figs, guava, citrus, and a wide range of vegetables, as well as many flats of garden starts. A top-bar-style beehive was added in 2006, supplementing the native pollinators and providing bee products to the owners.

Investing substantial personal assets, Chuck McDougal purchased the property and began designing Mountain Meadow Farm. He envisioned a place to "explore materials, methods, tools, systems and relationships that forward our quest for sustainable living in our place on the planet, and to actively share what we learn with others." Putting into practice an ongoing study of permaculture design, Chuck, with the support of his wife Denise, has created a homestead that contributes to research and development of sustainable human habitats for the high desert and mountains of the American Southwest.

Mountain Meadow Farm gathers and composts organic matter from sources around Flagstaff. Compost is used as the base for potting soil and garden and orchard fertilizer. Worm composting predominates. The redworms feed poultry and are shared with schools and other local foods projects. After market farming for several years, the increasingly productive landscape and bioshelter are now used to produce food for family and friends, while constantly testing production methods and crops. A CSA program rents half of the bioshelter's greenhouse in the spring to start seedlings — it served 70 members in 2010.

Acting as a resource for permaculture in the Flagstaff region, Chuck is assisting in the development of an urban farm at a more appropriate site. In addition to advocating and demonstrating local food system development, Mountain

Meadow Farm has worked with Flagstaff Foodlink's Youth Gardens Project, Master Gardeners, and students of public and private educational institutions.

In a climate with harsh extremes of temperature, moisture, wind, and intense sunshine, food production requires substantial shelter. In 2010, the farm celebrated six years of frost-free food production in its solar-heated bioshelter. Designed to produce farm animal products, fruit, vegetables, herbs and native plants for their personal consumption and sharing, this shelter has performed beautifully. Chuck first tested his ideas on bioshelter design in a smaller home-scale greenhouse system. This 10 by 20 foot structure provided fish and greens for his family. His success with this prototype led to further bioshelter study, including Three Sisters Farm's bioshelter. Chuck adapted what he learned in his research and from the management of his home-scale greenhouse to the design of the bioshelter for Mountain Meadow Farm.

Truly essential for sustainable high-desert homesteading, careful placement was important. Sited in a relatively flat and open area, the structure is oriented to true solar south. The bioshelter includes a 1,340-square foot greenhouse on the south side and animal housing, food processing and storage space on the north side. Animals include small flocks of chickens, turkeys and a pair of China geese. A large root cellar built into the north side provides cool storage. Also included are a potting and plant propagation area and a crop processing area with sinks, counters and refrigeration. There are also facilities to dry and store herbs, vegetables and fruits.

The bioshelter's passive solar design includes 12-inch-thick, grouted sand-block greenhouse walls; an insulated block foundation; a ground-level greenhouse gravel floor; ceramic tile on the balcony growing area; 1,500 gallons of water in storage tanks; the steel framework for the greenhouse portion; and raised growing beds. A solar hot water heating system provides some backup heat in cold weather. A newly installed, grid-tied, 1,500-watt photovoltaic system easily covers electrical needs.

The northern half of the bioshelter is built with a post-and-beam framework, in-filled and insulated with strawbales and

Inside the Mountain Meadow Bioshelter, located near Flagstaff AZ.

CHUCK McDOUGAL

CHUCK McDOUGAL

*Inside the Mountain Meadow
Bioshelter, located near Flagstaff AZ.*

finished with cement stucco. The greenhouse's triple-wall polycarbonate glazing system and framework was purchased from a commercial greenhouse supplier who worked with a local engineer to design the unique glazing and venting system.

Building Systems

Plants are irrigated with rainwater and snow-melt captured from the farm roofs. Cisterns with a total 15,000-gallon capacity hold this water until needed. Irrigation from the cisterns gets delivered with a solar electric-powered pump or by gravity feed. When the cisterns run low, city water is used to refill them, ensuring that a sizeable supply of water is always available. Indoor air circulation is enhanced with fans year round. Twin automated front and roof vents, windows and sliding glass doors allow for natural ventilation. On hot summer days, cooling is greatly assisted by limiting venting on the hot south side and opening up the screened windows and doors on the north side. When the ridge vent is wide open, the "chimney effect" draws in cooler, carbon dioxide-rich air from the north side/barn. Graywater from the building's crop processing areas is collected and used to supplement the irrigation of the landscape's fruiting trees and shrubs.

Mountain Meadow Farm is an ongoing project of permaculture design and subsistence living in the high desert and mountains of the North American Southwest. A new state-of-the-art green farm house is soon to be completed. All utilities and systems will be integrated with the bioshelter, further demonstrating to others in the region the value and beauty of bioshelter-assisted permaculture systems for life on a small farm.

PROFILE

Cape Cod Ark Bioshelter

A Tour of the New Alchemy Institute

Before building our bioshelter, we visited the New Alchemy Institute and spent two days studying their composting greenhouse and Cape Cod Ark bioshelter.

We arrived in late winter, on a crisp New England morning. The air was cold and the morning overcast. After letting the folks in the office know we were there, we were free to wander the property and examine the structures and the landscape. Even after years of reading about the Ark in New Alchemy's books and newsletters, we were astounded when we entered the Ark. The bioshelter was a time machine, transporting us from early March into the middle of June. The multi-level space was full of color. Many flowers were in bloom: purple and yellow pansies, red nasturtiums, white and pink snapdragons, yellow and orange calendula, and the striking bird-of-paradise plant. Still warm from stored solar heat of previous days, the contrast between the bioshelter garden with the brown and windblown landscape outside was striking. Seedlings were germinating in flats along the south knee wall. Frogs were swimming in a concrete pool. The growing beds, stone-lined terrace gardens and planters were abundantly planted with lettuces, chard, kale, bok choy and other greens. A 12-foot-tall loquat tree was heavy with ripening fruit. Water tanks were located throughout the building, including nine 700-gallon tanks in the entrance/aquaculture room. All were in good health and well tended.

We visited the other greenhouses, including the geodesic pillow dome and the composting greenhouse. These were productive and instructive, green and sheltered microclimates. Still, we were continually drawn back to the Ark. Its design created a frost-free microclimate — with no backup heat! The combination of water tanks, stone walls, gardens, pathways, flowers, vines and loquat tree was a true ecosystem. Insects buzzed about, ladybugs patrolled the plants and fish splashed in tanks. Our children, then aged two, five and eight were equally impressed with the Ark bioshelter. "Is our greenhouse going to be like this, dad?" five-year-old Christopher asked. Two-year-old Terra stooped to smell all the flowers, while Zack scrambled among the walkways and terraces. My wife Linda, too, was amazed and inspired by the possibilities of year-round gardening.

When the morning fog lifted, the building was flooded with sunlight. The colors brightened. The air was fresh and moist, smelling of humus, herbs and flowers. I had spent much

Composting greenhouse at The New Alchemy Institute in 1988.

DARRELL FREY

time in my youth discovering special woodland groves, where the combination of trees, water and landform created that sense of natural wholeness, greater than the sum of the parts, where the earth's energy seemed tangible and alive. The Ark was such a place.

Our goal on this first visit was to experience the space and to see firsthand what a bioshelter was. That afternoon and evening we met with the staff of New Alchemy and discussed our bioshelter design plans. Earle Barnhart and Hilde Maingay, who also had a consulting company called Great Works, gave us excellent advice for improving our design and management plans. Bruce Fulford, of Biothermal Associates, provided details we would need to complete our plans for incorporating compost chambers into our bioshelter.

The following day, we spent more time observing the site. I was happy to see that the building was not airtight. After 12 years, the reinforced plastic glazing (Kalwall) had lifted off its sill in several places, and the doors did not shut tightly. Yet the place was still frost proof with no backup heat. This was good news because our calculations had shown we would need backup heat in our bioshelter, but I wondered how well the building would hold up to the forces of entropy. The fact that the Cape Cod Ark was a little leaky gave me confidence. Also, the interior wood framing was enduring well. The floor plan was laid out to make the maximum use of sunlight. The distribution of thermal mass and the relationships between water tanks, stones and soil created microclimates within a microclimate. Highly fertile fish tank water helped maintain the healthy plants. The abundance of flowers promoted an ecological balance of pest and predator insects.

The Cape Cod Ark was designed as a research and demonstration facility. Areas for working and research were incorporated into the bioshelter. The New Alchemy Institute was founded in 1971 with the mission "to restore the lands, protect the seas and inform the earth's stewards." For the next 25 years, the Institute provided cutting-edge research, demonstrations and education in many aspects of ecological design. Organic gardening, permaculture design, aquaculture, alternative energy, forest farming,

Pond inside Ark bioshelter.

season extension, and bioshelter design were the primary areas of research. (This research is well documented in the *Journal of the New Alchemists*. Quarterly newsletters and various research reports are available at www.thegreencenter.net/.) For over two decades, thousands of visitors, volunteers, interns, students and neighbors have learned lessons of sustainable living at the New Alchemy Institute.

To the New Alchemists, the bioshelter was much more than a building. It was a metaphor for the newly emerging paradigm of sustainable, ecological design. It was the integration of horticulture, aquaculture and architecture, as explained by John Todd in his article in the *Journal of the New Alchemists* (No. 6, 1980). The Cape Cod Ark bioshelter was intended to "point the way toward a solar-based, year-round employment, creating agriculture for northern climates." The elimination of fossil-fuel heating and use of functional thermal mass demonstrated the possibilities for solar greenhouse design.

The original Cape Cod Ark was 90 feet east to west and 28 feet north to south, which gave 1,950 square feet of floor space. Built into a sloping hillside, the bioshelter is embraced on three sides by the earth. Whitewashed, curved concrete walls on either side of the bioshelter formed a "solar courtyard," that reflected light and protected the building from the cold north wind. Its south-facing walls and roof,

Ark Interior.

and east and west wall are glazed to allow sunlight to enter. The foundation is insulated with R-9 2-inch foam board. The north walls and roof are all insulated with R-24 fiberglass insulation. The 1,550-square foot glazed south roof rises at a 45-degree angle to meet the insulated north roof. There are several types of thermal mass material: 43 cubic yards of rocks in a rock box; more than 9,000 gallons of water in 14 solar algae tanks (a tank made of cylindrical fiberglass); and a 2,800-gallon, floor-level concrete fish pond. For several years, a blower had circulated air in the bioshelter through the rock box, storing heat in the day and releasing it at night. Eventually, the blower was turned off because it was not needed to keep the building productive.

In 1989, the building's glazing system was repaired. A steel framework was installed to replace the wooden rafters and support a double layer of thin-film, reinforced fiberglass glazing. The top vent was removed and the lower vents enlarged.

The Cape Cod Ark Today

In the 21st century, the Ark has become a home to people as well as a horticultural ecosystem. After the New Alchemy Institute ceased operations in 1992, the land was purchased by a co-housing co-operative. Former NAI researchers Hilde Maingay and Earle Barnhart have renovated the Ark bioshelter and attached their home to the north side. Earle describes the arrangement as "living with the ark." He explains that tending the building is like visiting a friend to see what is going on. Almost every day offers something new in the world of plants, fish and animals in the Ark.

The building needs very little upkeep. The plants are tended every few days and watered once or twice weekly as needed. Succession plantings of broccoli provide an almost daily supply year round. Dwarf lemon trees also produce all year. Flowers, climbing vines and many types of salad and cooking greens fill the growing spaces. All watering is done manually, as is the regulation of the passive ventilation. This bioshelter is an ecosystem that has been 30 years in the making. Beneficial insects and frogs are all established residents. The building's nine above-ground fish tanks and the floor-level cement pond are used to raise fish and provide nutrient-rich irrigation water.

Earle and Hilde replaced the glazing system with thermal pane glass and triple-walled polycarbonate. The rock box was removed and replaced with storage for 2,000 gallons of water. This water is pumped by a solar-powered pump through a solar hot water system.

No supplemental heat is used. The Ark bioshelter is totally heated by the solar energy passively and actively collected in the thermal mass. Similarly, excess heat is removed with natural, passive ventilation.

The 3,000-square foot house connects to the bioshelter through a glazed roofed dining room/sunspace. Every room in the house receives natural light. Tight-fitting doors and windows regulate the exchange of air between the house and bioshelter. This prevents excess heat and humidity from entering the house. According to Earle, "the best part of living with the Ark bioshelter is to be surprised by wildlife, sometimes small wildlife, intent on their busy lives, or just passing through. Hummingbirds dart in, look around, and dart out. Frogs travel on the soil, hunting bugs. Tiny predatory insects cruise through the parsley, looking for an aphid. The best part of living with a bioshelter is sharing the space with other interesting living beings."

New Alchemy Ark bioshelter transformed into a home.

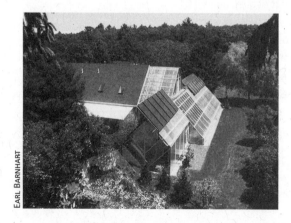

EARL BARNHART

CHAPTER 10

Bioshelter Management

B IOSHELTER MANAGEMENT HAS MUCH IN COMMON with standard greenhouse management. Both strive to provide a proper environment for plant productivity. A bioshelter is a departure from a standard greenhouse because it attempts to create and sustain a diverse organic system of plants, poultry, renewable energy, insects, gardeners and building. Such a system is complex, and it evolves as new species and varieties are added, seasons change and managers learn. After two decades of managing our bioshelter, we are still enthralled by the prospect of spending a day in a hot, sunny garden surrounded by snow and ice. (However, the need to feed the fire on long, cold winter evenings and feed the chickens each morning is a bit less enthralling. It's important to have a backup team to allow days off and travel.)

As stated in the previous chapters, we see bioshelters as a vital feature for any sustainable community in a temperate climate. Much of the value of a bioshelter is the use of the space for social and educational activities. At Three Sisters, weddings and other events have been held between the compost chambers and the planters, among the vines, and beside the fireplace.

Sunlit interior gathering spaces integrated with crop production — edible conservatories, so to speak — will be commonplace when local food systems are fully developed. Our bioshelter is an aesthetic environment as well as a production facility, so we try to keep the space neat and orderly as well as productive.

This chapter details the management of the bioshelter as the heart of the permaculture farm.

Managing the Indoor Ecosystem — An Overview

The primary objective in managing a bioshelter is crop production. Three Sisters Farm is a commercial enterprise. At the same time, we have always had a goal of furthering bioshelter research. Ecological management of an indoor polyculture system is a fascinating study. Our bioshelter has been developing as an indoor ecosystem since 1989. Over the years, our seasonal crop mixes have varied, and perennial plants have been rearranged many times. We have not always managed the building simply for profit. At times we have just experimented with crops or taken a break from intensive management to pursue other activities (such as writing this book). We have given space over to collections of unusual ornamental plants. Linda loves working in the sunny bioshelter — getting her hands in the soil and growing healthy plants. My own fascination with the natural world and garden ecology has been engaged by the various components involved in the site's development. Our mutual belief in creating new possibilities for abundant and ecological landscapes has kept us interested all these years.

Many aspects of the bioshelter's management have been consistent through the years. Seasonal planting schedules, soil management, harvest schedules, and daily maintenance are well-established patterns. Production of salad greens, herbs, herbal bouquets and edible flowers has been ongoing. Every day since 1989, someone has tended the chickens, watered the plants and managed the heating or ventilation.

Daily Management

Daily care of the bioshelter begins with the chickens. Fresh hay or straw is strewn over the morning's chicken droppings. This makes a cleaner space for the chickens and reduces ammonia evaporation, conserving nitrogen and keeping down the odor. It is surprising how well this works. A well-tended chicken coop smells like fresh straw instead of ammonia. In winter months, when chickens have less access to the outside, extra straw is added, and rock phosphate or crushed limestone is occasionally added to absorb nitrogen.

Bedding is removed and composted several times in the summer and in late winter. After the old bedding is removed, a base of sawdust is sprinkled with rock lime then and covered with straw to begin the process again.

The chickens' water is refreshed and feed troughs are filled with a daily ration. In all but the coldest weather, chickens are allowed access to an outdoor yard and forage areas. Eggs are collected two or three times daily. Additional feed may be given to the chickens at midday. As we work in the gardens and bioshelter, wheelbarrow loads of weeds and plant trimming are given to the chickens. Chapter 12 provides a more in-depth discussion of the role of the poultry on the farm.

Heating and Cooling

A fire is maintained in the woodstove during the winter months on cloudy days when the temperature is below 30°. When the outside temperature falls below 20°, we also heat the 600-gallon water tank to act as a radiator. A submerged aluminum "Snorkel Stove" provides highly efficient water heating. The pre-warmed water is used to irrigate garden beds as needed. Two fans near the wood stove and water tank blow the air from the center of the building toward the east and west ends on cold nights.

During spring, summer and fall months, roof vents are opened to allow passive cooling. As discussed in the previous chapter, we have multiple options for ventilation, depending on the season and weather. An electric exhaust fan is used as needed on summer afternoons to keep building temperatures below 100° on the second floor.

Plants require daily monitoring for water needs, health and insect infestations. Most plants are watered daily and fertilized with liquid seaweed; they get fish emulsion weekly during the spring through fall growing season. Less fertilizer is applied in the cold and darker winter season.

Ongoing Building Maintenance

Besides regular building maintenance, such as painting and window washing, the bioshelter has some unique needs. Most of the wood-paneled walls, floor joists, trusses and wood framing for planters are painted every few years with a mix of beeswax dissolved in turpentine (See Chapter 9 for the wax treatment "recipe"). This soaks into the wood and provides a waterproof finish. Floors and decks inside are painted as needed with linseed oil stain. Extra care is taken to keep windows and doors in good repair to keep out drafts.

The building exhaust fan, the active heat recovery blowers, and several air circulation fans are kept clean and in good repair. The faucets and plumbing in the bioshelter and associated hoses, watering wands and nozzles are also kept in good repair to prevent leaks.

Wood stoves and chimney pipes require proper care for fire safety. The dampers and doors of our wood stove are maintained to keep an airtight seal. Chimneys are kept clean by proper fire management and timely sweeping.

We have found wooden planters are not cost effective because they rot too quickly. Most of our planters are heavy recycled plastic barrels. Still, we have found it necessary to have some wood in contact with soil at the ends of the planter beds to allow them to be emptied if we want to change soil or access air ducts under the beds. These wax-coated hardwood boards are occasionally replaced, maybe every eight or nine years.

The first floor of the bioshelter is covered with crushed gravel over packed subsoil. The gravel is raked as needed to keep it clean of plant residue. Every four or five years, the gravel is removed and used to fill in potholes in the driveway. Fresh gravel is brought in to cover the floor.

Plant Growth Factors

Factors affecting plant growth include fertility, temperature, light, moisture, air circulation, CO_2 levels, insect pests and disease. All these factors interrelate. A few examples: in a cold or dry soil, nutrients may not be available; in low light, plants cannot metabolize nutrients; insects and diseases attack weakened plants. The goal of managing a bioshelter is to maintain a balance of growth factors to encourage productive and healthy plants.

The main factor we cannot control in our bioshelter is the light level. We decided we wanted to use the absolute minimum of energy-intensive artificial light. The only artificial lights we use are the florescent grow lights we use to get seedlings off to a good start. Therefore, during the short, cloudy days between early December and mid-January, we keep the building cool and take a break from the busy schedule we keep the rest of the year. If the sun happens to shine more than usual, we appreciate the extra harvest income.

During the summer, many of the planters and beds in the bioshelter are allowed to lay fallow or we do succession plantings of buckwheat as green manure. Perennial plants (ornamentals, herbs, flower and tomatoes) are maintained year round in the bioshelter. In the summer, the annual crops most susceptible to flea beetle damage — brassicas, eggplants and salad mix cutting trays — are grown inside. Flea beetles rarely enter the bioshelter, but they render these crops unmarketable in our outdoor beds.

Microclimates within Microclimates

The bioshelter's plantings are arranged to take advantage of the building's various microclimates. Heat-loving figs, tomatoes and nasturtiums are grown in planters and beds on the warmer second floor. Taller shrubs and vines are located on the north end of the beds to mimic a forest edge and prevent them from shading other crops. Shorter perennial herbs are planted in the narrow divide between the deep beds. This is done to divide annual planting beds and slow down the migration of any pests that might try to take up residence.

Salad crops are grown in the cooler beds on the first floor all year and in the second floor planters all winter. Houseplants and ornamentals are located throughout the building but are concentrated at the ends of the building, which are shaded in the morning or evening.

Inside Three Sisters bioshelter in the spring

In August and early September, we begin preparing and planting for fall and winter production in the bioshelter. Around this time, soil tests are done to check for nutrient levels and pH. Mineral levels are high after years of organic management, so we usually do not need to add much more than a few inches of compost each fall.

Pests

At the beginning of the fall season, we monitor the bioshelter for pests. Aphids, thrips, whitefly and spider mites are likely lurking among the perennials and self-seeded crops in the beds. Sow bugs (also known as pill bugs) are ever present. While usually a valued part of the decomposition process, sow bugs can be a major consumer of crops when populations get out of control. To get the season off to a good start, we try to get the insects in balance early. Bowls of beer are set around the beds to trap sow bugs by the hundreds as well as any slugs that may have come in with plants or flats from the outside. (Sows bug and slugs are a big treat for the chickens!)

However, one must be careful when trying to control pests. Colonies of aphids and whiteflies usually support beneficial predators as well. Before we spray insecticidal soap, we look for these predators, collect and relocate them. Aphid predators include the bright red gall midge larvae, syrphid fly larvae, tiny parasitic wasps and ladybugs. These all come into the building from the surrounding gardens. We encourage them by planting insectary plants, such as tansy and fennel, just outside the building.

The parasitic wasp, *Encarsia formosa,* consumes whitefly. If their numbers are low, we purchase a few thousand larvae. But they are usually present in sufficient numbers to keep the whitefly in check. Occasionally, a summer crop of tomatoes or other host plant will become host to vast numbers of whiteflies. When that happens, we resort to vacuuming and then spraying soap before releasing more *Encarsia.*

Ladybugs and their larvae are major predators of aphids. Unfortunately, they only reproduce one generation in the greenhouse. For about eight years, we bought ladybug larvae by the thousands from California. But we didn't feel right about it. Besides the unknown effect on California ecosystems, we thought we might be introducing new ladybug diseases or parasites into our gardens. Then, in the mid 1990s, an Asian ladybug introduced by the USDA began to appear in our area. We do not know how they are affecting the local native ladybugs, but the Asian ladybugs are now here in abundance. On warm Indian summer days (a warm period after the first killing frost), we leave the doors and windows open, and they come into our bioshelter by the hundreds seeking a sheltered place to hibernate. We no longer have to purchase ladybugs. We do order predators for thrips and spider mites each fall and again in late winter.

These pests are difficult to see until after damage has been done (the presence of spider mites is only obvious once an infestation strips leaves of their outer surface), so if we see signs of them, we add predators.

Even before the advent of the Asian ladybug, our use of imported predators was greatly reduced by the appearance of the insect-consuming fungus *Beauveria bassiana* in the bioshelter and surrounding gardens (see Chapter 6). This fungus spends the warm months in the soil. When the air cools in the fall, we find the fungus infecting aphid colonies in the gardens. We collect a dozen or so "fuzzy" aphids (already infected with the fungus) in a deli tray and allow them to spore. Spores are mixed with distilled water and sprayed throughout the bioshelter. New seedlings are given an application of the spores all winter and spring. The fungus becomes active in the bioshelter in November, just as predators go into dormancy. Some years, when temperatures are cool and humidity is high, the *Beauveria* fungus virtually eliminates aphids, whiteflies and thrips until March — when it gets too warm for the fungus. But by then, the predators are active again.

But every year is different. The winter of 2006-2007 lasted so long that the *Beauveria* fungus remained active into May. Other winters have been too sunny and warm for the fungus to be completely effective. When that happens, we spray more with insecticidal soap and rinse plants more often. On the brighter side, though, predatory insects are active earlier in those sunny winters.

In the fall of 2002 we had no Indian summer — and the ladybugs did not invade the bioshelter. The early onset of winter that year greatly reduced the population of Asian ladybugs in our region in 2003. In the fall of 2004, they came back in smaller numbers. Since then, their numbers have stayed level; they seem to have reached a more balanced relationship with the local environment.

Habitat

A profusion of flowers are planted throughout the building to provide nectar and pollen for the adult parasitic wasps, syrphid flies and ladybugs. The New Alchemy Institute's publications discussed establishing such "biological islands" of habitat in a greenhouse. So we have included plantings of pansy, alyssum, scented geraniums, calendula, nasturtium, nicotiana and snapdragons — all of which are good habitat in the winter

DARRELL FREY

Edible pansy flowers and lambsquarters at Three Sisters Bioshelter.

greenhouse. Plants that tend to be pest free, such as rosemary, thyme or sage, are placed at key spots in the continuous deep beds to minimize pest migration through the beds.

A large two-story bioshelter like ours contains many zones, each with a different combination of average daily light, temperature ranges and humidity. Plants are located in the appropriate zone to maximize healthy productivity. The second floor tends to be hotter, so tomatoes go in the planters on the second floor and hang down to be harvested from the first floor.

Pollinators and Other Fauna

Each February, carpenter bees (a wood-boring bumblebee look-alike) emerge from hibernation in the bioshelter ceiling and pollinate our tomatoes. A few weeks later, adult syrphid flies appear and pollinate as well. By mid-March, syrphid fly larvae are busy consuming insect pests. Syrphid flies are active pollinators and their larvae are active predators until late fall. Ground beetles (Carabidae family) are plentiful in the bioshelter beds and planters. These predators must be eating something, but we are not sure what. They probably consume various soil fauna (hopefully, pill bugs). They are active from March through October.

Toads often find their way into the bioshelter, patrolling the floor under planters and among the plants in the deep beds. Usually we put them outside where they are safe from the high traffic in the building. Nevertheless, every winter we find one or two toads nestled in the soil of the bioshelter beds, or in the gravel floors. Their spring song, a lovely, soft trilling, is a unique entertainment in March.

Spiders are common in the bioshelter. Although their contribution to pest control is hard to quantify, spiders occupy many niches in the building. Black jumping spiders usually become active in February and can be seen patrolling for insect prey throughout the building. Several other spiders weave their webs along window frames and in odd corners of the building, catching flies, aphids, whitefly and cabbageworm butterflies. One interesting, long-legged spider we have yet to identify survives by eating other spiders.

In the spring and summer, hummingbirds come and go through open doors as they please to drink from the nasturtiums. Swallows also visit the bioshelter at their leisure. Most other birds need help finding their way out.

Disease

Our bioshelter harbors only two persistent diseases: mosaic virus and powdery mildew.

Mosaic virus causes stunted and misshapen leaves. Aphids carry the virus from one plant to another. The disease seems most prevalent when plants are stressed by cold soil. We try to plant seeds certified "virus-free" in the bioshelter, and we aggressively root out diseased plants before the aphids can spread it around. Chickweed, a valuable wild edible, is especially susceptible to mosaic virus.

The other disease in the bioshelter is powdery mildew. Plants affected develop a white fungal bloom when low light and poor air circulation allow the fungus to grow. Control is a combination of removing affected leaves, avoiding getting leaves wet when watering, and using fans to keep air moving. Ornamental plants and the non-edible leaves of plants like pansy or tomato can be sprayed with sulfur water to kill the powdery mildew fungus. When the sun shines brightly for at least three days a week, it has an antifungal effect on the soil surface, greatly reducing disease problems.

Densely planted salad greens, such as arugula, lettuces, mizuna and tatsoi, may succumb to mold in cold, cloudy weather. This is best prevented by planting less thickly in winter and not watering on cloudy days. Usually the crops end up thinning themselves. Some die, but many plants remain. The dead leaves and plants between the survivors are cleaned out and the remaining plants generally thrive.

Crop Management

Healthy crops begin with healthy soil. Beds are monitored and tested for fertility and fertilized as needed. Compost is added to most beds

DARRELL FREY

Interior plantings on the first floor at Three Sisters Bioshelter.

and planters each year. Green manure crops are occasionally planted as well.

The bioshelter beds are managed for high productivity by continuous cropping from September through early June. When salad crops start to bolt, they are removed and replaced with a new crop. Lettuce, endive and other salad plants are seeded every week most of the year to ensure a steady supply of seedlings. We are always trying new salad ingredients, and we vary the mix seasonally.

Lettuces and other salad plants are closely interplanted in the planters and beds. Self-seeded edible crops germinate among the transplants. Different types of self-seeded plants appear at different times of the year. Miner's lettuce and mache begin to grow in October as soil cools; amaranth appears in February as soil temperatures increase. In March, as the soil gets even warmer, purslane appears. Chickweed and lambsquarters germinate all year. Tango lettuce, a green oak-leaf variety with deeply serrated leaves, in particular, grows well in the bioshelter; we have let it self-sow for many years. The overall result can be a beautiful, multi-layered mix of many different plants all growing together. A 6-square-foot planter can accommodate lettuce, amaranth, lambsquarters, wood sorrel and a tomato or nasturtium vine.

Many crops are direct seeded. In warmer months, we seed full or partial beds of lettuce, spinach, mizuna, tatsoi, red mustard, pea shoots and other salad greens. We often direct seed a crop of cutting greens between mature transplants, to save time on the rotation. After the new seeds are established, we harvest the previous crop.

Arugula is always direct seeded thickly to provide weekly cuttings. A small handful of seeds will cover about 25 square feet. Seeds are scattered on a prepared bed and raked in with a hand rake. The beds are watered deeply every day until the seeds germinate and get established. This builds up a reserve of water in the soil.

Seeding Cutting Trays

Cutting greens can be seeded into trays and flats and harvested at any size — from 1-inch baby greens to 6- or 8-inch mature leaves. Kales, mustards, kyona mizuna, tatsoi, spinach, pea shoots, shungiku, lettuce, arugula, and just about any edible leaf can be grown for cutting greens. Each week, we seed the equivalent of about ten flats (about 20 square feet) that will be harvested four or five weeks later.

We sometimes start by soaking used trays in hydrogen peroxide bleach and water to kill plant pathogens. Often, they simply need to be washed off with a hard spray from a garden hose and dried in the sun. The flats are filled with potting mix to a depth of no more than two inches and watered thoroughly. A small handful of seeds — approximately one to two tablespoons — is spread evenly over the soil surface and covered with a thin layer of potting mix. This is lightly watered and the trays are placed in a fully sunlit section on the second floor. Once they germinate, trays of cutting greens need to be kept well watered. Because the soil is shallow, they dry out easily. As they mature, watering is lighter to prevent flattening the seedlings and causing them to rot. Cutting greens can be planted the same way in the greenhouse and garden beds, but growing in trays makes planting and harvesting easier.

The greens are harvested with a sharp knife or harvest scissors. Greens grown in flats usually yield only one harvest, sometimes two. Harvest size is based on customer's preference.

Because they are planted densely and only harvested once, cutting greens grown in flats can be expensive in terms of seed costs and potting mix. They are priced accordingly. We recycle the spent potting mix for potting up other plants. Or, in the winter, when there is a smaller supply of green plant food available, the spent flats are dumped in the chicken yard or poultry room, so the birds can eat the remaining stems and roots

Second Floor planters and shelves at Three Sisters Bioshelter.

Gardener as Ecosystem Manager

Humans are indispensible caretakers in the bioshelter ecosystem. The building systems require a minimum of an hour or two of daily care.

Fires have to be fed in winter, fans and blowers turned on and off, vents opened on sunny days, and plants have to be watered. The bioshelter managers monitor the system dynamics and decide what, where and when to plant — and which self-seeded plants to keep for desirable traits. (Interesting variations in leaf shape, such as oak leaf lambsquarters or round-leaved sheep sorrel are allowed to re-seed.)

Humans can also be vectors for insect pests. Care must be taken not to spread spider mites and aphids when working among plants. It is best to wash one's hands and brush off clothes after working with infested plants.

Although we spend a lot of time in the bioshelter (we often spend evenings around the fire), it is not designed as a home. Plants like a wider temperature range than we do. Nighttime temperatures can drop into the low 40s; daytime temperatures can rise into the 90s. Integrating living spaces into a bioshelter would require special considerations for humidity and temperature control.

Along with the poultry and insects, we are also consumers in this ecosystem. In addition to the food and aesthetic yields we take from the bioshelter, we "harvest" lessons in solar energy dynamics and ecology when we give tours to school groups and other visitors.

A bioshelter ecosystem is a dynamic place. From the ecology of the soil to the building's role in the larger farm system, and from the daily work schedule to the seasonal cycles, a complex web of energy, nutrients and organisms interact as a whole system — and we are its directors.

Bioshelter Management through the Seasons

The four seasons are generally adequate for describing the climate of northwest Pennsylvania. However, in a bioshelter we add a fifth: *winter/spring*. Each of the five are distinct periods in the indoor garden. Each has a different crop mix, different pests and control methods, different maintenance schedules and different solar dynamics. Understanding these differences and working with the cycles of the year is the key to successful year-round gardening.

Summer

Summer in the bioshelter is from June until early September. Summer is partly a fallow time in the bioshelter because most crops are grown outdoors. However, a number of potted plants are maintained in the greenhouse: perennial herbs, insectary plants, some early-crop tomatoes, peppers, eggplants and the last of the winter nasturtiums. Some ornamental plants, including the scented geraniums and some edible flowers, remain indoors. Summer crops grown in the bioshelter include basil, arugula and baby greens planted in trays. Most years, we seed buckwheat in fallow beds to serve as green manure.

Building Management: Summer

Ventilation is a daily concern in the summer. Most screened vents and windows are left wide open all day (except in stormy weather, when they are closed to keep out rain and prevent wind damage). The south wall windows are opened to allow ventilation up under the glazing and out the roof vents. The east wall exhaust fan is used each day from 1 p.m. until 4 p.m. unless the day is cloudy.

The rear barn is a major source of ventilation air in summer. The exterior sliding door is kept closed. Air drawn in through the east barn door is cooled as it passes through the shaded barn. This cooled air is drawn from the barn into the greenhouse section of the bioshelter.

The thermal mass of the concrete blocks and soil that helps hold heat in the winter also moderates temperature in summer. The floors and concrete block beds are sprayed with water in extremely hot weather and the evaporating water helps cool the building. Roof vents are left open overnight during hot weather to allow the building to cool down at night.

Through June, our main focus is getting seedlings and other plantings into the outdoor gardens. In early July, the interior growing beds and large planters are cleaned of crop residue and planted with buckwheat. After four to six weeks of growth, the buckwheat is chopped into the beds. Most years, a second crop is planted in many of the beds. When in flower, buckwheat provides important forage for beneficial insects. The organic matter then feeds the soil ecosystem and helps keep minerals actively cycling.

In early September, several inches of compost are chopped into the beds and planters. Based on soil test results, lime, rock phosphate and greensand (described in more detail below) are added as needed along with the compost.

Seedlings for outdoor plantings are started in the bioshelter under the grow lights through the summer. After four weeks under lights, seedlings are moved outdoors to the grow frames to "harden off."

Daily chores in the bioshelter include watering, opening and closing vents, feeding and watering the poultry, and collecting their eggs three times each day, morning, noon and late afternoon.

Insects in Summer

Insect pests are rarely a problem in the bioshelter spring through fall. The predators previously discussed are active and provide effective control. An application or two of soap spray may be needed to keep thrips, spider mites and whiteflies in check. However, certain crops grown inside can create problems. Several years we tried growing gourds in the bioshelter, and they were quickly infested with whiteflies and thrips. We had to remove the plants early — before we were able to harvest any gourds — to prevent a winter infestation. When brought under control in the spring, though, whiteflies are rarely a problem in summer.

Fall

From September through late November, crops are established in the bioshelter for the long winter season. Fall is distinguished by the shortening days and the lower angle of the sun. Light penetrates further into the building each week from July through November. For a month after the autumnal equinox, the combination of day length and solar angle fill the building with light and heat. In fact, in October the bioshelter receives more daily solar gain than in summer. After mid-November the shortening days bring diminishing light and diminishing returns in crop production.

Plantings in the fall are scheduled for a weekly harvest of salad, herbs, edible flowers and cooking greens from bioshelter, cold frames, grow tunnels and gardens (of hardy outdoor crops) until late November and then bi-weekly harvest through the winter solstice and beyond.

Building Management: Fall

In late November, we prepare the bioshelter and season extenders for winter. Roof vents, some windows, and the exhaust fan vent are sealed with plastic sheeting. Door and window weather stripping are checked and repaired as needed. Thermal mass is assessed and upgraded where possible. Kindling is gathered for winter fires. Windows are washed and faucet washers replaced as needed.

A main goal of building management in the fall is to allow the active storage of heat without overheating the bioshelter. This requires daily hands-on management of blowers, the exhaust fan and windows. The blowers are run to draw warm ceiling air down through the ductwork and send it into the rock storage under the deep beds.

We do not exhaust heat from the building until the second floor exceeds 90°. Then, second-floor windows are opened at each end of the bioshelter to allow prevailing winds to provide flow-through ventilation. On days of full sun in September and early October, we may run the exhaust fan for an hour or two in midday. By early November, the exhaust fan is not needed, so we disconnect the fan motor and seal the opening until April. Sometime in the middle of November, we begin to burn wood in the fireplace in the evening. When the temperature drops below 30°, we burn the stove through the night.

At some point in late November, usually when the first cold, windy snowstorm is approaching, we put clear plastic vapor barriers over the roof vents and windows. (But we leave two of the vents and one window on each end uncovered until winter hits hard. This allows us to ventilate if necessary.) When the first storm hits, we begin to heat the central water tank in the evening for overnight radiant heating of the building. During extreme weather, both the water tank stove and the regular wood stove are fired.

The fall bioshelter garden is most productive when planted in early September. We usually begin weekly succession plantings of lettuce and greens the second week of September and continue until mid-November. We start them in row flats, which have twenty shallow rows per flat, as well as in six packs. They are put under the grow lights and then moved into the beds and planters throughout the building. Spinach, beets (red and golden) and chard are planted for regular cuttings all winter.

The cold frames attached to the bioshelter's south wall are planted early each fall. The cold frame lids are replaced before the first frost and are opened and closed daily to allow for light and ventilation. Plantings include pansy and perennial miniature roses for edible flowers, cooking greens, and various salad crops. Other plants that survive the winter in the cold frames include mache, miner's lettuce and kale (all of which are self-sown) plus chard, lettuces, arugula, spinach and the wild edibles, chickweed and ox-eye daisy.

Insects in Fall

Many insects go dormant in the fall as the days grow shorter and the weather cools. On warm Indian summer days in October and November, the ladybugs seek shelter. Hundreds enter the bioshelter through open doors and windows. They will hibernate in cool corners until spring. Most of the predatory insects are far less active in the fall. Aphids and whiteflies take advantage of the lack of predators and begin to multiply. We keep them under control with soap spray, rinsing or even vacuuming extreme infestations with a lightweight, handheld vacuum cleaner. When the *Beauveria* fungus appears in late October, we begin to culture the infected aphids and spray the spores on pest-prone crops.

Winter

Winter in the bioshelter lasts from early December until mid- or late January. Some years, it is longer, lasting into February. After six to eight weeks of very short days, the days begin to be noticeably longer, and the increased solar gain is evident in increased plant growth. During this winter period, harvest is dependent on weather. Sunny days bring plant growth; on cloudy days, plants remain semi-dormant. Harvest yields can vary greatly from winter to winter. Harvest is best done after a period of sunny days.

Once the weather is reliably sunny, new seedlings are started under lights to prepare for winter/spring plantings. We also start herb seeds and cuttings in winter.

Building Management: Winter

Sunny days in late December, January and February bring temperature extremes inside — temperatures can easily exceed 100° — so passive ventilation is often necessary. With full sun, the bioshelter will be 75–80° on the first floor and 90–95° on the second floor. Opening two second-floor windows on the east and two windows on the west (12 square feet on each side) is usually adequate to prevent overheating.

Ventilation is regulated more closely on days with partial sun. During these days, the active solar storage system is used. The building is allowed to heat up as much as possible to "charge" the thermal mass with heat. The active heat recovery fans are turned on at midday to move heat from the ceiling to the rock storage under the deep growing beds to maintain moderate temperatures at night. Some days, second floor windows have to be opened for short periods to reduce humidity.

The building gains little energy from the sun on overcast and cloudy weather, so supplemental heat is required. With outdoor temperatures in the low 30s, a fire in the woodstove is necessary in the evening and at night. In periods of below-freezing weather with no sun, a fire is maintained morning through night. The 600-gallon water tank is also kept warm or hot, as needed. The goal is to maintain interior temperatures at a minimum of 40° at night and 60° during the day.

Cold frames and tunnels are rarely harvested in winter. Growth is slow, and they are often inaccessible because they are covered with snow. In sunny weather, the frames may need a little ventilation; if not ventilated, they overheat and kill plants. Pansies, rosemary and hardy greens are occasionally harvested from the cold frames in good weather.

Insects in Winter

Insect pests and disease are at their peak in winter. Plants need regular inspection — early detection and control is the best prevention against outbreaks. Although some insects are dormant, aphids certainly are not. Their numbers increase rapidly when plants are stressed by low light and low temperatures. Applications of soap spray and washing infested leaves cuts down their numbers. As discussed above and in Chapter 6, applying spores of the *Beauveria* fungus further reduces aphid outbreaks.

Whiteflies can begin to reproduce and be active in late winter. This is a good time to release the whitefly predator wasp, *Encarsia formosa*. We also use sticky traps and spray insecticidal soap to control whitefly.

Winter/Spring

Winter/spring begins in mid-January or early February and lasts through the spring equinox. Each week, the longer days and the increasing, but still low, solar angle (inclination) flood the bioshelter with sunlight and heat. Plant health increases and beneficial insects become more active. Periods of cloudy weather can dampen the progression, but, day by day, the winter/spring garden unfolds.

Building Management: Winter/Spring

Increasing sunny days recharge the bioshelter's thermal mass as the days grow longer. Gradually, the average soil temperature rises, as evidenced by germinating wild edibles such as lambsquarters and amaranth. We need to burn the fire less often and begin to ventilate more on sunny days.

In early winter/spring, the bioshelter is often the center for our evening activities. Books are read, plans are discussed and friends are entertained around the fire.

Winter/spring chores include doing inventories of seeds and tools, planning for the coming year, placing seed orders, and cleaning out the poultry room bedding.

February brings the return to weekly harvest. Tomatoes begin to flower heavily. Fall-planted arugula and tatsoi flowers are ready to be used in the salad mix. Bolted lettuce is fed to chickens, although a few lettuce plants may be allowed to set seed.

By early March, the bioshelter has a lush tangle of tomato plants and nasturtium vines climbing to or hanging from the second floor beds. The growing beds and shelves are thick with salad greens. The entire building is liberally splashed with the brilliant colors of pink snapdragon, scarlet pineapple sage, yellow and red nasturtium flowers and golden tomatoes. As the season progresses, a succession of wild edibles and self-seeded salad crops begin to germinate and grow. Among these are miner's lettuce, chickweed, wood sorrel, sheep sorrel and lambsquarters. Winter lettuces are replaced with fresh seedlings. Many new plants are seeded and production increases with each weekly harvest. In winter/spring cuttings are taken to propagate many herbs and geraniums.

Native wildflower seeds were sown in row flats in late December and stratified in the refrigerator. These are germinated under grow lights in January. Beginning in February, the seedlings that will go into the cold frames and tunnels are started on top of the compost chambers, under grow lights. The compost provides bottom heat for the germinating seedlings from mid-February until the end of March. (For more on the management of the compost chambers, see Chapter 11.)

The cold frames and tunnels also become productive in winter/spring. The protected environment and lengthening days combine to encourage plant growth in these unheated structures. Ventilation is crucial to success because a cold frame can quickly overheat.

In late February or early March, we begin to plant seedlings for many early spring and summer crops. Tomatoes, peppers, broccoli, cabbage, and many other crops are seeded. (For timing, see the planting calendar chart in Chapter 8.)

Insects in Winter/Spring

Many insects that were dormant in the bioshelter in winter start to become active in winter/spring. Whiteflies begin to emerge and reproduce, just ahead of their predator, the tiny parasitic wasp, *Encarsia formosa*. Occasional use of insecticidal soap slows the whitefly's progress, allowing the wasp to catch up as the days grow warmer. Fungus gnats are sometimes quite numerous in the bioshelter, but seem to do little harm. In recent years, there have been fewer of them, probably due to some predator eating their tiny, soil-dwelling larvae. When excessive populations do hatch, they are easily captured on yellow sticky traps.

The *Beauveria* fungus becomes less active as the building warms up. We usually collect and spray the spores on newly emerging seedlings through winter/spring to help plants get a pest-free start.

Aphids that hid out under leaves through the coldest weather begin to grow wings, mate and lay eggs. They, too, are attracted to the sticky traps, but so are their wasp predators. So it is best to limit use of the traps to major outbreaks. Winged aphids (and their predators) are attracted to the south-facing windows, where they are easily wiped off with a damp mop or cloth. The common house spider also captures them by the hundreds in window-frame webs.

Ladybugs and other aphid predators are slow to emerge. The best aphid control in winter/spring is a vigilant eye for outbreaks. Infested plants can be washed with water and the bugs crushed with fingers. Insecticidal soap is useful for larger infestations. Occasionally, if the aphids are out of control, a crop will need to be pulled and fed to the chickens or composted. We are careful to watch for wasps and mummified aphids containing wasp larvae before spraying and removing plants.

Spring!

Spring in the bioshelter lasts from the spring equinox until late May. The building gets noticeably shadier in the northern half due to increasing solar angle. But, because of the lengthening day, the building grows ever warmer. Soil temperatures rise and new types of wild edibles germinate in the bioshelter beds. Summer weeds like purslane, wood sorrel and amaranth add color, flavor and nutrition to the salad crop. Many seedlings and cuttings are started for the outdoor gardens and for sale. Every available space is crammed with flats, trays and planters. There is continuous rotation of flats as seedlings nurtured on the compost chambers are put under the grow lights on the second floor and then taken outside to harden off in protected frames.

Building Management: Spring

In spring, ventilation is as important as heating. Generally, the longer days and increased solar gain provide sufficient heat gain to keep the building warm through the night, so supplemental heating is only necessary during late cold spells. The average temperature of the building and the soil increases as the day length increases.

Ventilation is achieved passively. Plastic vent seals are removed and roof vents are opened daily. East and west windows are also open daily. By mid-May, the exhaust fan in the east gable is needed because second floor temperatures can exceed 90°.

The final batches in the compost chamber provide heat through March. After that, the building is warm enough to germinate seeds, and the soil is warmed enough by the sun for good growth.

Insects in Spring

A sure sign of spring in the bioshelter is the emergence of wood-boring carpenter bees from their hibernation in the bioshelter ceiling. They are a welcome sight because they pollinate our tomatoes. Each year, these helpful members of our community awake and become active four or more weeks earlier than they would outside. We try to keep the window screens closed so they will not escape outdoors too early and freeze or starve. But they seem to know their way around quite well, and they stay inside, gathering pollen and nectar for their brood from the snapdragons, nasturtiums, tomatoes and other blossoms.

Another welcome spring resident is the pollinating syrphid fly and its aphid-eating larvae. Gall midge larvae also begin to appear and devour aphids in the bioshelter, but they are more active in the summer.

Other Aspects of Bioshelter Management
The Potting Room

The potting room is the center of activity in the greenhouse. Here we store small tools, potting supplies and equipment, seeds, catalogues, reference books, and planting records — and it's where we pot plants. All farm workers should be thoroughly familiar with the potting room. The duties here include mixing potting soil, starting seeds in row flats, and transplanting seedlings to six-pack inserts and pots.

The potting room is stocked with all the supplies needed for mixing our organic potting soil. Recipes are posted above the workbench. Everyone on the workforce will be called to mix potting soil at one time or another. The posted recipe, easily accessible ingredients, and an organized and equipped workspace help this task go smoothly. All the equipment needed for mixing soil is stored in the potting room: a wheelbarrow, a ¼-inch screen tray, buckets, scoops and a short-handled, flat shovel.

The actual potting area has a workbench where flats are filled with the potting soil and then moved to a table for pre-moistening. Then they are moved again to a potting bench near the grow lights. The watering table and potting benches are set up at counter height to reduce the need for bending and stooping.

Several types of flats and seeding trays are necessary, including 20-row flats for starting seedlings, flats with drainage holes, and various sizes of flat inserts. We most commonly use 60- and 72-cell per flat inserts for herbs and vegetable seedlings. Other sizes include 48- and 32-cell per flat inserts, which are used for herb cuttings and a few other crops. Plant markers, pens, pencils and markers are also kept handy.

Properly storing seeds saves money and aggravation. A cool, shaded, dry location is best. As a rule, the sum of temperature and humidity should not exceed 100. This means that if the temperature is 70°, the humidity should be less than 30 percent. Conversely, if the humidity is high, the temperature needs to be lower to preserve seeds. Bulk seeds are sometimes kept in a refrigerator. Mice love to eat seeds, so rodent-proof containers are required. Coffee cans work well. Some of ours have lasted for nearly two decades in the potting room. Heavy snap-top plastic containers also work well. We have on occasion found a black, smooth-skin caterpillar eating our seeds, so it is good to check your stored seeds every so often.

Records should be kept of all seeds started. In addition to being valuable in crop management, the records are needed to comply with the requirements of organic certification. An inspector will want to see seed purchase records and receipts, planting records, harvest records and sales receipts. This way, an organic crop is traced from seed company to market.

We plan our annual gardens in January so there is time to submit the plans along with our organic certification application. The plans detail crop rotations for each garden and any soil-building amendments.

Writing down what you seed each week helps to keep track of what you are growing. When a plant label or two falls out of a row flat of 20 varieties of heirloom tomatoes, records are an invaluable backup. Poor germination and other problems can be quickly traced to a specific seed pack. Records of what was planted in previous years also help when you are planning next year's garden.

We produce up to 1,000 seedlings a week between February and mid-September. Seedling production is a critical job on the farm and therefore is closely monitored by the farm manager to ensure that workers understand the processes involved. Not everyone has the patience to transplant seedlings for several hours at a time. If a worker does not like the work, it may get done too slowly or poorly. We usually assign that person more active tasks.

The potting room has a number of reference books relating to seedling production and plant care. Workers are encouraged to consult them as needed.

From a Tiny Seed: Starting Seedlings

"Though I do not believe that a plant will spring up where no seed has been, I have great faith in a seed. Convince me that you have a seed there and I am prepared to expect wonders."

—Henry D. Thoreau, *Faith in a Seed*

Planting a succession of seedlings is a key strategy for intensive market gardening. As soon as one crop is finished, the next crop needs to be ready to go into that bed. We seed row flats every week from February until September each year, and at least once a month from October through January. Seedlings are started in a potting soil enriched with minerals, kelp meal and compost (more on this, below).

Seedlings for transplants are started in a protected environment to ensure a high yield of plants. They are started outdoors in a sheltered hot bed or cold frame or inside, under lights. Once germinated, seedlings require proper light (in terms of both quality and duration), ventilation, water, and fertilizer to develop with health and vigor.

For the home gardener, simplicity is best. In the 1930s, Grandmother Florence Decker Frey started her hardier crop seedlings in simple cold frames set over garden beds. Her potting soil mix was mostly well-manured garden soil that she sterilized in the wood cook stove's oven. She used it for her tomato, pepper and flower seedlings and germinated them on a sunny windowsill. After germination, she transplanted into a hot bed. Seedlings were usually transplanted to the garden directly from the hot beds or cold frames — without pots or flats. She got such good results that she was able to market hundreds of seedlings each year to her rural neighbors.

For the commercial grower, there are a number of good reasons to start seeds indoors and under lights. During the winter and spring months, the temperature and available sunlight is too unpredictable to be reliable. For the home gardener, a little flexibility is possible, but a commercial grower requires a higher degree of certainty.

A commercial grower does not have time to coddle marginally healthy plants; they need to be vigorous from the start. Grow lights are an important tool that give plants a good start by providing steady light without the drying effects of the sun and wind. Protected from the extremes of weather and the munching pests found in the garden, a higher percentage of seeds planted in the greenhouse survive to healthy maturity. This can be of critical importance with expensive organic and heirloom seeds.

Getting a head start on the season can make a major difference in the success of a market garden. Early and late basils are more valuable to customers (especially restaurants) than the main season basil crop. And having crops ready early can help establish season-long customers for weekly harvests of parsley, cilantro and dill.

The productivity of a market garden is dependent on always having transplants ready to go into the garden. As soon as the first outdoor crop of spring lettuce is harvested in mid-May, peppers and tomato seedlings should be ready to go into the bed. Provided there is enough room to repot plants as they mature, seedlings started indoors can be already flowering or even fruiting when placed in the garden.

Starting seedlings indoors also helps guarantee healthy plants. Because they are in the center of Zone 1, they can receive daily care. Properly tended seedlings become vigorous transplants. Any weak, diseased, or otherwise undesirable plants are discarded.

Soil Mix

The goal of creating potting soil is to have a good medium for plant growth that is not so expensive that it eats into profits. A good medium has the capacity to hold water and yet drain well enough to leave room for air between soil particles. Good potting soil should be high in humus because it has the capacity to hold and release nutrients to the

Practical Floriculture

Writing at the end of the 19th century in *Practical Floriculture,* New York City horticulturalist Peter Henderson describes a potting soil based on two ingredients: compost and pasture sod. His compost was made either from spent brewery hops or horse manure. Occasionally, fully decayed leaf mold was used in place of compost. The compost materials were mixed with cut pasture sod. The sod was usually a loam soil with cropped grass and roots. Henderson recommended two parts sod to one part hops or manure. When the pasture soils were clay, he recommended more compost; when sandy, he used a little less to balance drainage and weight. After the compost and sod was composted and aged one year in a pile, the potting soil mix was ready for use. With a few exceptions,

such as ferns and azaleas, this mix — prepared in heaps that "loomed up like a miniature mountain" — was used to propagate hundreds of thousands of flower and vegetable seedlings each year.

While Henderson's advice is sound, one should question the sustainability of harvesting pasture sod. Perhaps the intensive use of horses and livestock and a rotational harvest kept the pasture soil renewed. When growing seedlings for use only on the farm, garden soil can be a good part of the mix. When growing plants for sale, though, a potting mix without the soil is probably more ecological. There are a number of issues when buying potting mix supplies, and the answers to the issues are rarely straightforward.

plants — without being *too* rich. A pH between 6 and 7 is desirable. A lightweight potting soil makes the work of mixing and handling easier.

A working farm operates best with standard procedures. They help in maintaining stocks of supplies and managing labor. Potting mix recipes vary depending on crops, the farmer's preferences and available materials, but having a standard potting recipe is important, especially in the busy spring planting season. However, finding a perfect blend of local materials can be challenging.

Sustainable Soil Mix

Short-term planning includes choosing ingredients from among the materials available to you — with an understanding of the pros and cons of each. Long-term planning may mean setting up areas to age compost, old straw, wood chips, leaves, or other sources of organic materials. The goal is to find or create a stable humus material to serve as the base of your mix.

Minerals

If a garden soil has a full range of minerals at adequate levels, it can serve as a source of minerals for a potting mix. But over the long term, selling your topsoil along with your plants is not sustainable. The alternative is purchasing bags of the necessary minerals: limestone, greensand, rock phosphate and other rock-based minerals. But these, by necessity, involve heavy equipment to mine, crush, bag and transport them. The good news is a little goes a long way, and eventually the soil of a well-managed farm will build to have adequate mineral levels. As plants and seedlings are placed in the garden beds, the high-mineral potting mix gets incorporated into the garden soil where it will continue to feed future crops.

Recycling Soil

We use our potting soil to grow cutting greens in shallow trays. If not given to the chickens as mentioned above, the soil mix and any roots it contains are piled in a bin outside for a year. This is later used for potting up nursery plants and container-grown plants in the bioshelter. After a year of aging outdoors, the minerals in the mix are more available to the potted plants than they are in freshly mixed potting soil. We mix the "used" soil half-and-half with well-aged compost.

Peat Moss

Peat is an organic material that is high in stabilized humus and formed in generally anaerobic conditions in wetlands over decades or centuries. Commercially available peat includes reed-sedge peat and sphagnum peat moss. Reed-sedge peat is the most common peat harvested in the US. The sustainability of reed-sedge peat harvesting is questionable because it has an environmental impact on wetlands. In the 1980s, a farmer near us excavated several acres of swamp; in the process he created a pond and was able to sell the dried and shredded peat. The material was coarse, formed from marsh grasses, reeds, cattails, and probably leaves and other debris blown in from surrounding forests for thousands of years. It made an excellent garden material. But the enterprise was closed by the state department of environmental protection because it violated wetland protection laws. There was a reasonable concern that such harvests, if allowed, would degrade wetland habitats regionally.

Sphagnum peat moss, usually from Canada's extensive peatlands, is considered more sustainable because the sphagnum moss begins to regenerate several years after harvest — and because Canada has a lot of it. Because it is lightweight and has the ability to hold water, air, and nutrients, peat moss is a valuable soil mix ingredient. It is generally acidic, so rock lime is used with it to adjust the pH upward.

Peat moss can be difficult to wet when it is completely dry. Because of this, many commercial potting mixes include wetting agents. These wetting agents are *not* acceptable for organic production, so be sure to use untreated natural peat moss if you seek organic certification. When in doubt, contact the manufacturer before buying.

Other Materials

Leaf mold (well-decomposed leaves) can easily replace peat moss in soil mixes. I have also known greenhouse managers to mix 5-year-old leaf mold half-and-half with peat moss and get excellent results. The main problem with leaf mold is the need to know the source. Leaves collected in towns and cities are an unknown quantity. Because they may contain pesticide residue, plastics or other litter, organic standards prohibit the use of municipal leaf compost on organic farms. This, of course, is an individual choice in the search for sustainability, but if you choose to use municipal leaf mold in your potting mix, you will jeopardize organic certification. We gather leaves from yards we know to be pesticide free and use them directly for garden mulch rather than age them to use as leaf mold.

Coir is the shredded fiber of coconut husks. Its look and performance is very similar to sphagnum peat although it is not as acidic as peat moss and it persists longer in the soil. It is shipped in highly compressed blocks, but it still may be more expensive than bulky bales of peat moss. Because coconut fiber is a by-product of crop production, its use is considered by some to be more sustainable. It is certainly being promoted as a sustainable replacement for peat moss; suppliers often claim to have an organic product. But again, certified organic farms are required to document such claims, so you may need to contact the supplier for details.

Kenaf fiber is made from the shredded stalks of the kenaf plant, *Hibiscus cannabinus*. This relative of hollyhock can be grown in the southern half of the US. It is used for making paper and adding bulk and drainage to peat moss-based soil mixes. We have no experience with kenaf. As with other ingredients, organic certification requires the farmer to document the source of the product.

Compost

Well-prepared compost is an important addition to potting soil. Our potting soil compost is a mix of already composted horse manure, sawdust and straw that has aged three to five years. It has the consistency of a heavy reed-sedge peat. Compost provides most of the soluble nitrogen, phosphorus and potassium needed by young seedlings, and it increases the moisture-holding capacity of the soil. Compost destined for potting soil should be tested to determine its nutrient content, and the soil mix should be supplemented with organic fertilizer as needed.

Mushroom Compost

Mushroom compost is readily available in our region because there is extensive local button mushroom production. Mushroom farmers compost a mix of horse manure, minerals and other plant materials before using it to grow mushrooms; it is re-composted and aged after the mushrooms are harvested. We often use bagged mushroom compost instead of our own compost for vegetable seedling we want to sell. We do so because our own compost occasionally sprouts wild edibles and a few other weeds, and we do not want to sell a weedy product. We also use mushroom compost when our own piles are not mature enough. Again, the grower should check the source for compliance with organic standards.

Sand

Coarse sand is added to soil mix to improve drainage. We buy it very inexpensively by the ton from a local sand and gravel mine. It is heavy and adds no nutrients to the mix, but it can keep a soil mix aerated.

Perlite and Vermiculite

Perlite is a heat-processed volcanic rock used to provide drainage and improve the water-holding capacity of a mix. Generally it is used as a lightweight replacement for sand. The fine dust of perlite should not be inhaled. It is best to wear a mask and pre-wet perlite before handling it.

For years, vermiculite (a naturally occurring mineral) was a main ingredient in our soil mix. The combination of its relatively low cost, light weight, ability to hold water and nutrients and promote good drainage made it ideal for our potting soils. We bought only horticultural-grade vermiculite to avoid the asbestos fibers found in some grades of vermiculite. However, an EPA study determined that even horticultural-grade vermiculite can contain minute amounts of asbestos fibers. Inhalation of carcinogenic asbestos fibers can cause debilitating lung disease. The EPA began in 2000 (see www.epa.gov/asbestos/pubs/vermfacts.pdf) to advise consumers to use the following methods to minimize the danger in using vermiculite:

- Use vermiculite outdoors or in a well-ventilated area.
- Keep vermiculite damp while using it to reduce the amount of dust created.
- Avoid bringing dust from vermiculite into the home on clothing.
- Use premixed potting soil, which usually contains more moisture and less vermiculite than a pure vermiculite product and is less likely to generate dust.
- Use other soil additives such as peat, sawdust, perlite, or bark.

Since the EPA's announcement of its recommendations, the vermiculite industry has countered by testing and documenting asbestos-free vermiculite sources. Our garden supplier offers a South African vermiculite that tested asbestos free. We highly recommend that readers conduct up-to-date research on any products they use. Talk to both the supplier and the manufacturer and request product data sheets that you can show your customers and the organic inspector upon request. As with perlite, wear a mask when handling vermiculite.

Pulverized Limestone

We use a locally mined and crushed dolomite limestone that is mostly calcium and magnesium. This rock, formed from ancient sea deposits, helps balance the pH of the peat moss and provides important plant nutrients.

Greensand

Greensand is also an ancient sea deposit, consisting of iron potassium silicate. It is 5–6 percent potassium, and contains up to 30 other trace minerals. Greensand increases the water-holding capacity of potting mix, and its minerals are slowly released as soil acids and organisms make it available to plants. Adding it to potting soil puts the minerals where they are needed, at the base of the growing plant in the garden bed. Greensand is a good provider of potassium, which is important in photosynthesis, helps plants resist disease and is used by plants for strong growth and fruiting.

Rock Phosphate

Rock phosphate is pulverized phosphate rock (primarily calcium phosphate). We use both rock phosphate and colloidal phosphate on the farm. We apply rock phosphate to the garden beds directly, but do not generally use it in the potting mix because it is not as fine or available to plants. Colloidal phosphate is a finer grade of rock phosphate in a clay matrix. Both contain many other minerals as well.

For a potting mix, colloidal phosphate is better because it is more available to plants. Rock phosphate does not readily dissolve into the soil's chemistry. Both need the biochemistry of plants and microorganisms to absorb it and put it into the nutrient cycle. As with the other minerals, putting rock phosphate into the potting mix makes phosphorus directly available to the maturing plant, promoting strong, healthy plant growth, flowering and fruiting.

Bone Meal

Steamed bone meal provides highly available calcium phosphate and nitrogen. A by-product of the meat processing industry, steamed bone meal is made by cooking and grinding bones of slaughtered animals. With 30 percent phosphate and 2 percent nitrogen, this ingredient can offer immediate benefits to young plants. We especially like to use it in a soil mix for flowering plants. When mad cow disease appeared in England, we suspended our use of bone meal for several years. We have since learned that our concern was ungrounded. There is virtually no chance of plants transmitting prions, the rogue protein responsible for the disease. In England, gardeners are cautioned to wear masks when using bone meal. But this is a good precaution to take when mixing almost any potting soil because the light, fibrous materials used in the mixes are easily inhaled.

Kelp Meal

Kelp meal is dried and ground sea kelp. Because it grows suspended in the ocean, it has a wide range of minerals, including 2 percent potassium, 2 percent nitrogen and 2.5 percent calcium. We like to add kelp to the potting mix because it provides trace minerals and root-promoting hormones, giving seedlings a strong start. One should shop for sustainably harvested kelp that comes from relatively clean waters.

Wood Ash

Wood ashes contain up to 2 percent phosphate and 10 percent potassium. Ashes are very alkaline and highly soluble, so they can damage young plants. We do not use wood ashes directly in the potting mix. We do mix hardwood ashes from our bioshelter woodstoves into our compost piles before the last turning. This allows the compost to absorb the nutrients and balance the pH of the ashes.

Ideally we would like to move to a potting soil mix based primarily on leaf mold and compost, plus the minerals and kelp. Each year, we collect about 60 large garbage bags of leaves that we know to be pesticide and litter free, but that isn't enough for all our potting soil needs. Because we want to remain certified organic, we can't use municipal leaves. So, a soil mix based on leaf mold would require us to add a yard service to our operation. In the meantime, we will continue to use our standard potting soil mix.

Three Sisters Potting Mix

Linda researched the options and developed a standard potting soil mix that has served us well for 20 years. We use it for transplants and repotting plants. The variation we use for starting new seedlings is also given below.

Three Sisters' Standard Potting Soil Recipe:

We mix directly in a wheelbarrow. We start by sifting two five-gallon buckets of peat or sphagnum moss through a ¼-inch mesh heavy wire screen mounted in a wooden frame. To the peat moss, which is generally acidic, we add a third of a quart (one and one third cups) of pulverized rock lime to balance the pH. (We periodically test the finished soil mix to confirm this is the right amount of lime to use with a particular supply of peat moss.) After the lime, we add two cups of greensand, two cups of either rock phosphate or steamed bone meal, and a cup of kelp meal. A cup of bagged organic fertilizer supplements the fertility of the compost. These ingredients are mixed thoroughly in the wheelbarrow using a short-handled, flat shovel. Next, we add five gallons of sifted compost and two and a half gallons of coarse sand and we mix thoroughly again. Finally, we add one five-gallon bucket of either vermiculite or perlite and mix a third

Biochar

As this manuscript is being prepared for printing, we are testing charcoal as a potting soil ingredient. Charcoal has the potential to replace vermiculite and perlite. It generally has a neutral pH, is lightweight, and can likely improve drainage while holding moisture and nutrients. Because charcoal is extremely porous, it can provide habitat for millions of soil microorganisms in fertile soil.

Many soil researchers are beginning to investigate the use of charcoal, also known as *biochar,* as a soil amendment in temperate climates. Biochar has been found to be a major component of *Terra Preta,* a highly fertile soil created by pre-Columbian civilizations in the Amazon basin. Biochar, in addition to its ability to enhance soil productivity by storing moisture and nutrients, is an extremely stable form of carbon. Biochar can persist in the soil for centuries. Charcoal can be made on the farm with simple equipment. On a small scale, charcoal screened from wood ashes can be rinsed, crushed and used in soils. Soaking the biochar in fish emulsion and liquid seaweed, or in a liquid compost extract (compost tea), can help charge it with nutrients before it is added to soil and soil mixes.

time. This recipe makes about 23 gallons of potting soil (a wheel-barrow full). The average flat of six-pack inserts requires about one gallon of potting soil.

A Few Cautions

Gloves and a dust mask should be worn while mixing. To prevent damping-off pathogens from multiplying in the mix, keep the mix as dry as possible until ready to use and don't mix a whole season's worth all at once. (We mix no more than a week's supply at a time.) Pre-moistening the ingredients reduces dust, but it makes mixing more laborious. However, vermiculite and perlite do need to be pre-moistened to reduce dust.

Seed Starter Mix

When preparing a potting mix for starting seedlings, we use the standard potting mix without the compost or fertilizer (so, just the peat, lime, greensand, kelp meal and rock phosphate or steamed bone meal). We also add extra seaweed (kelp meal). Compost is not added because it often contains microbial organisms that cause seeds to dampen off. Also the compost can bring in weed seeds. If you want to use compost in a seeding mix, it should be sterilized first to kill weed seeds and disease organisms. For instructions on how to sterilize soil, visit www.colostate.edu/Depts/CoopExt/4DMG/Soil/sterile.htm.)

Artificial Light

Grow lights are expensive to purchase and maintain and, of course, they require electricity, so judicious planning is needed to get the most value out of them. The number of grow lights required depends on production goals. A set of four fixtures (each with two 4-foot florescent tubes) covers 16 square feet of space, which is enough space for eight standard flats. Depending on the number of cells per flat (generally 60 or 72), this set of grow lights can produce 480 to 576 seedling every three weeks.

Twelve to 16 hours daily of light is best for seedlings. Our racks of grow lights are located above our compost chambers along the north greenhouse wall. As previously noted, the compost provides the seedlings with bottom heat. Most seedlings can be rotated out from under the lights and into direct sunlight two to three weeks after transplanting.

Grow lights, when used, can and should be set up in the least sunny spaces of the greenhouse. This leaves the more valuable, naturally lit space for crop production.

If air temperatures are below 60° when florescent lights are turned on, fixtures with *cold temperature ballasts* are required. Otherwise, the life span of the florescent tubes will be considerably shortened. To get a good balance of light, we use one full spectrum grow tube and one cool white light tube per two-tube fixture. Many growers use cool white bulbs alone. Cool white light will produce healthy seedlings and are less expensive. But we think the closer we can get to sunlight, the better the plants will be — but the differences are hard to quantify. Certainly, mature plants and flowering plants are healthier with the mix of spectrums.

Bulbs need to be replaced at least every two years, more often if used continuously. Hang lights no more than 6 to 8 inches above plants, but allow space to water plants without hitting lights with water wand. We keep sets of lights at different heights so plants can be moved as they grow.

Light Racks

Our two-bulb fixtures are fastened to wooden frames. Four fixtures are evenly spaced on the frame, or rack, to provide 16 square feet of lighted space. The top of the frame is covered with aluminum sheeting for reflection. Metal eye screw hangers go through the sheeting into the wooden frame. The rack of lights is suspended on adjustable wires from the ceiling. A rope and pulley system is used to raise and lower the lights to adjust for different plant heights and to change bulbs. All four fixtures are wired to a waterproof junction box attached to the rack. A single cord connects the rack to a power strip. Each rack of eight 40-watt tubes requires 320 watts. Two to three racks are plugged into a power strip, and then to a timer, which is plugged into the wall outlet. The timer is set for 14 hours, morning to evening. A typical flat measures 2 square feet. Our five main

Plants under growlights.

EMILY BRAGONIER

light racks together give us 80 square feet of lighted area, which holds 40 flats. They use 1,600 watts of electricity.

Temperature

Germination temperature varies for each type of seed. Tomatoes will germinate and sprout in six days at 72°, but can take several weeks at 55°. A temperature range between 65° and 75° will give the best results for most vegetables started indoors. Quick and even growth is the goal. If the soil is too cool, seedlings can take too long to germinate and grow, which exposes them to a higher risk of disease or invasion by aphids and other pests. After they are a few weeks old, seedlings can tolerate a wider range of temperatures.

Good air movement around seedlings helps prevent damping-off diseases and promotes strong stems. A small fan blowing a slight breeze past the seedlings helps them develop sturdy stems and also helps dry the daily watering off the leaves and soil surface, greatly reducing the risk of infection.

Row Flats

Starting seedlings in a row flats, rather than six packs saves space and time. Row flats are the same size as a standard flat, roughly 11 inches by 23 inches, and have 20 shallow rows. Seeding soil mix is added and then pre-moistened. We seed most crops heavily, anywhere from 30 to 70 seeds per row. Tomatoes, peppers, basils and other plants subject to dampening off are seeded at about 30 seeds per row. We plant enough lettuce seeds to get flats of 60 to 72 plants per row. Other crops are seeded about 40 seeds per row.

Crops started in row flats include tomatoes, peppers, cabbage and its relatives, herbs, (excluding dill, cilantro and chervil, which are only direct seeded), lettuce, endive, radicchio, pac choy and bok choy, fennel, parsley, wildflowers and other flowers. Root crops and other direct-seeded plants like corn, beans and many of the greens are rarely started in flats.

After the seedlings have germinated and developed a bit, they are transplanted from the row flats into six-pack inserts. When a seed germinates, the first leaves are the cotyledons, which emerge from the seed. These are followed by the "true leaves." Almost every garden book says to wait until the true leaves have developed. We rarely wait this long. For one thing, we are in a hurry to keep the succession of plants moving. The amount of transplanting on a market garden farm requires expediency and good timing. Also, we have found seedlings can get leggy and develop disease if left too long in the row flats. The denser the seeds are planted in the row flat, the sooner one should transplant. More expensive seeds, such as heirloom tomatoes, are planted more thinly in the row flats and so may be allowed to develop more before transplanting.

Flats are prepared three or four at a time to keep a good workflow. They are filled with soil, placed on a screened table and watered thoroughly. It takes several doses with the hose before the mix is wet through. Once saturated and drained, they are ready for the seedlings.

Seedlings are tougher than you might think, but handle them with care until you develop a good transplanting technique. We have a small tool, a block of wood with six 3/8-inch wooden pegs, that we use to form holes in the soil in the six packs. Some workers prefer to use their finger to make a hole. Push the seedlings roots into the soil and lightly press the soil down around it. Most seedlings can be planted a little deeper than they were in the row flat. This will help make the roots stronger and anchor the seedling to minimize stress.

After flats are planted, they are gently watered with a diluted liquid seaweed solution to encourage rooting. Seaweed is rich in trace minerals and provides hormones that encourage root formation. Seaweed solution is made by adding granulated or liquid kelp to water. Follow directions on the label of the brand you use.

Inserts can be reused many times if handled with care. We wash flats and inserts with a mix of water and hydrogen peroxide bleach to disinfect them and then store them in a dry bin until needed. As mentioned above, we usually use six packs sized for 60 or 70 plants per flat for most crops. Tomatoes, peppers, cabbages and other tall and long-season crops are planted 60 per flat. Lettuce is planted 72 per flat. Many herb plants are planted in larger inserts, with 32 to 48 plants per flat.

Soil blocks and paper pots can be used instead of plastic inserts. We use paper pots to get a head start on squash and melons because they do not transplant as well as other crops. We have found soil blocks to be tedious to make and not as efficient as six packs for most plants. If too soft, they crumble and if too dense, they are hard to transplant into. They do work well for cuttings and for crops we may start in plug

Plug Flats

Plug flats are flats that have any number of holes on a single sheet, from 20 or 30 to several hundred. Seedlings started in plug flats are then direct planted into the garden. The more plugs per tray, the more watering they need and the more tedious it is to plant seeds into them. The trade-off is the savings in transplanting time. Plants that do well in plug flats include kale, chard, and onions. Kale and chard are planted with one seed per plug. Scallions can be planted with six to ten seeds per plug. When transplanted to the garden, they grow in bunches. This makes harvest for market a little faster.

flats. We line the bottom of the flat with paper before putting in the soil blocks. This helps them remain moist as the plants develop. The choice between using soil blocks and plastic flats depends mostly on the scale of the farm. Certainly, the reduced use of plastic is an important consideration. But on an intensive market farm, the extra difficulty of working with soil blocks can add to production costs.

Potting Up

Potting up is transplanting seedlings to larger pots as they mature. Many crops, we never pot up. Most are simply planted from six packs or plug flats directly into garden beds. Those that are potted up — tomatoes, peppers, eggplants, herbs and flowers — are plants we either maintain in pots or ones we want to get bigger before planting or selling. When putting plants into larger pots, we use a mix with more compost and sand, which gives more fertility and drainage. We maintain piles of recycled potting mix from our cutting green trays (described above). The mix is rich in mineral nutrients, and the minerals are more available after being piled for a year because the soil community has begun to act on the rock phosphate, lime and greensand. Recycled potting mix is combined with equal parts compost and a couple shovels full of sand in a wheelbarrow. Most of our wildflowers and herbs receive this mixture in the final potting.

Watering

Watering seedlings is a daily chore. Plants in six packs must be watered enough to thoroughly soak the soil, but not wash away nutrients. A garden hose sprinkler nozzle with a wide spray pattern gives good results. It is important to be gentle; we let the water spray upward and fall on the plants. This prevents damaging stems and leaves with too hard a spray. It takes a little practice to develop a technique. When watering a lot of flats, it is best to go over each tray several times, allowing the water to soak in each time. Later, check the flats and you can see where you may be missing cells or not watering enough. Often, the top half of a pot or cell can be wet but the bottom remains dry. This stresses plants and stunts their growth, so be sure to water thoroughly. Watering should be done early in the day, before seedlings get too hot and wilt. This allows them time to dry after watering, preventing disease.

Watering the deep beds and planters is a similar process. A gentle spray, directed upward and allowed to fall on tender seedlings will minimize damage.

Before planting a bed with seeds or transplants, the soil should be thoroughly moist. And the surface needs to stay moist for the seeds to germinate. Initial watering of newly planted beds should be deep enough to build a reserve of water deep in the soil. When the seeds germinate, they will grow deeper roots. If only the surface is wet, the roots will be shallow and the seedlings more prone to wilting. The plants will be weaker and more likely to mat down and die when watered.

We water the bioshelter pots, planters and beds heavily at least twice a week. A 25-gallon planter, with a soil saturation of 25 percent, will hold about 8 gallons of water. Providing this amount of water takes about a minute for each planter. So, 50 planters require nearly one hour. Then there are the pots and deep beds. When we are too busy to water heavily (harvest days), a light watering is ok for a day or two, but after that the reserve of water begins to be depleted.

Because thorough watering is so time consuming, it is a good time to monitor plant growth, pull weeds, and look for pests. Aphids can be washed off of leaves while watering. Wetting leaves is not desirable in cloudy weather, but on sunny days adding humidity to the leaves and washing insects off onto the soil will encourage *Beauveria* fungus to infect the insects.

Feeding Seedlings

Potted plants and seedling flats need to be fed at least once each week. A solution of fish emulsion and seaweed diluted according to label instructions will keep plants growing well. A hose siphon is the perfect tool for large-scale fertilizing of seedlings and crops. This is a fitting attached to the faucet that draws a small amount of liquid fertilizer into the stream as the water flows past. We mix the seaweed and fish emulsion concentrate in a bucket with water at a rate of about one cup of concentrate to three gallons of water. The siphon further dilutes this to about ten to one.

Many minerals are absorbed directly by leaves, so we use this solution for potted plants and foliar feeding of our seedlings. For several days after foliar application, we make sure to wet the leaves to help the plant absorb residual fertilizer. As a rule, it is best to not fertilize crops in the week or two before harvest. They are not likely to benefit, and you want to avoid fertilizer residue on mature crops. When used properly, foliar feeding maintains healthy plant growth and plant vigor. In addition to plant nutrients, foliar feeding with the seaweed adds plant growth hormones.

We have found it useful to occasionally fertilize plants with nettle tea. Nettles and other plant teas can provide nutrients and other "bioactive plant compounds" (see Diver, "Notes on Compost Tea," in the Resource List for more information). We use nettle tea to fortify and rejuvenate potted plants in the spring. Chamomile tea has been suggested as a remedy that strengthens plants against mildew and fungus. Comfrey tea is also thought to invigorate plants. We prepare plant teas by soaking fresh or dried leaves in a bucket or barrel for about a week. This liquid is then applied (diluted with a hose siphon) to both foliage and soil. For food safety reasons, we do not use these teas on leafy greens and roots near harvest time.

Hardening

Seedlings require hardening off before being placed in the garden. The harsh effects of direct sunlight can damage and kill tender seedlings. Sudden exposure to wind and temperature extremes also can be detrimental. After their time under grow lights, we rotate plants to sunny benches on the bioshelter's second floor in the winter and late spring and to outdoor frames in summer. The outdoor frames have removable plastic sheeting covers. The covers are closed at night in cold weather and when rainstorms threaten. When we are heavy into spring seedling production, after a week or two in the outdoor frames, we move flats to a more exposed holding area to await transplant or sale. Shade-loving plants are stored under the kiwi arbor.

Starting Seeds in Hot Beds and Cold Frames

Seeds can be planted directly into cold frames or hot beds for later transplanting into the garden, as Grandmother Frey did. For best results, seeds should be sown thinly or the seedlings thinned to several inches apart to allow the plants room to develop. A sturdy root system is needed for successful transplanting. Crowding the seedlings may make them leggy and will reduce the root ball when transplanted, increasing transplant shock. It is best to thin by cutting unwanted seedlings, to prevent disturbing the roots of the remaining plants.

Planting Out

Seedlings are ready to plant into the garden when the roots are developed enough to hold the soil together in the cell. If they are transplanted too soon, the soil will crumble away and the plant may struggle to recover. If left in the six packs too long, seedlings lose vigor from nutrient depletion and root crowding.

Potted Plants

Plants maintained in pots need a richer and lighter mix than seedlings. The key to success with hanging baskets and other potted plants is a regular fertilization schedule. Some plants spend many years in a pot. Allow an inch or so of space from the soil surface to the top of the pot. This space allows thorough watering and the occasional top dressing of compost or organic fertilizer.

Cuttings

Propagating plants from cuttings allows you to produce clones of favorite or unusual plants. In some cases, it is the only way to produce more of a particular plant. Variegated varieties of herbs, for instance, can often *only* be propagated by taking cuttings.

Cuttings are best taken in early morning or at dusk. Fill flats of six packs with seeding soil mix and water with seaweed solution. Cuttings are a 3- to 4-inch section of plant stem, including leaves. Branch tips are best, but not necessary. Cut the stem on a diagonal to get a larger exposed edge. New roots grow from the cambium layer. Remove bottom leaves. Insert cutting 2 inches into potting mix. Water again with seaweed solution. Cuttings need 16 hours a day under the grow lights for several weeks. Make sure they get good ventilation. Keep moist, but not soggy. For most plants, roots will begin to develop in two to four weeks. The plants can then be moved to a spot with natural light in the greenhouse or hardened off before moving them outside.

Plants easily propagated by cuttings include scented geraniums, roses, nasturtiums, mint, rosemary, sage, oregano, thyme, lavender, African blue basil, and many perennial flowers.

Other Plants

We give a lot of bioshelter space to ornamental "house plants." These plants are grown in the proper soil mix for each plant and have varied and often special needs for watering and fertility. Some of these we sell or use in bouquets, but many are for our own enjoyment.

Clivia, amaryllis, freesia, and snapdragons all bloom in our large beds and planters. Baskets of Boston fern, spider plants, wandering Jew and petunias, to name a few, hang in baskets around the building. These grow well in shadier places. Many of these plants help keep household air fresh and thus are a valuable item to supply to offices and homes. We have a wide range of cactus and succulents as well. These are grown in the east end of the bioshelter, on the barrels that line the wall just outside the poultry room.

Edible Perennials in the Bioshelter

We have a number of edible perennial plants in the bioshelter. Large rosemary plants and lemon verbena have permanent locations in the deep beds and planters. Jujube (Chinese date), brown turkey figs, smaller rosemary plants and bay laurel are kept in large pots. Scented germaniums, mints and other herbs (sages, thyme, and oregano) for culinary use and bouquets are planted in pots and in the deep beds and planters throughout the building. Tomato plants can remain productive for several years, producing crops several times a year.

Countless Bioshelter Possibilities

Our bioshelter management strategies have evolved. We are primarily guided by the need to provide crops for our customers, especially salad crops and herbs, and we manage plantings to maintain habitat for populations of beneficial insects. But we also work to maintain an aesthetically appealing environment.

As there are many possible design strategies for bioshelters, there are countless possibilities for planting schemes and management plans. Each farm has its own mix of crops. Each gardener has a personal aesthetic and favorite plants. Annual cycles vary from farm to farm. As bioshelters become more commonplace in the sustainable future, each one will be a uniquely productive environment.

CHAPTER 11

Compost and Biothermal Resources

Things cannot be destroyed once and for all.
When what is above is completely split apart,
it returns below.

— Hexagram 24: Return, *The I Ching*

THE COMPOSTING SYSTEMS AT THREE SISTERS FARM are a dynamic
part of the bioshelter's design and mission. Nutrients in the form
of agriculture "waste" and other organic materials are pollutants when
not cycled back to the soil, so we feel it is important to re-introduce
a complete nutrient cycle to the farm. A permaculture market garden
farm can require a lot of compost to maintain high yields. This chapter looks at the management of compost from the permaculture design
perspective.

Harvesting Compost's Multiple Yields

A compost pile is more than just a pile of manure and plant residue. As myriad microorganisms process and convert organic
matter into finished compost, they release heat (biothermal
energy), carbon dioxide, and ammonia. We can capture these
products and put them to work.

Heat

Heat from compost comes in two forms: The air itself is warmed
by the contained pile, and heat is also present in the large volume of water vapor released in the composting process. (That
there are large amounts is quite obvious on cold mornings, when
the pile can be seen steaming.) The heat required to turn water

Compost chambers provide supplemen-
tary heat and CO_2 to the deep beds at
Three Sisters Bioshelter.

DARRELL FREY

347

to vapor can be thought of as being "stored" in the water vapor. This stored heat is released when the vapor condenses back into water. (For more on water's heat storage capacity, see Chapter 5.) In the bioshelter's compost heating system, the water vapor within the compost bin is blown through ductwork into rock storage under the deep beds. When the warm vapor hits the cooler rocks, it condenses and releases its heat energy, which is absorbed by the rocks and then transferred slowly to the growing beds. The heated air from the compost chamber also contributes to warming the deep beds.

CO$_2$ Enrichment

Carbon dioxide gas is essential to photosynthesis. Plants combine CO_2 and water to create sugar, releasing oxygen in the process. In cold weather, when ventilation and air circulation is low, plants can use up all the available CO_2 in a greenhouse. Carbon dioxide enrichment is a greenhouse production method used to boost plant health and growth that has been in use for decades.

Normal CO_2 levels in the atmosphere are currently about 390 parts per million (ppm). Increasing the levels to 800 to 1,000 ppm is common in commercial greenhouses. Commercial greenhouses commonly burn fossil fuel to release the CO_2 — which is something we want to avoid in a permaculture bioshelter. Recovery of CO_2 from the composting process uses CO_2 currently in the geological cycle rather than releasing ancient CO_2 and contributing to climate change.

Capturing Ammonia Using Biofilters

Anaerobic decomposition creates methane (this is what happens in a digester), which is a flammable and potentially explosive gas. Therefore, it is important to manage decomposition in compost chambers to be *aerobic* and *ventilated* to prevent methane production. Unfortunately, one product of *aerobic* composting is ammonia gas — and excess ammonia is not wanted in the soil because it can build to toxic levels. For a bioshelter composting system to work correctly, the ammonia gas has to be removed before the compost "exhaust" reaches the growing beds. This is done by passing the ammonia-filled air in the composting bin through a *biofilter* as it leaves the compost chamber. This sounds very technical, but in reality, a biofilter is simply a tray of peat moss-based potting soil. The tray needs to be checked daily. When it becomes saturated with ammonia and moisture (about twice a week), the potting soil is removed and replaced. The ammonia-saturated potting soil is then mixed back into the compost, where it will feed microorganisms and re-enter the nutrient cycle.

We have found that our potting soil mix is an excellent media to use in the biofilter. The peat is coarse enough to allow the air to move through it, while being absorbent enough to soak up ammonia vapor. The vermiculite, rock phosphate and lime in the mix also absorb nitrogen. The compost in the mix may inoculate the mix with beneficial microorganisms that process nitrogen into less volatile forms. The reduction of ammonia in the airstream coming from the compost helps prevent excess nitrogen from building up in the soil.

Legal Issues

Many state and local governments regulate composting operations. A first step in planning a nutrient cycling/composting operation is to investigate all applicable regulations and other legal constraints (to get an idea of what some of the issues involved are, see Appendix A at the end of this chapter for regulations in the state of Pennsylvania).

Traditional Uses of Biothermal Energy

Compost energy has been used in *hot beds* for millennia. A hot bed is a cold frame that has a layer of hot compost topped with a layer of soil. The hot compost provides bottom heat for germinating seeds and young plants.

A hundred years ago, hot beds were common on farms and market gardens and in greenhouse production. They deserve a revival. Writing in the American Garden Calendar, Bernard McMahon discusses the use of "bark stoves," or bins of composting tree bark to heat greenhouses in the early 19th century.

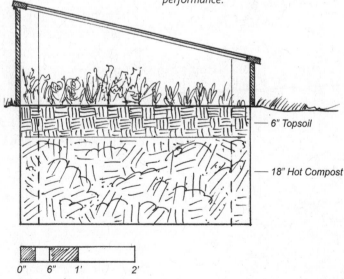

A hot bed utilizes biothermal energy to warm soil, enhancing season extender performance.

— 6" Topsoil

— 18" Hot Compost

0"   6"   1'   2'

Basic recipes call for hot manure or other compost mix with a high ratio of nitrogen to carbon. The material is generally pre-composted for ten days. This stabilizes the heat, and — if done thoroughly — such pre-composting kills pathogens and weed seeds. The compost is then added to a pit or frame and tamped to a depth of 18 inches.

The compost will heat up again after being placed in the bed. When a hot bed is intended for producing early spring seedlings, 6 inches of soil is placed on top of the compost. When the soil temperature drops below 100°, seeds can be planted in the soil.

When rooting cuttings or nurturing potted plants, the layer of hot compost is topped with 6 inches of sawdust and the pots are placed into the sawdust.

Hot beds are generally set up in March for early spring crops and in April for early summer crops. A hot bed can also be used to extend the fall growing season.

The compost will provide warmth to a bed for over a month. Hot bed performance can be controlled by manipulating the carbon-to-nitrogen ratio. For hotter compost that lasts several weeks to a month, use hot manure with only a little bedding. For a "warm" compost over a longer period (one to two months) use more carbon (in the form of straw, sawdust, wood chips, bark, etc.). Be sure the compost has enough nitrogen for a hot pre-composting.

Note that hot beds need *daily* monitoring. When the sun shines, they must be ventilated or the compost will overheat. In cold but sunny weather, vents need to be open but plants in the beds should be covered with row cover to protect them from the wind.

Hot Beds and Food Safety

Health concerns: due to the potential for illness from *E. coli* bacteria and other pathogens found in manure, organic farming regulation prohibit growing food in un-composted manure. Manure must be thoroughly composted *before* being applied to the soil. Otherwise, crops should not be harvested from soil treated with manure for at least 90 days. Therefore, hot beds should be used wisely. A thorough hot pre-composting of the hot bed material should, in theory, yield a compost that is safe for crop production. However, it is best to not take chances until you master the art of composting. One should use extreme care when growing root crops, such as radishes, carrots and beets in hot beds because they can incorporate pathogens from the manure. Leafy greens in contact with the soil can likewise harbor pathogens. Because of this, we have chosen not to grow any of these crops in hot beds. We find hotbeds to be most useful for starting seedlings for transplant into the garden.

Any systems that integrate composting into the growing environment must assiduously follow practices that keep raw manures away from the growing crops. Contained composting, dedicated compost yards and tools dedicated solely to the composting operation are strategies to keep composting and food production separate.

Compost and the NOP

The USDA's National Organic Program (NOP) has created a series of guidelines to define requirements for "organic" food and their guidelines extend to compost that comes into contact with food. Organic certification requires detailed records be kept for each batch of compost, including material sources, temperatures and management. An organic certification inspector does an annual inspection to verify compliance. Below is the NOP's definition of an acceptable compost process that removes pathogens from food.

Darrell Frey explains windrow composting to a tour group.

Compost. The product of a managed process through which microorganisms break down plant and animal materials into more available forms suitable for application to the soil. Compost must be produced through a process that combines plant and animal materials with an initial C:N ratio of between 25:1 and 40:1. Producers using an in-vessel or static aerated pile system must maintain the composting materials at a temperature between 131°F and 170°F for 3 days. Producers using a windrow system must maintain the composting materials at a temperature between

131°F and 170°F for 15 days, during which time, the materials must be turned a minimum of five times.

— USDA

Record Keeping

Organic standards and good farm management practices require diligent record keeping of compost operations. Records should document the source of raw materials, the date they arrive on the farm, temperature data and turning dates and times. Each batch of compost must be documented.

Material Handling

Material handling is a major consideration when managing compost. Making transportation of materials *un* challenging is the key to making this system work. If the manure source is off the farm, it must be loaded in a truck or wagon, then transported to the compost chambers and loaded in. After chamber composting, the finished product has to be moved to windrow piles for final composting. (A windrow is long, narrow row of organic material mounded to promote thorough composting.) Each step in the journey — from the stable or barn to the garden — requires work, equipment and safety measures.

The expense of hauling manure or other materials for composting from off farm includes the cost of owning a wagon or truck and the labor involved in loading and unloading. Because a truck or wagon is required for many other farm chores, this cost is spread over a number of tasks. However, the labor involved in loading manure from horse stables can be prohibitive. One solution is for a market gardener to offer stable cleaning services for a fee, with tools and equipment dedicated to the task. Our solution is to keep our farm dump truck parked at a stable. In exchange for hauling away the stable bedding, the stable managers load wheelbarrows full of manure (mixed with sawdust, hay and straw) on the truck. The load is often topped off with either a front-end loader or by hand with a pitchfork.

Unloading manure is much easier with the dump truck. We rake the materials down from the tilted truck bed into carts. Then we fork from the carts into the compost chambers. This requires about one hour of labor per loading.

Managing contained compost includes monitoring moisture, aeration and temperature. The pile must be moist enough when loaded to begin a thorough composting. Aeration tends to dry the pile, so well water is added via a sprinkler as needed to maintain a moist pile. Our system's blower operates several hours each morning, then again in the afternoon and evening, for a total of six to ten hours daily. Temperature is monitored daily to assure proper temperatures are attained for each load of manure and bedding.

The piles begin to decompose, settle and compact after a week to ten days. Then they are usually turned and loosened and watered if necessary to ensure thorough composting. After the raw materials are composted, the weight and volume is greatly reduced. Generally, the volume of finished compost is one third the volume of the raw materials.

Biothermal Resources

Jean Pain, a French gardener and inventor, has developed some interesting systems to recover heat from composting wood chips. One of them is a hot water system consisting of coils of plastic water line buried in a compost pile. A 20-foot diameter pile of woodchips heats water for up to 18 months. His work illustrates the potential for creative application of biothermal energy. Pain directly influenced the development of the composting greenhouse at the New Alchemy Institute.

New Alchemy's Composting Greenhouse

The composting greenhouse at New Alchemy Institute was the inspiration for our own system. Their design was very simple: A 48-foot by 12-foot pipe-framed polytunnel was glazed with double-skin, inflated, 6 ml greenhouse plastic. The structure was oriented with the framed end walls facing east and west. This basic polytunnel was equipped with a series of wood-framed compost bins (collectively referred to as the "compost chamber") along almost the entire length of the north side.

Composting greenhouse at The New Alchemy Institute in 1988.

DARRELL FREY

The compost chamber performed several functions that moderated the greenhouse environment. It helped insulate the structure on the north side and provided biothermal energy to warm the soil in the ground-level garden bed on that half of the tunnel. The top of the

Sewage Sludge: A Matter of Scale

As Joe Jenkins has thoroughly examined in the *Humanure Handbook,* the trouble with composted sewage sludge is simple. People dump nasty stuff down their drains. Many household wastes get flushed. Paints, cleansers, and other potentially toxic household wastes are dumped down the drain. Pharmaceutical residues in our wastes further complicate the issue. Many towns and cities have storm water drains combined with the sewage system. So anything spilled on the road or left along the curb gets mixed with the sewage. Many commercial and industrial facilities use municipal sewage systems to get rid of waste products.

In theory, composted sewage sludge should be a fine fertilizer, completing the final step in the nutrient cycle. In practice, though, it is impossible to predict what will be in the finished product. While composted sewage sludge is tested for "acceptable levels" of heavy metals and other potentially harmful chemicals, it has been deemed unacceptable by the organic industry and the USDA for use on certified organic farms.

In simple terms, use of composted sewage sludge often means bringing home someone else's problems. In a saner world, we would separate storm drains from the sewage system, educate everyone not to contaminate their wastewater with toxins, and develop more sophisticated testing of sewage-based compost. Until we have these safer methods of managing our urban and suburban human nutrient cycle, composted sewage sludge will remain unacceptable for organic food production.

However, the home-scale use of Jenkins' humanure composting system, with appropriate management and monitoring, is a safe and proven method to complete the nutrient cycle. Other composting toilet systems have also been proven effective and safe.

Note: If you choose to do humanure composting, remember that when in doubt about the finished product of composting toilets, you can perform a secondary thermophilic composting. Simply layer materials removed from a composting toilet with coarse organic material and added nitrogen (such as organic fertilizer, urine, animal manure or fish emulsion) if needed, to create a hot compost. Monitor and turn the pile to ensure the material has adequate exposure to the proper temperatures for the proper amount of time. Jenkins states that a temperature of 122° for 24 hours is sufficient to kill pathogens.

JOSEPH JENKINS

Joe Jenkins managing humanure compost bins.

compost chamber provided 150 square feet of space for growing potted plants and seedlings. Heated air, rich in growth-promoting CO_2, was pulled from the compost chambers through a biofilter and blown through a network of perforated flexible plastic drainpipes buried in the garden bed.

Supplemental CO_2 generated by the decomposition of the compost materials promoted crop production during the cold winter months. They used horse manure and sawdust stable bedding from local stables as their raw material. After one month, the composted material was moved to an outdoor bin to finish composting and fresh material was added to the bins.

Many Possibilities

There are other innovative ways to harvest the multiple yields of compost. As Bruce Fulford explains in his evaluation of New Alchemy's research (see Resource List), the costs of material handling and the benefits of recovering heat, CO_2 and nitrogen must balance economically. An intensive market garden that requires both early spring seedling production and tons of compost is one place where the costs do balance.

Our friend Jack Stupka developed a simple method of using biothermal energy for starting his garden seedlings. A box made of wooden pallets, measuring about 4 feet by 8 feet is filled with the bedding accumulated in his chicken coop all winter. After an initial composting, seedling trays are germinated on top of boards placed over the composting manure. A sheet of clear plastic helps hold in the heat. Seedling grow fast and strong in the frame, with the heat and CO_2 from the compost aiding their growth. On very cold nights, an old blanket is placed over the box for extra insulation.

On a larger scale, roll-off dumpsters filled with compost materials and covered can act as containment vessels. Some farms have set up a piping system to aerate the pile and recover heat and CO_2, which can be directed to an adjoining greenhouse or to cold frames. Compost bins and aerated compost piles also can be located near these season extenders to make the transfer more efficient. Farms with livestock can set up material flow pathways and schedule composting of animal bedding to ease the process of composting and compost resource recovery.

Compost systems that do not use animal manure are more easily incorporated into a greenhouse. A loose garden soil enriched with organic matter will emit CO_2 as it respires. Making compost from bales of straw or hay materials inside a greenhouse will also release appreciable amounts of CO_2. A CO_2-generating compost bin that uses only vegetable matter would probably not generate enough ammonia to require a biofilter.

Composting Practices at Three Sisters Farm
The Indoor System

Two compost chambers are located along the outside north wall of the bioshelter. Each chamber is made of concrete block, with approximately 11.5 cubic yards of interior space. Each compost chamber has four evenly spaced doors (30 inches by 42 inches) that provide access to the chambers from the barn area. Approximately 2.5 cubic yards of

Vermiculture

In *vermi* composting, the redworm (*Eisenia foetida*) is used to turn food wastes into useful products. We have not set up vermicomposting systems at the bioshelter because our chickens eat all the food waste we generate. On farms with no poultry to consume farm waste, vermicomposting can be an important source of fertilizer and potting soil compost. One option we have studied is feeding the spent trays of cutting greens to worms. They would benefit from the rock minerals in the potting soil and actually make the minerals more available to plants. Worm castings could then be added to soil mix for subsequent cutting green production. The soil would be continuously cycled from trays to worms.

We have also considered adding worms to our compost chambers in the spring (when we are no longer interested in having the compost produce heat) and allowing the worms to finish the compost in the compost bins. This would alleviate the need to move it by hand.

As mentioned earlier, we have not yet integrated aquaculture into our farm, but vermiculture bins could be a big part of aquaculture. (They would also serve as thermal mass.) Catfish are said to be able to consume worms as 20 percent of their diet.

Working with Christine McHenry-Glenn of the PRC (and in consultation with staff at the PRC and the Pittsburgh Children's Museum), I designed a worm bin for the Pittsburgh Children's Museum as a project of Three Sisters Permaculture Design. The bin is designed for year-round, outdoor use. The 20-cubic-foot bin is constructed of 2-foot by 4-foot pine lumber, with 2-foot by 6-foot lumber for the lid. Both the main box and the lid are insulated with 3 ½ inches of EPS board (expanded polystyrene board) and have an 1/8-inch luan wood interior and a 3/8-inch plywood exterior. The inside of the bin, including the underside of the lid, is lined with butyl rubber roofing. The exterior of the box and lid are ☞

raw compost can be loaded into each of the four chamber sections. Total chamber capacity is 10 cubic yards because there has to be about 1.5 cubic yards of airspace above the compost. Because there are two chambers, total compost capacity in the bioshelter is 20 cubic yards.

Our practice is to put our poultry bedding in first. So it goes onto the wooden planks at the base of the chambers. The bedding is a mix of straw or hay and chicken manure. This provides a coarse bottom layer that allows good aeration because it does not contain sawdust, which tends to mat down. We then fork in the horse manure and straw/sawdust bedding material.

As mentioned earlier, stable bedding is loaded into a pickup truck (3 cubic yard capacity) at a local stable. We unload the truck into a garden cart and load the compost chambers using shovels and pitchforks. At the end of a 28-day cycle, the compost is manually removed back into the cart and unloaded into windrows outside for further composting.

finished with PVC foam board (weatherproof plastic) and sealant. The lid is sloped to shed rain. An acrylic mirror is attached to the underside of the lid to allow the kids to see the bin contents when the lid is opened. A 1 ¼-inch sink drain at the bottom allows excess water to drain out.

A grid of plastic wood supports the worm bedding and allows castings to drop to a collection space at the bottom of the bin. A hinged clean-out door provides access so castings can be collected.

The bin is ventilated by 2-inch soffit louvers and a 4-inch solar-powered vent fan. Supplemental heat in winter is provided by 24 feet of heating cable lining the bin's interior wall. The bin is held off the ground by 4-inch by 4-inch plastic lumber boards. Hinges, handles and other hardware allow the lid and clean-out door to be opened and latched.

This design was adapted from a previous design by the staff of the Felician Sisters Convent in Coraopolis PA, which was in turn an adaptation of a design from UC Berkeley's Vermitopia Bin.

Design of the Worm Bin at The Pittsburgh Children's Museum.

One of the four chamber sections is loaded each week. When we use the compost chambers to full capacity, both chambers are filled. Compost residence time in the chamber is about 28 days — enough time to adequately exhaust its heat and CO_2 resources. A soil thermometer is used to ensure adequate temperatures are reached and maintained. Blowers at the end of each chamber pull the exhaust fumes from the compost and ventilate the chicken's room through the biofilter located on the intake side of the blower to reduce ammonia in the air stream. Biofilters are changed when ammonia odor is detected. Spent peat filters are returned to the compost chambers to recycle nutrients contained in the peat filter.

The blowers direct the exhaust to rock storage areas beneath the raised growing beds in the bioshelter. The warm, moist air condenses onto the rocks and soil, giving up its stored heat. Any remaining ammonia in the exhaust is converted to nitrates by bacteria in the soil, which are taken up by the plant roots as a nutrient. The CO_2 in the exhaust passes upward through the soil in the beds, to be taken up by the plants. The CO_2 is in the right place because a plant takes in gases through its *stomata* (pores) on the underside of its leaves.

As described previously, the tops of the compost chambers are used to grow seedlings, which benefit from the bottom heat (as well as the CO_2 the compost provides to the greenhouse).

Compost removed from the bins is piled in windrows, covered with leaf mulch and soil, and left to finish composting and mature.

Compost chambers are used during the heating season, which runs from January to sometime in April. However, in recent years, we stopped using the compost chambers in December and early January due to the concern about potential generation of excessive ammonia, leading to increased nitrate levels in lettuce leaves (more on this, below).

Indoor System: Heat Generation

Calculations (see Appendix B at the end of this book) show that the maximum heat-generating capacity of the compost chamber is 20 million Btus per month, or 120 million Btus over the 6-month heating season. The calculations demonstrate the potential magnitude of compost heat generation relative to other heat sources/losses in the bioshelter. These calculations assume the compost chambers are used at maximum capacity.

Taking December as a typical cold-weather month, compost heat can contribute as much as 35 percent of the solar gain entering the bioshelter through the glazing. Composting manure and bedding has the potential to generate more than three times the amount of annual heat losses from the bioshelter and three times more heat than three cords of hardwood burned in a stove. In reality, only half of the compost chamber's capacity has been used in recent years. Even at half of the above-stated heat contributions, the compost chambers represent a significant source of heat in the bioshelter. (Body heat from the chickens also contributes about 18,000 Btus per month — which is 1,000 times less than the amount provided by compost heat.)

Indoor System: Carbon Dioxide Generation

CO_2 is as essential to plant growth as water, light levels, soil nutrients, and temperature. Plants average about 50 percent carbon by dry weight, deriving most of this carbon from atmospheric CO_2 via photosynthesis. So it is necessary to pay attention to CO_2 levels in a greenhouse. Greenhouses designed to cut energy costs by reducing air infiltration can suffer from low CO_2 concentrations. Without CO_2 supplementation, plants can deplete concentrations to photosynthesis-limiting levels within a few hours following sunrise.

CO_2 enrichment in many commercial greenhouses commonly produces a 20 to 30 percent yield increase when levels are maintained at three to five times ambient concentrations of CO_2 (roughly 1,200 to 2,000 ppm). When operating at half of maximum capacity (typical operation), our compost chambers can increase CO_2 ambient levels in the greenhouse air to the range of 125 to 285 ppm above ambient levels (or 515-675 ppm, which is roughly 1.5 to 1.75 times ambient concentration).

Note that more is not necessarily better with CO_2 supplementation. Optimal enrichment levels for many crops appear to be much less than three times ambient CO_2 levels. Some suppression of photosynthesis occurs at levels in excess of 1,000 ppm. And damage to leaf tissues in some crops can occur at levels ranging from 800 to 1,600 ppm CO_2.

Our main goal with recovering the CO_2 from the compost and directing it to the growing beds is to assure levels in the bioshelter do not drop below ambient levels in the well-sealed building. Excessive CO_2 enrichment, in my opinion, is similar to using excess fertilizer, too much of a good thing can cause harm. And because we grow such a wide range of plants, we can not be sure all would react well to the high levels of CO_2. We have noticed that bioshelter plantings do respond well to the increase of heat and CO_2 when our compost chambers are in operation. We estimate the average increase in CO_2 levels in our system to be close to 60% — which is about 624 ppm (vs. the 1,200 to 2,000 in commercial operations).

Problems of Existing Systems

Loading and unloading the compost chambers is labor intensive. There are several possibilities to reduce the current labor-intensive practices. One possible retrofit involves moveable carts that can be rolled in/out of the chamber sections. This method would minimize the number of times that the compost would need to be manually handled.

A potential problem with winter-grown greens is the accumulation of nitrates. Nitrate in food can convert to nitrite in the body. Nitrite is considered a carcinogen and can interfere with oxygen uptake in the bloodstream. Compost gas contains ammonia that is converted to nitrate when the gas is discharged into the soil beds. Plant roots take up the nitrates. This is a normal process. But during cool, cloudy weather some plants take up more nitrates than they can process, so nitrate levels can exceed what is considered safe. We have found that nitrate concentrations build up in lettuce leaves during winter months — when daylight hours are shortest and little photosynthesis is taking place. In a study done by graduate student Elizabeth Affleck, testing of our mixed salad greens revealed that nitrate levels can be as high as 900 ppm (900 mg/L) during December when the compost operation is in full operation. Ideally, nitrate levels in the lettuce should be below 600 ppm (600 mg/L). Affleck's study showed that the problem can be controlled using several measures: careful use of fertilizer to reduce soil nitrates, selecting greens that are naturally lower in nitrates, and harvesting late on a sunny day. In the second December of the two-year study, we did not use the compost chambers, and nitrate levels were below 300 mg/L.

Some researchers suggest that the health risk of nitrates may be less than the health risks of not receiving the high food value of leafy greens. Regardless, pregnant women and nursing mothers are cautioned to avoid excess nitrate consumption.

Further research is needed to establish the best method of nitrate management in the composting greenhouse during extended conditions of low light levels. For the moment, our answer to the problem is to simply do minimal harvesting in the darkest weeks of winter — our production is at its low point anyway. However, a strategy needs to be found that allows a farm to process compost and recover heat and CO_2 while controlling nitrate buildup.

There are numerous biofilters that can oxidize ammonia (NH_4) to nitrate. Vermiculite and zeolite can bind NH_4 in soil and zeolite can strip gaseous ammonia directly from the air stream as it passes through the biofilter. Increasing the carbon content of the soil and keeping the pH of the soil below neutral (pH 7) enhances capture and assimilation of gaseous ammonia (NH_3) by the soil. The filters work because gaseous ammonia (NH_3) in the compost exhaust converts to ionized ammonium (NH_4) when it is dissolved in liquid (the water condensing from humidity coming out of the compost pile), and then the ammonium is biologically oxidized to nitrite (NO_2) by the bacteria *Nitrobacter*. The NO_2 is then further oxidized to nitrate (NO_3) by the bacteria *Nitrosomonas*. These biological reactions readily occur in the warm, moist soil (or similar filter material) through which the compost exhaust is passed. Therefore, the answer may be as simple as placing a layer of soil over the top of a compost pile at the start of its decomposition cycle. This would filter the ammonia from the compost exhaust before it enters the building, reducing the need to change the peat filter in the bioshelter.

Other Ways to Capture Compost's Yields for Indoor Use

There are some other methods for capturing compost's yields. Air can be recovered from aerated windrows, also known as aerated static piles (ASP). These are covered piles with a manifold of perforated piping at the bottom of the pile that allows air to reach all sections of the pile. This eliminates the need to turn the piles. The NOP regulations reduce the number of required monitoring days from 15 for a windrow to 3 for ASP systems. Because the temperature of the pile must be recorded to ensure initial thermophilic decomposition, a 3-day process significantly reduces record keeping requirements.

A further modification to ASP piping is using fans to actively draw the heat/exhaust into a bioshelter's deep beds. The result would be similar to our existing system: heat, ammonia and CO_2 would be discharged into the base of the soil beds. This modification would eliminate all of the labor associated with the loading/unloading of indoor compost chambers.

Another possibility is an air-to-water heat exchange system. An air-to-water heat exchange system uses plastic waterlines layered in the compost pile to collect heat. The water is either used directly or pumped to storage. While less efficient, this system does eliminate ammonia buildup. Heated water could be introduced into the greenhouse in multiple ways: direct watering into beds, radiant pipes or radiators, or as a water pre-treatment system for the kitchen hot water heater. Such a system would require extensive design and calculation work prior to implementation.

The Outdoor System

For most of the year, we compost in outdoor windrows. We used to bring in manure and stable bedding in a one ton dump truck. Currently the farm received a dumpster of 25 cubic yards (cy) of raw stable bedding as needed. This includes horse manure, hay, straw and sawdust.

Agricultural composting is regulated at the state level, for us by the Pennsylvania Department of Environmental Protection. They have worksheets available that define which systems do and do not require permitting (based on source of compost supply, sizing, location and the desire to sell the product). Three Sisters receives a compost mix consisting of horse manure/urine and straw/sawdust. These materials fall into the category of farm products and off-farm co-products, respectively. As long as the sawdust is clean, unpainted and untreated, it is acceptable for use. Other amendments such as leaves, grass clippings and mushroom compost are exempt from permit consideration. The farm products used meet the additional criteria of disturbing less than five acres and can be made to fit within the

Gourds planted on maturing compost shelter the windrow, conserving moisture and yielding a crop.

DARRELL FREY

necessary location constraints, and therefore do not require permitting. However, *sale* of finished compost does require licensing and product registration.

We do not compost food wastes, which require compliance to stricter regulations. (See the sidebar below, "Restoring the Nutrient Cycle.")

The USDA National Resources Conservation Service (NRCS) requires a Nutrient Management Plan for systems that exceed a defined quantity (weight) of animals. Because our compost material comes from private stables rather than "for profit" farms, Three Sisters Farm is exempted from consideration.

Our windrows each require about 300 square feet of space (10 feet by 30 feet by 4.5 feet). Windrows are shaped to have flat tops and steep sides to absorb rain or irrigation, as needed. Windrows are mulched with straw to hold heat. The first 15 days is the thermophilic stage, when temperatures reach 140° to 170°. After this phase, the compost naturally transitions into a mesophilic stage (111° to 122°) for up to two months.

After the first two weeks, we turn the windrows by hand to aerate the compost. Turning is done by pulling the base of the pile outward, raking the top down and then piling what was on the base to the top.

After 6 months, the compost will have reduced in size to about one third of its original volume. Windrows are then moved with a skid loader to another location on the farm where they are allowed to age further. It ages from one to two years before we consider it to be finished compost. The compost is generally used within two years.

Over time, the finished windrows become host to an array of useful, wild plants (e.g., purslane and lambsquarter). Perennial plantings of roses, berries and sod surrounding the windrows capture nutrient runoff and grow well in the rich soil that accumulates on the edges. Patches of burdock that thrive around the compost yard are harvested for chicken food.

We have up to eight windrows of raw compost at any given time, each measuring 10 feet by 30 feet (by 4.5 feet). We have capacity to store 22 rows of 6-month and older compost that are 10 feet by 10 feet (by 4.5 feet). The farm's gardens require about 75 cubic yards of finished compost when all gardens are planted to full capacity (see calculations in Appendix B at the end of this book). Since there is a significant reduction in volume (about two thirds) from the raw to finished product, the farm requires about 225 cubic yards of delivered raw compost annually. (Note that the semi-finished material can be moved and combined into piles because it has already been subjected to the required temperatures.)

Compost Sales

As regional recycling centers, farms can handle specific materials from trusted households and restaurants (e.g., food wastes, leaves, pine needles, newspapers, cardboard). Given sufficient space, knowledge of the processes involved, licensing and product registration, compost is a marketable product.

Restoring the Nutrient Cycle

Many European and early American market farms were fertilized by the cycling of horse manure, brewery waste and other sources of organic matter. Our challenge in the 21st century is to restore a nutrient cycle to our food systems by recovering organic material from the waste stream for composting and reuse in gardens and landscapes. Much work is being done around the country to reconnect local farms with local customers. Reconnecting the nutrient cycle is the next step.

A farm needs to either have enough land to allow fallow seasons and green manures to replenish soil organic matter or access to sufficient compost and mulch to keep soil healthy and productive. As described elsewhere, poultry provide a simple means of managing nutrients on-site. However, a small farm may need to purchase bedding materials and feed grains. Urban gardeners no longer have access to as much horse manure as 19th century market gardeners. Similarly, livestock is not as readily available for suburban and urban market gardens.

One solution is for the farm to develop functional connections with sources of organic waste: grocers, restaurants and food processors, chemical-free yard care services, livestock farms, horse stables, and zoos. Waste material can be supplemented with bagged organic fertilizer of course, but economically and ecologically, humus-rich compost is preferred for long-term soil improvement.

Food Waste = Urban Nutrients

Every day, many tons of food is brought into cities and tons of food waste is generated. Most of this "waste" goes to landfills. The EPA estimates that 10 percent of wastes entering landfills is compostable organic matter. A 2006 study of food waste in Pittsburgh examined the volume of compostable organic matter generated by small, medium- and large-scale purveyors and processors. This study, funded by the Pennsylvania Department of Environmental Protection and conducted by the Pittsburgh office of the non-profit Pennsylvania Resource Council, is titled "Food Waste Collection and Composting Feasibility Study." The following discussion is largely based on this study.

The study classifies food waste as either *pre-consumer* or *post-consumer* material. Pre-consumer food ☞

Outdoor System Requirements

- Large spaces for aging compost piles
- Means of solving aesthetic problems if uncovered compost piles are located next to public access areas
- Significant labor or expensive equipment to manually turn windrows as required by regulations
- Means to capture nutrient-rich leachate (raspberries, roses, mulberries and fruit trees at the low end of any new compost areas will gather excess nutrients that will run off from the compost piles)
- Truck access and driveway maintenance

waste includes any organic material generated by wholesale produce distributors, grocers, processors, cafeterias and restaurants up to and including food preparation. Post-consumer food waste is what comes off plates and trays in cafeterias and restaurants. Small-scale food waste generators include restaurants, grocers and cafeterias that produce up to 200 pounds or less of waste a day. Medium-sized food waste generators include processors and purveyors that produce 500–1,000 pounds per day. Large-scale food waste generators (primarily food processing facilities) produce over 1,000 pounds per day.

Following are some considerations in the recovery of food wastes and their return to the nutrient cycle:

- Time and space required to separate and store food waste
- Costs to collect and transport food waste
- Management of the composting facilities
- Availability of high carbon materials to balance high nitrogen food waste
- Regulations, permitting and other legal issues

The main choices for composting food wastes are in-vessel composting or windrow composting. (Vermicomposting was not addressed in this particular study but is discussed below.) In-vessel composting facilities are expensive to build and operate but offer better assurance of rodent control and minimize the amount of turning that the compost requires. All composting generates some odor, so biofilters may be needed to operate these facilities in urban or suburban areas. As demonstrated by the compost chambers at Three Sisters Farm, in-vessel composting systems can be integrated into greenhouses, providing thermal mass, biothermal heat and CO_2.

Windrow composting of food wastes requires less infrastructure. Facilities do need to be large enough to maintain piles and store materials. Equipment is needed to turn the piles and move the finished compost. As with in-vessel systems, additional carbonaceous material may be required. The windrow system is better suited to areas where the potential smell and drainage can be controlled. Vigilant rodent control is also required.

Both systems require space to allow the compost to cure after the initial high-temperature composting ☞

Composting windrows need to be located so that a truck can continue to dump off the materials. Generally, the dump area must allow for the 30-foot windrow, plus turning radius and the length of the truck.

period. They also need adequate monitoring of moisture, ventilation, and drainage to safely and efficiently process food waste into useable compost.

High-carbon materials needed for the composting process can be supplied from waste paper, shredded cardboard, or straw. Municipal leaves and wood chips can also be used (if organic certification is not a concern). Sawdust (no treated wood, please) is a possible source of carbon, but the finished compost may need extra curing time to finish.

Of special concern are meat by-products in food waste. The high protein and fat of food wastes containing meat is especially attractive to rodents. Systems that include these products may need to be contained to prevent rodents.

The Pennsylvania Resource Council's Pittsburgh office has also worked to promote vermicomposting and backyard composting of food waste for small-scale businesses and residents in Allegheny County. A 127-cubic-foot worm bin was installed at the Felician Sisters Convent in Coraopolis Pennsylvania. This system processes the convent and school cafeteria's food waste into high-quality compost. Another system is in place at the Children's Museum of Pittsburgh. This 40-cubic-foot system processes food wastes from the museum cafeteria. Museum visitors have the chance to observe the feeding of the worms every afternoon. In both these systems, redworms feast on the food waste and transform it into highly fertile vermicompost.

Working in association with the Allegheny County's Health Department and the City of Pittsburgh's Environmental Services Division, the Pennsylvania Resources Council (PRC) has distributed over 6,000 home-scale backyard compost bins to Allegheny County residents. Homeowners sign up for a workshop, and for a small fee receive the plastic compost bin and training on how to manage the system. A similar program has distributed over 500 home-scale worm bins. In discussing the report, PRC staffer Corinne Ogrodnik estimated that the average household produces 360 pounds of food and yard waste per year. This means that the 6,000 bins that have been distributed have the potential to remove over 2,000,000 pounds of food and yard wastes from landfills each year in Allegheny County.

Interesting Waste Stream Facts:

- About one quarter to one third of the waste stream, by weight, is organic waste, like food, leaves, and grass.
- Nationwide, leaves and yard trimmings amount to approximately 28 million tons (or 13.4 percent) of the waste stream.
- Each year, the average person creates about 360 pounds of food and yard waste.
- One cubic yard of leaves weighs approximately 200–250 pounds.
- One cubic yard of grass weighs approximately 350-400 pounds.
- 27 percent of all food produced in North America is wasted.

CHAPTER 12

Chickens in the Greenhouse

No matter how small or large your farm, you should keep hens.
The outlay for good stock, equipment, and feed, and the amount
of time and labor to manage them properly are small in compari-
son with the advantages.

— M. G. KAINS, *FIVE ACRES AND INDEPENDENCE*

CHICKENS HAVE A SPECIAL PLACE in the permaculture farm eco-system because they offer multiple products and services. They consume a wide range of plants and crop residue, snails, slugs and bugs, and they produce manure for the compost pile. A laying hen can produce up to 150 pounds of manure a year. In a rotational pasture system, much of this is applied directly to the soil. When accumulated in the chicken house, manure and bedding is the basis of high-quality compost.

Wild birds play a similar role in the natural ecosystem. Birds gather insects, seeds and fruits and leave their droppings scattered about, returning nitrogen to the landscape in a highly soluble form.

Mollison's *Introduction to Permaculture* uses the chicken to illustrate the process of analyzing design elements. Each element in the system has specific needs, yields and characteristics. Recognizing what these are allows a designer to provide for the needs of the plant, animal or structure, and to connect all the elements in the system to one another.

Birds need food, water, small stones and grit, shelter and habitat that include space, sun, shade, and air circulation. Yields include meat, eggs, CO_2, body heat, feathers and manure. General characteristics include behavior such as foraging habits, scratching in the dirt, flying and fighting; breed-specific characteristics include climate tolerances, color, efficiency of egg laying and weight gain, and whether they are docile or aggressive.

Food Security

Chickens are an important element in the search for food security. Their low cost and the minimal care they require make poultry a good source of low-cost protein.

Because they can transform food and garden wastes into meat, eggs and fertilizer, and they will clear an area of most weeds and bugs if fenced at the right densities, a small flock is a valuable addition to a garden system. The sustainable community of the future will again welcome the backyard chicken flock as an important part of the neighborhood's food system. At the moment, though, many towns and cities limit flock size and ban roosters.

Demand and Economics

Free-range, organically raised poultry and eggs are in high demand. It's not surprising — an understanding of the modern industrial production systems for these foods can ruin one's appetite for them. A growing number of consumers believe birds that have a more natural existence, with fresh air, sunlight, exercise, and lots of variety in their diet will be healthier and produce a more wholesome food. However, relative to large-scale production facilities, it is much more expensive to raise chickens in such an environment. Organic feed prices are high and so is labor. However, with careful budgeting, producing and selling organic, free-roaming chickens and their eggs can be profitable. Room to grow your own feed is helpful, but not necessarily available on a small farm. Designing and planning for a harvest of *multiple yields* from poultry (or any other livestock) can help balance

Poultry flock at Three Sisters Farm.

LEO GLENN

the costs and benefits. Having the ability to supplement feed with plant residue and pasture will also help increase profitability.

Finding the Right Bird

For the purpose of this discussion, chickens fall into two categories, meat birds and egg layers. When raising chickens in the greenhouse, we are mainly concerned with egg layers. And, as already discussed, the birds are managed as a part of the building's ecosystem.

Egg-laying breeds may not grow as fast or as plump as birds bred for meat production (although many older breeds are "dual use"). A common homestead practice is to buy twice as many unsexed (not separated by gender) chickens as you need for the egg flock, and then butcher the extra roosters for the freezer at about 10–12 weeks of age. Because egg laying declines after a year, it is also common to purchase new chicks each year and butcher the year-old hens. When properly timed, the supply of eggs can remain steady and the farm freezer kept stocked with birds.

At Three Sisters Farm, the focus is primarily on keeping layer hens. For those wishing to develop systems for raising meat birds, please see sidebar, "Design: Raising Meat Birds at Three Sisters Farm," in which I describe a plan we've designed, but not yet implemented, for a rotational meat-bird system. I highly recommend the reader consult Joe Salatin's writing on livestock and poultry production when planning an ecological husbandry system. *Pastured Poultry Profits* is a good one to start with.

Other Birds

Other poultry and birds can be integrated into a farm system if it is properly designed. Geese and turkeys will forage for some of their food, but they are relatively large and can be imposing to young children. Geese and ducks have been successfully used on farms as weed and pest eaters. Careful matching of poultry variety with the farm's cropping system is important. As with all livestock, garden fencing and rotational pastures are required to get the most benefit from integrating them into a farm or garden.

Quail can be raised for eggs as well as meat. Pigeons are mainly raised for their meat. Both would integrate well into a permaculture farm. Other game birds may fit on larger properties.

Food Safety

Uncomposted Manures

Of particular concern is the need to follow organic standards when applying un-composted manures to soil. This means waiting 180 days to harvest crops in direct contact with the soil, and 90 days for above-ground vegetables.

When we use chickens to remove grasses from blocks of garden beds, we follow with a winter cover crop to capture nutrients. The following year we till it in and the beds are ready for crops.

Eggs

Eggs are a fine food. Despite the question of how our bodies metabolize cholesterol in our food, eggs are a complete protein food and are convenient and easy to use. They are low in calories and saturated fat and are high in vitamin A and riboflavin. They also contain significant amounts of calcium, phosphorus, and potassium. Eggs are a good source of choline (important in our nervous and muscular systems) and lutein (helps keep our vision strong and prevent cataracts). Egg yolks are the source of the fat and cholesterol in eggs, as well as the vitamin A and most of the calcium and phosphorus. Egg whites have neither cholesterol nor fat, but they have most of the potassium in an egg. Nutritionists increasingly are focusing more on the need to increase consumption of omega-3 fatty acids to control cholesterol than on the need to eliminate dietary cholesterol. Daily consumption of eggs may not be wise, but a vegetable-filled omelet or two and a few quick egg breakfasts a week is fine for most people.

Eggs do have the potential to harbor *Salmonella* bacteria if they come from infected hens. *Salmonella* is one of the most common causes of food poisoning; raw and undercooked eggs and poultry are the most common source. Cooking eggs will kill *Salmonella* bacteria, so handling and eating raw eggs is not recommended. Proper handling and cleaning of eggs is also important. When ordering new chickens, order birds that are certified free of *Salmonella* pathogens.

Eggs should be collected at least once a day, though three times is preferable. We collect any eggs in the nests at feeding time each morning, around noon and again in the late afternoon or early evening. Adding fresh hay to the nest boxes daily helps keep eggs clean. Keeping a deep mulch in the chicken yard reduces soiling of the eggs by the birds' muddy feet (plus, microorganisms in the mulch provide poultry with vitamin B-12). Any eggs soiled with mud or chicken manure are gently washed with a 10 percent mixture of hydrogen peroxide and water and wiped with a paper towel. Badly soiled eggs are crushed and put back into the chicken's daily feed or added to the compost.

Eggs need to be kept refrigerated. Cartons should be new and clean and should be marked with the date the eggs were laid. We have never had enough eggs to satisfy demand, so have not had to store eggs for long periods of time.

Each state is governed by both state and federal food and agricultural regulations. Any farmer selling eggs needs to keep updated on these regulations.

Breeds

The choice of breed is an important decision. Many breeds have been developed in the past two hundred years. Each has distinctive characteristics.

Virtually all chickens will need some amount of daily grain rations to be productive. But some breeds are excellent at foraging and scratching for bugs, slugs and seeds in mulch and litter. This can reduce grain rations by up to 30 percent. Chickens not as good at foraging are more dependent on their daily grain rations. Game birds are known for their aggression, while other breeds are more docile. (Once, while helping a

---

Pennsylvania Department of Agriculture Regulations Regarding Eggs:

"If an egg producer has fewer than 3,200 laying hens, sells eggs within five days from the date of lay, and sells eggs predominantly within a 100-mile radius of their production or processing facility, then the following summary of regulations will apply when selling eggs to the consumer. 1. All eggs must be maintained at 60°F or less from the time of gathering to the time of sale. This also applies to eggs sold at farmer markets or at roadside stands. 2. Each carton, flat, or container of eggs must be labeled with the producer's name and address, date of lay, statement of identity (eggs), net contents (in 3/16-inch-high letters), and "Keep Refrigerated." 3. If you do not weigh the eggs or if they are of mixed size, and you do not wish to assign a grade, they must be labeled as unclassified. You also must remove dirty, leaker, or loss eggs."

(from "Agricultural Alternatives: Small-Scale Egg Production [Organic and Nonorganic]," Penn State College of Agricultural Sciences, 1999.)

DARRELL FREY

Barred Rock Hens at Three Sisters Farm.

friend catch a game rooster, it stabbed me in the shin with its long, sharp spur. It hurt for weeks. I have also seen young children cornered by a rooster guarding his territory. It is best to choose more docile birds on a small farm.)

When choosing among the various breeds it is important to acknowledge your goals and management plans. A few layer hens foraging about a woodlot and forest garden would need to be a sturdy breed with good foraging habits. A larger flock of confined birds should be a gentler breed.

We highly encourage all small farmers to adopt a variety or two of rare breeds of whatever animals they raise. The American Livestock Breeds Conservancy has prepared a conservation priority list for farm livestock in need of special efforts to preserve breeds (see albc-usa.org/). These are breeds that are no longer produced on a large scale. Rare breeds are listed as *Critical, Threatened, Watch,* or *Recovering.* "Critical" varieties of chickens may number fewer than 500 birds in the US. Threatened varieties have fewer than 1,000 breeding birds; those on the watch list, fewer than 5,000. Recovering breeds are breeds that were once listed but are now more widely raised.

Early in our permaculture studies, we worked with friends to develop a sweet corn-and-chicken rotation. Their 50 birds foraged and cleared out weeds in a yard measuring 50 by 30 feet, while an adjoining yard was planted with sweet corn. Each year, the flock was rotated to the other yard. We noticed that of their mixed flock of birds, the Black Australorps were the best natural foragers. They love to scratch and dig for bugs and worms. So we tried a flock of Black Australorps in the bioshelter. Although they were great layers, we found they were a little too active for the confines of our poultry yards. For our second breed, we chose the Barred Rock variety; they are a little more docile and better suited to the space provided.

Barred Rock is considered a recovering breed. According to the American Livestock Breeds Conservancy, the breed was developed around 1869 and was the most popular American breed until the 1940s. Known as a gentle bird, they are a hardy, dual-use bird. We replenish our flock with new chicks every four or five years, and allow most to grow old. We generally keep one or two roosters. We have eaten some of the chickens to thin the flock a bit; some have been killed by young dogs not yet trained; and most eventually die of natural causes. When they are several years old, the hens slow down their laying but are still quite active at converting weeds and garden waste to compost.

Recipe:

Occasionally we make a variation of Coq au Vin; it's a very old method for cooking old hens and roosters.

The birds are killed, cleaned, skinned, and cut and allowed to sit overnight in the refrigerator. The chicken is then covered in water with a cup or two of red wine and slowly simmered in a stewing pot. After a long, slow stewing, the meat is a little tenderer and the broth is dark and richly flavored. We usually cut the cooked chicken into small pieces, and return it to the pot with seasonal vegetables for a hearty stew, or added it to pasta or grain dishes.

Economics

The economics of raising chickens is an important consideration, though the issues are not always straightforward. Writing in the 1930s classic *Five Acres and Independence: A Handbook for Small Farm Management*, Maurice Grenville Kains suggests that a small market garden farm should always have a flock of chickens, but no more than 100 layers. More than this, Kains explains, requires so much attention that other aspects of the farm may be neglected.

A small-scale intensive farm will need to purchase much of the feed required for the birds and probably straw or hay for bedding. While good use of rotational pastures can reduce feed costs up to 20 percent (or even more in an intensively designed and managed system), the current high costs and the difficulty of locating a steady supply of organic feed can make the net income from egg production negligible. (This has been our experience so far at Three Sisters Farm.) However, farm fresh eggs bring in customers who almost always buy other farm products and who keep coming back for years. (For more about profit in raising birds, see the sidebar, "Design: Raising Meat Birds at Three Sisters Farm.")

Chickens' ability to transform farm waste, weeds and insects into eggs, meat and fertilizer, though hard to quantify in dollars, makes poultry a valuable asset on the farm. As discussed in previous chapters, locating the poultry room in the bioshelter allows us to harvest not just eggs and meat, but also heat and CO_2 from the chickens.

Poultry Habitat

Chickens require at least 2½ square feet of floor space per bird. Our poultry room is 175 square feet, so in theory we could house 70 birds. We prefer to allow 3 square feet (or more) per bird, making the maximum flock size 50 birds. For most of the year, except in extreme weather, they have access to at least 200 square feet of outdoor space as well.

Body Heat and CO_2

As previously discussed, our poultry room is ventilated by a blower. The blower pulls the warm, CO_2-enriched air from the poultry room through the east compost chamber and sends it directly to the deep growing beds. The daily application of fresh straw over the droppings keeps the air fresh and prevents the nitrogen from evaporating as ammonia. The poultry room is sealed from the rest of the bioshelter with a black plastic vapor barrier. The sealed room and ventilation further reduce any chance of foul odor emanating from the room.

Daily Care

We start the chicken's day with whole grain scratch. They also get large handfuls of comfrey and burdock leaves each morning from April through October. Tossing this to the outdoor yard gets the chickens all out of the poultry room. While they eat, we can spread straw over the floor and fill their indoor feed trough with egg mash and refresh their water. Chickens need their watering cans refilled each day. Fresh straw is added to the nest boxes as needed. Eggs are gathered in the early morning, around noon, and again in the late afternoon.

The poultry room has two doors, an entrance from the bioshelter's greenhouse and an entrance from the storage barn. Feed and straw are stored in the barn, and the bedding is usually cleaned out through this door. The birds also have a small door leading to the outside chicken yard.

Generally, one nest box is required for every four layers. Our ten boxes have served our flock, which has averaged about 40 layers. Roosts are provided because the birds prefer to sleep off the floor. Altogether, we have 40 feet of roost length, allowing about 10 to 12 inches per bird. The roosts are set up to allow for easy cleanout between them.

Chickens are never allowed in the greenhouse proper and are almost never allowed to leave their yards to wander the gardens. Their habit of scratching in the mulch and eating crops makes them as unwelcome in the gardens as woodchucks and rabbits.

Outdoor access is required for healthy birds. They are self-regulating. They go outdoors in the early morning to hunt for the worms and slugs attracted to the deep mulch in the chicken yard; at noon they retreat inside or seek shade in their forage yard to escape the hot sun. In the late afternoon and evening, they come back out to forage weeds added during the day.

During the summer and fall, the chickens are rotated between two larger pastures connected to the main yard. Each pasture measures about 500 square feet. The south poultry area includes a shady grove of black locust trees with an understory of comfrey. The east poultry area includes a hop vine trellised along the south fence for shade. Comfrey is planted along the north edge of the fence for toss-over feed. When one pasture is occupied by the chickens, the other grows up with grasses and edible weeds.

Life Cycle

Chickens are purchased as day-old birds. These arrive (shipped overnight) from the hatchery having never been fed. This is very important for an organic flock. If they have ever had non-organic rations, they will not be acceptable for organic production of eggs or meat. (We

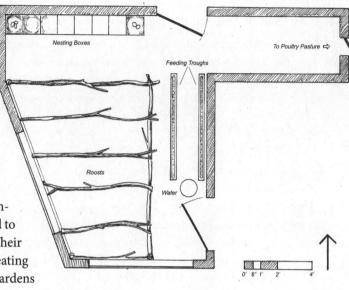

Poultry room in Three Sisters Bioshelter.

Poultry Yard Design and Rotating Yards

Writing in an introduction to Rodale Press's condensed version of Charles's Darwin's *The Formation of Vegetable Mould through the Action of Earthworms,* J.I. Rodale describes the poultry system at the Rodale Homestead, in Emmaus, Pennsylvania. Sixty birds had access from the poultry house to two outdoor forage yards. Each yard measured 300 square feet. Chickens were allowed access to one yard for a week and then rotated to the other yard the second week. During the week without chickens, the earthworms moved in to feast on chicken dropping. The next week, the chickens would scratch for and feast on the earthworms. Rodale assisted the chickens by turning the soil in the beginning. Later he discovered that a deep mulch would keep the soil moist and attract more worms — and make the chickens' worm hunting easier.

occasionally incubate our own eggs to renew our flock. It's a 21-day process during which the incubator temperature need to be monitored and the eggs turned twice daily.)

Young birds need to be kept warm with a brooder lamp. For their first two weeks, we keep the chicks in a large box heated by a hanging lamp. When they outgrow the box, we move them to a brood cage in the chicken room. They need to be protected from cats, rats and weasels and anything else that will eat a small bird (including older chickens!). Young chickens will fly out of their yard, so they need to have the first ten wing feathers of one wing trimmed to prevent this. This does not hurt the chicken — it's like a haircut. After they are a year old, they are less likely to fly over the 4-foot fence, so the wing feather trim is unnecessary.

Protection

Chickens need to be well fenced to keep them where you want them. The fence needs to be strong enough to prevent wandering dogs, foxes and coyotes from getting in and attacking the birds. Raccoons and weasels will kill chickens, and both can kill many in one night. A good fence and a door on the entrance to the henhouse will help. A well-trained farm dog is also a good defense against henhouse raiders.

Control

Many farms, especially Amish farms in our region, have a small flock of chickens roaming the place and gleaning what they find to supplement their rations. This is fine on an isolated rural farm, but does not always work for a small-scale intensive farm. Chickens love to scratch in mulch, garden beds, compost piles and any other place they might find something to eat. They can quickly make a mess of a well-tended landscape and will eat prized plants. As with any animal, controlled access to pastures, garden spaces and forage yards is the best way to integrate them into a system. Good fencing is a start.

Feed

Egg-producing chickens are usually fed two types of feed: scratch grain (includes cracked corn and another grain, usually oats) and layer mash.

Purchased feed is probably difficult to avoid for a small-scale farm. The equipment and space required to grow and process a mix of grains is beyond the budget of most small farms. Storage bins for a year's supply of feed would also be required.

Organically grown poultry feed is expensive and often difficult to come by. We recommend small farms develop relationships with regional organic grain growers to assure a steady supply of grain. Grains begin to lose vitamins after they are ground into feed, so finding a source for fresh-milled feed is important.

The mix of feeds should provide the proper ration of carbohydrates, protein and vitamins and minerals. Vitamins and minerals can come mostly from pulverized rock minerals added to the feed and green matter fed to the birds. Most egg mash has a protein content of 16 to 20 percent.

Our flock has done well for many years on purchased scratch grains and a homemade mix of milled grains. We purchase excess flour from an organic mill and mix the various types, including whole wheat, oats, soy, spelts, rice and rye flours. To 300 pounds of mixed milled grains, we add several cups each of rock lime, rock phosphate, greensand and kelp meal. (These are our potting soil ingredients as well, so they are always handy.) We also add extra kelp and organic barley grass meal in the winter. Feed is supplemented daily with garden weeds and comfrey spring through fall, and with weeds and cutting greens trays in the winter.

We do not analyze each batch of feed. We rely on our blend of many grains and soy flour to provide proper nutrition. It seems to be working. Our birds are healthy and productive into their old age.

Forage

The primary considerations for planning a permaculture poultry systems are described by Susana Kaye Lein in her article, "A Self-Forage

System for Chickens," in *Permaculture Activist* magazine. Chickens eat a wide range of plants. They will almost certainly eat any plant people eat and many more we will not. Their forage area needs to provide a diversity of food plants and areas to scratch about for worms and insects, as well as small stones and grit for their gizzard. Given suitable habitat, stocking density, and the proper plant food at the right time, chickens can thrive in a forage system. When possible, poultry forage incorporated into forest gardens and orchards encourages the birds to fertilize the trees and shrubs as they forage. Poultry pastured with other livestock will hunt insects and fly larvae in manure and glean wasted feed. The chickens' habit of scratching in the mulch and leaf litter for insects, worms, slugs, seeds and other foods is an important behavior to account for when planning a forage system.

Laying hens require a diet that is 16 percent protein. Meat birds require 20 percent protein for fast growth. Insects and worms are the best source of protein, followed by seeds and grains, and protein-rich forage plants. Plants also provide carbohydrates, vitamins and minerals. Layers, especially, need calcium. Forage areas should be given the standard minerals, rock phosphate and crushed limestone as needed for proper soil pH. Limestone will interact with the soil's organic matter to make calcium available to the plants, and thus to the chickens that eat the plants.

Many of the wild edible plants we use at Three Sisters Farm are also our chickens' favorite foods. Chickweed, yellow dock, lambsquarters, amaranth, alfalfa, dandelion, quackgrass, and galinsoga are all relished by the birds. These are cut and fed to the chickens as we weed the gardens. These plants also self-seed in the forage yards between rotations. Burdock is allowed to grow along the tree line edge adjoining the compost yard. Comfrey is planted around the chicken yard to catch nutrient runoff. These are cut and fed to the chickens, especially on days we are not weeding the gardens. As the summer progresses, the main chicken yard develops a deep mulch that is prime forage ground for worms.

Seed crops such as sunflowers, amaranth and buckwheat can be purposely introduced into pastures for even more varied forage. Beware, though. Plants that chickens do *not* eat can take over a forage yard. Sweet Annie and goldenrod are two plants they avoid, so it is best to keep an eye out for those and do some hand weeding if necessary.

When we get large populations of slugs or snails, we gather them and feed them to the chickens, who can't get enough of them. Pill bugs trapped in beer traps are also eagerly consumed.

Moveable Henhouses, Chickens in Polytunnels and Chicken Tractors

We have not had space or time to tend chickens in moveable shelters, but some farmers put them to excellent use. There are a number of variations. Generally, a moveable chicken shelter is moved daily to give the birds fresh grass. An Amish farmer we know uses a team of horses to move his henhouse seasonally. This allows the birds to forage and manure his fallow fields. Gardens and poultry pastures are rotated annually.

On other farms, hens winter in polytunnels or are allowed seasonal access to tunnels to clean up crops and add manure. Poultry can do a great job cleaning up weeds and pests and fertilizing a hoop house. However, from a food safety perspective, I repeat: *Follow organic guidelines.* Wait 180 days after direct application of manure to harvest any crops in contact with the soil; 90 days for crops not in contact with the soil. This is necessary to prevent pathogens from the manure contaminating the crop.

A chicken tractor is a small moveable coop designed to cover a garden bed. Chickens are left in one spot for several days to a week to eat weeds and crop residue and dig up the soil.

Design and management strategies should allow for fresh feed and water each day. There are many system options to experiment with.

Zone two orchard at the Robert A. Macoskey Center for Sustainable Systems Research and Education at Slippery Rock University, with movable poultry yard.

JOHN WHEELER

Design: Raising Meat Birds at Three Sisters Farm

During a permaculture course at Three Sisters Farm led by Susana Lein, a permaculturist and poultry forage researcher, participants developed a design plan for us to raise meat birds in a pasture system. The accompanying map illustrates the plan.

The design course participants were given the task of designing a poultry production system for several hundred meat birds. We first needed a design for a warm and predator-proof brooder room for the chicks. The participants recommended a new seasonal henhouse be constructed on the east end of the existing bale-shelter greenhouse. The layer flock would be moved there each spring, providing them with fresh pasture and allowing the bioshelter's poultry room to be used as a brooder house for purchased chicks. The layer flock will eat the fallen mulberries and clean up and fertilize the bale-shelter after spring production is over.

After the layer hens move out, the poultry room will be cleaned of bedding, hosed down, allowed to dry, and strewn with fresh straw bedding. Cages will house the day-old, purchased birds until they are three or four weeks old. When old enough to be let outside, they will have access to the forage area adjoining the bioshelter.

At about six to eight weeks old, the chicks will be moved through a series of eight fenced paddocks. Every two adjoining paddocks will have a small shelter to provide cover from rain. Paddocks will each be 750 square feet; the size will be adjustable, however, because we will be using temporary fencing.

Day-old chicks will be ordered May 1. By late June, the chicks will have been transferred to the larger paddock areas. On July 1, more chicks will be ordered and put in the brooder.

This method will produce two saleable flocks per year. This size of the flock can be adjusted up or down to meet market demand. After the second flock moves into the paddock areas, the egg layers can be shifted back to the bioshelter anytime. As is, the system is sized to produce up to 250 birds twice a year. They would be ready to be butchered and sold August 1 and October 1.

Grain Production

The paddocks will be planted each fall with winter grains, such as wheat, oats and rye, and undersown with field peas, mustards and other fast-growing spring forage in March. These will be cut for the birds to pick through. As the first batch of birds move through the paddocks, other, faster growing summer grains, such as millet and buckwheat, and sunflowers will be sown for the second batch of birds. Supplemental grains and plant food will be given to the birds as needed to keep them growing fast.

Trees, Shrubs, Vines, and Cover Crops for Forage

The course participants created a list of perennial plants to grow in and around the paddocks for supplemental feed. As in any good permaculture design, plants that could serve multiple functions were selected. Some fruit in early summer (May–July), others in late summer (July–September). Size at maturity, shading capabilities, and nutritional value were also criteria taken into consideration. Small trees and shrubs will provide the birds shade from the midday sun as well as forage. Comfrey will be included in these paddocks because it is a fast-growing, nutritious feed that the chickens love.

The plans call for the following plantings (see map): *Elaeagnus* (Goumi) plants that produce berries; a long hedge of Siberian pea shrub (*Caragana arborescens*), a legume with high-protein pea pods; black locust ☞

trees; and honey locust trees. Pasture plantings in the plan include clover, alfalfa, wheat, buckwheat and millet, amaranth, lambsquarters, dandelion, kale, and sunflowers.

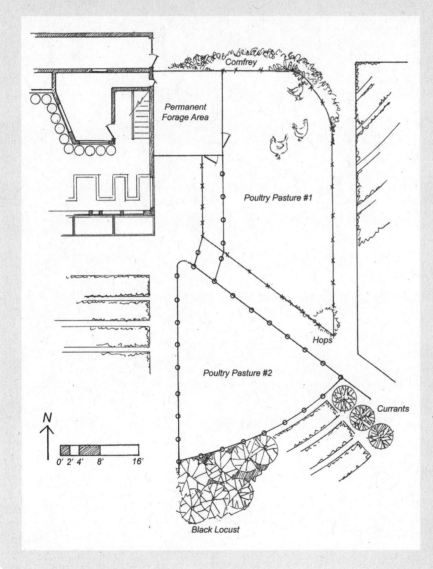

Poultry forage areas adjoining Three Sisters Bioshelter.

THANKS TO SUSANA LEIN, JAY TATARA, CRAIG KALLOCH, DIANE TRETTER, BEN DEVERIES, MARY ANGELINE DEKE RITTER AND ERIN WEBB

CHAPTER 13

Permaculture for Wetlands

W ETLANDS ARE AN IMPORTANT PART OF THE NATURAL ENVI-
RONMENT. They benefit the agricultural landscape by adding
ecological diversity and stability, recharging aquifers, enhancing edge
habitat and functions, and attracting a diversity of wildlife. The capacity
of wetland ecosystems to absorb nutrients and remove pathogens from
water makes them useful in treating wastewater.

The distinctive combination of soil and shallow water make wetlands
among the most productive landforms that exist (they incorporate a
high ratio of solar energy as related to their size). Yet, they are generally
an untapped resource, inappropriate to the needs of conventional agriculture. In fact, conventional agriculture has tended to regard wetlands
as a problem rather than a resource.

As with anything in permaculture, it is important to first understand a system before designing for it. So when considering the wetland's
role in a permaculture landscape, the first step is observation. What
are the natural patterns and processes? What are the wetland's primary functions? Only after questions like these are answered, can it
be determined how to integrate a wetland into a larger agricultural
system.

Restoration and creation of wetlands, rain gardens, and ponds not
only re-engage natural processes in exchange for artificial agriculture,
they can provide a small-scale farm with valuable harvests.

This chapter explores the issues related to ponds, rain gardens and constructed wetlands in the sustainable landscape. A wetland species list is included.

Lost Wetlands

Of the estimated original 127 million acres of wetland in the US, 52 million acres have been drained, filled, or flooded for development of various kinds. Lost wetlands mean lost habitat for plants and animals and lost functionality of wetland hydrology and ecology.

Throughout northwestern Pennsylvania, thousands of miles of drainpipe drain thousands of acres of wet soil and thousands of ephemeral pools. Only in 1983 did state and federal regulations begin to restrict the ways people can change the basic characteristics of wet soils and wetlands. Wetlands can still be used for agricultural and forestry purposes, but no new drainage (or fill) can be installed without permits. Only drains installed in fields prior to 1983 can legally be repaired.

Care of the Earth

While regulations may allow the cultivation of wetlands, with so many wetlands having already been altered, we must be careful before making changes to existing natural systems. We know much more about ecosystems than we used to — enough to recognize the need for humility and respect when dealing with natural systems. To take the example of Pennsylvania's wetlands in particular, consider that they have been evolving a mix of species and functioning as habitat since the region's glacial ice sheets receded between 10,000 and 12,000 years ago. Further south, wetlands may be far older. Many of these slowly evolved habitats and accompanying species have become damaged, endangered or extinct in the name of progress. Of major concern when dealing with wetlands is the preservation of rare or endangered species and habitats.

Simply having natural habitat on the permaculture landscapes contributes to the biodiversity, and thus the ecological health of the system. Natural wetlands are especially important as breeding grounds for amphibians which then disperse across the landscape to prey upon insects. As part of a productive, small-scale farm we can try to improve the productivity of these natural systems in order to harvest their products. For example, we can introduce more productive varieties of naturally occurring wetland plants such as swamp rosehips and swamp blueberries.

Classes of Wetlands

Natural wetlands are classified by three components: water, soils and plants. Wetland soils are *hydric* soils, which are soils that stay wet long enough each year to be low in oxygen. Soils low in oxygen tend to have low levels of minerals (carbon, nitrogen, iron, manganese and sulfur, in particular) chemically bound to soil particles, which means they can only support plants that do not need to get such minerals from soil. Wetlands are often "muck soils" — soils that are wet and high in organic matter.

Swamps are areas of standing water in which trees and shrubs grow. Although people often associate swamps with more tropical climates, this is incorrect. There are many swamps in northwestern Pennsylvania.

Marshes are shallow-water grasslands that contain a variety of grasses, sedges, reeds, cattails, and many other forbs (flowering plants not in the grass family) adapted to the sunny, wet habitat with standing water.

Fens are a type of peat land. Peat is a deep accumulation of dead plant material. Fens are generally high in pH; the water in a fen comes from seeping or flowing groundwater.

Bogs are peat lands that get most of their water from rain. They are often old lakes or ponds that have filled with accumulated organic matter. They tend to have low pH and be low in nutrients.

Vernal pools are seasonal wetlands. In northwest Pennsylvania, vernal pools are generally wet or flooded in the spring and dry during the summer. They are important habitat for turtles, frogs, salamanders and birds.

Plants

Wetland plants are generally classed according to their environment and/or their position within it: swamp, marsh, bog or shoreline; and submerged, free-floating or emergent (growing partially in the water and partially above it).

Swamp and marsh plants grow in water one foot deep or less. These include cattails, reeds, iris, calamus, sedges, rushes, other grasses, horsetails, smartweeds, mints, ferns, forbs, and shrubs. At the edges of a swamp, a succession moves from herbs (groundnuts, nut sedges, mints, monarda, angelica) to shrubs (elder, alder, willow, viburnum, swamp rose, blueberry) to trees (pepperidge, swamp white oak, ash, sycamore, wild allspice).

Bog plants range from herbs to trees and include cranberries, bog rosemary, rushes, sphagnum and other mosses, groundnuts, lichen, alders, bayberry, swamp rose, skunk cabbage, monkshood, winterberry, chokeberry, hemlock, birch and ash.

Lake and ponds offer shoreline plant habitat for emergent plants, such as bulrush, spike rush and cattails, and floating-leafed plants such as pond weed. Submerged plants such as coon tail and water milfoil grow near the shoreline. But the shoreline is only one pond habitat. On a pond's surface there may be free-floating duckweeds, algae, micro-organisms and insects. In shallow water (1 to 5 feet deep), there are emergent plants and plants with floating leaves, such as lotus, water lilies, pondweeds and arrowheads. These plants can form dense stands and provide valuable habitat and food for fish, crustaceans, amphibians, insects and microorganisms. Relatively few plants grow in water over 5 feet deep, but some do, including wild celery, chara (also called muskgrass or skunkweed) and elodea. A pond bottom, though, is not lifeless; worms, snails, clams, crayfish, turtles and fish forage a complex system of detritus, algae and microorganisms.

Other shoreline habitats are found on stream banks and in riparian zones (the floodplain along rivers, lakes and streams). Here, wetland plant systems are often dominated by trees and shrubs such as alder, willow, elder and various forbs. Stream plants grow along the shoreline, in quiet, protected areas and in shallow water where silt can collect. Many of the same plants that grow on a pond edge grow on stream banks. Factors limiting plant growth in streams include flow rate, oxygen level, temperature, bed composition, amount of organic matter, water clarity, pH and pollution levels.

Some plants are only at home in or near cold spring water. Among these are watercress, calamus, cowslips, some mints, peppercress, saxifrage and Jacob's ladder.

Basic Design Procedures

As with any landform, basic design procedures should be followed. Foremost among these is observation. Determine the nature of the wetland. Does the water level, clarity, oxygen content or algae population fluctuate? What species are present and what interactions can be observed? List existing functions and products. Avoid making major changes suddenly. Develop existing resources before adding new elements to the system. Introduce new species or varieties that can provide multiple functions and products. For example, a stand of willow on the southern edge of a pond can shade and cool the water preventing excess algae or duckweed growth, while also improving habitat for insects, frogs and birds. Willow provides food or tea (its young leaves are rich in vitamin C), medicinal bark and buds, pollen and basketry material.

Design for Wetlands

Whether designing with a pond, swamp, marsh, stream bank, ditch or just a bit of wet ground, wetlands and wetland plants can provide many products and perform valuable functions in the total design. These products include food for humans and animals, fiber, paper, thatching, craft materials and medicinal plants. Functions include bee forage, water purification, nitrogen fixation, biomass accumulation, and erosion control. Wetlands provide habitat and food for fish, waterfowl and other birds, furbearers, larger animals, amphibians and insects.

Regulations and good ecology dictate that the character of a wetland cannot be changed by drainage. However, wetlands can be managed for improved productivity. Permits may be required for extensive changes.

Rain Gardens

A rain garden is a garden or landscape feature that retains or directs rainwater or manages infiltration into the ground. Many urban and suburban areas are promoting rain gardens as a means of keeping storm water out of sewage systems. This is an excellent use of resources. Cities and towns often have "combined" sewage systems that capture both wastewater and rain runoff. Even a small amount of rain can cause a municipal sewage plant to overflow and discharge untreated sewage into streams and rivers and basements.

A rain garden is designed and sized to capture the rain coming off a roof or parking lot. They may be underlain with gravel beds to increase retention. Other systems include cisterns that collect rain and slowly release it into the garden. The amount of water collected during the

DARRELL FREY

Joe Pye weed thrives in a rain garden at Three Sisters Farm.

Arrowhead's starchy tubers are eaten by humans and wildlife.

growing season determines the size of slow-release cisterns and in-ground catchment basins.

Rain gardens provide the opportunity to add useful wetland plants to the landscape. The type of plants used in a rain garden depends on how quickly water drains from them into the ground beneath. They can be as simple as a swale built on a contour line, or they may be drip-irrigated, lowered beds of any shape. Regional species that are both beautiful and useful include cardinal flower, great blue lobelia, iris, swamp milkweed and butterfly weed, jewelweed, ironweed, Joe Pye weed and wild bergamot.

A rain garden at Three Sisters Farm collects rain runoff from the driveway. The wet spot was seeded by nature and maintains itself. Each year the succession of flowers, jewelweed, milkweed, Joe Pye weed, and ironweed provides a full summer of bee and butterfly forage.

Springs

Sunny, open springs on a gradual slope can be developed into a series of pools that are habitat for watercress, calamus, cowslips, mints, peppercress, and other food plants. (Note that because these are rather delicate ecosystems, hand weeding may be necessary to keep weeds in check.)

The development of springs for livestock or household water supplies is, of course, an ancient practice. We designed a spring development project for livestock water for a farm in Butler County, Pennsylvania. Our plan called for a small pond to be dug in order to increase the volume of collected water; from the pond, water traveled through gravity-fed waterlines to rotational cattle pasture paddocks. Another project at this farm was to direct seasonal runoff water into swales between garden beds to provide natural irrigation to watercress, mints and other water-loving crops.

Constructed Wetlands

Constructed wetlands use the biological functions of microorganisms, wetland plants and the environment to filter and remove nutrients from wastewater. Aquatic and wetland plants and microorganisms, such as duckweed, cattails, watercress, rushes, algae and bacteria have the ability to purify wastewater and remove heavy metals from water and soil. Systems can be designed for black water, which includes toilet water, or for gray water, which does not contain water from toilets. Constructed wetlands are also used to treat runoff from mining operations.

There are two main types of constructed wetlands: surface flow and subsurface flow systems. *Surface flow systems* use microorganisms in the water to reduce pathogens and use wetland plants to remove nutrients from the water. *Subsurface flow systems* rely on the complex system of plant roots and microorganisms in the soil to do the same job.

There is a third type that is truly constructed: The Living Machine system is a trademarked system that uses aquatic ecosystems to treat wastewater as it passes through a series of translucent tanks and small subsurface wetlands. You can learn more about this system at www.livingmachines.com/.

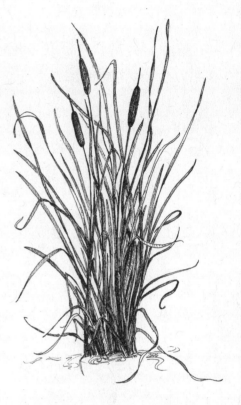

Cattails provide food, fiber and craft materials.

The Constructed Wetland at Slippery Rock

We installed a constructed wetland at Slippery Rock University's Robert A. Macoskey Center for Sustainable Systems Research and Education. The objective was to treat gray water from the center's office building. This project was part of a test of these systems being conducted by the Pennsylvania Department of Environmental Protection (PA DEP).

We began with a pond liner of local clay supplemented with bentonite clay. Many constructed wetlands are designed with liners made of heavy vinyl or EPDM

Gray Water

Caution: For crops that are intended for sale, great care must be taken to avoid contact between the crops and untreated gray water. Because of health concerns, we do not recommend the use of gray water for commercial production gardens. However, grey water passed through constructed wetlands *can* provide habitat for many living beings; such systems can also supply craft materials, such as reeds for basket making. Using subsurface gray water to irrigate tree crops is a possibility, but more research is needed before all health concerns are met.

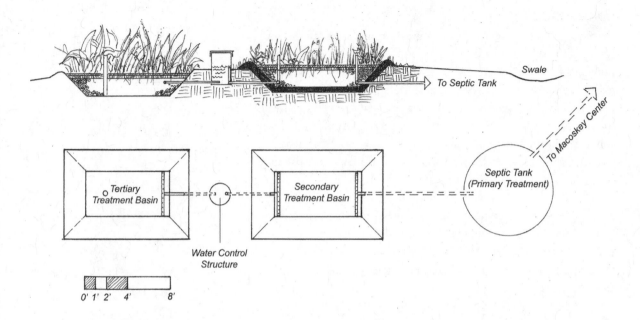

Tertiary
Treatment Basin

Secondary
Treatment Basin

Septic Tank
(Primary Treatment)

To Septic Tank

Swale

To Macoskey Center

Water Control
Structure

0' 1' 2' 4'   8'

The constructed wetland for waste water treatment at the Robert A. Macoskey Center for Sustainable Systems Research and Education, at Slippery Rock University treats grey water from the center's main building. Water leaving the center's sinks, showers and clothes washer enters a septic tank. From there it passes through a gravel basin planted in wetland plants and on to a second basin where the water returns to the ground."

rubber, but these require extensive excavation if they need to be repaired after a rodent or muskrat digs through them. We chose clay because we wanted the wetland to be a geological feature rather than a technological feature. Bentonite clay was added to the local clay because it swells when it absorbs water, so it is self-sealing for small leaks. Larger leaks can be easily repaired by adding more clay.

System Design

The constructed wetland system has five components:

1. a 1,000-gallon septic tank
2. a 4 x 8 foot lined basin
3. a water-level control structure
4. a 4 x 8 foot unlined basin
5. a monitoring stand pipe

Septic Tank

The septic tank receives gray water from the office building. The 1,000-gallon capacity provides a 10-day retention of water, which allows the small amount of suspended solids and organic matter entering the system to settle and microorganisms to begin to breakdown nutrients. The tank is dual-chambered and includes a sediment trap. Tank chambers are accessible for monitoring through sealed and covered risers, tubes leading from the buried tank to the ground's surface.

First Basin

Water flows into the first basin from the septic tank through a 2-inch diameter PVC pipe. Water is distributed in the basin through a 4-foot long, 2-inch diameter PVC header pipe set in 2–4 inches of washed gravel across the upslope end of the basin. The header pipe was drilled with 3/8-inch holes, 6 inches apart on the top, bottom and sides of the pipe. Cleanout ports are attached to each end of header pipes.

The bottom of the 2-foot-deep basin (covering 48 square feet) is lined with clay so it will hold water. To make sure the local clay would have sufficient water-holding capacity when compacted we did a small test plot before construction. The on-site clay was augmented with clay from a gravel pit and sealed by tamping bentonite clay into the top 6 inches of the basin sides and berms (two pounds of bentonite per square foot). A nylon mesh and straw landscape cloth was placed over the clay lining to stabilize it. The basin was then filled with 18 inches of ½–3/8-inch gravel.

Wetland plants from other wetlands at the Macoskey Center were planted in the gravel and topped with a layer of 2 inches of soil and 3 inches of mulch. Water levels are maintained at 2 inches below the gravel surface. Berms surrounding the basin rise at least 6 inches above gravel level to prevent overflow. These are landscaped to prevent rain runoff from entering the basin. A small spillway leads from the first basin to the second in the case of a major storm event.

Water Control Structure

The water level in the first basin is controlled as needed with the covered control structure placed in between the basins. Water leaves the first basin through a 2-inch header pipe (with vertical cleanouts), and gravity flows through a 2-inch pipe to a flexible pipe chained to the inside of the control structure. The water level in the first basin is adjusted by raising and lowering the flexible pipe. Water leaves the control structure through a 2-inch pipe to the second basin and is distributed through a 4-foot long header pipe. This 2-inch PVC header pipe lies horizontally across the basin and is drilled with 3/8-inch holes spaced 6 inches apart on the top, bottom and sides of the pipe. This header also includes vertical cleanout pipes.

Second Basin

The second basin is identical to the first, with the exception that the second basin is not lined and has only an inlet header (2-inch). The second basin serves as the drain field and is constructed above existing soil fragipan. This basin has a 4-inch monitoring pipe with 1/4-inch holes below the surface of the water to allow water to flow through.

Monitoring Standpipe

Outside the second basin is a capped, slotted, 3-inch schedule 40 PVC monitoring standpipe. The pipe is set 10 feet northeast (down slope) of the basin and set to the depth of the fragipan layer. The top several inches of the space outside of the pipe were back filled with sand and sealed with several inches of bentonite clay.

Planting the Basins

The first basin is planted with emergent wetland plants already present near the site. These include bulrushes, sedges sensitive fern, water horehound and arrowheads. These plants provide secondary treatment, "scrubbing" the wastewater of nutrients and pathogens. The second basin will be planted with a mix of seasonal wetland plants. These plants provide a tertiary treatment or "polishing" of the wastewater. Trees and shrubs are kept away from the site. Berms and areas around the monitor pipes are mowed to allow easy access.

Early results have shown that the system is working well at the Macoskey Center, but similar wetlands in the region have tested high in biological

Beaver Dams and Water Quality

Beaver, widely reintroduced early in the 20th century, have made a major comeback in the last several decades. The beaver's dams are again providing habitat for many species and improving water quality in the streams they inhabit. Beaver dams have been shown to reduce the impact of acid rain in the Appalachians by increasing pH of the streams. Many landowners believe beavers to be a nuisance because of tree damage and flooded fields. But from an ecological point of view their value exceeds the potential damage. Therefore many states have regulations protecting beavers and their dams.

oxygen demand (BOD). BOD is related to nutrient levels. Larger systems tend to allow absorption of nutrients.

Ponds

Pond Design

Pond design needs careful consideration for safety, efficiency and productivity. Ponds are usually built for multiple uses, including recreation (swimming and fishing), irrigation, wildlife habitat, fire safety, and aesthetics. In the past, and likely in the future, ponds were a source of ice, which was stored in an icehouse for summer use.

Zack Frey, Chris Frey, Terra Frey and Travis Rostron harvest pond weeds for compost.

Pond construction may require permits. Check with the local extension office for permits and design services. A pond has several components and special design considerations.

Pond components:

- water source
- clay soil base
- dam
- standpipe drain
- overflow floodway
- sloped banks

The area that drains into a pond will dictate its size. (As described in more detail below, on our farm we had a spring feeding the pond, but also at least five acres of land that drained heavy rains and snowmelt toward it.) When a pond is excavated, topsoil is set aside to finish the dam and for other uses.

The dam should be wide and high enough to hold the water and not be undermined by muskrats. As the dam is built up from material excavated to create the pond, it is compacted with the bulldozer. The standpipe is set upright in the pond and runs under the dam to drain the pond. The height of the standpipe determines the water level. A standpipe can clog and needs to be accessible, at least by boat, for cleaning as needed.

Because water overflowing a dam can quickly erode and undermine it, a pond's design needs to include a spillway large enough to accommodate overflow during extreme weather. The overflow spillway is a wide swale that leads water safely around the dam. Generally, a spillway should be no more than a foot higher than the intended water level and kept grassy and free of trees and debris that could impede water flow.

The sides of the pond will tend to slump back to the angle natural for the material being used. Generally, this will not be more than 2 or 3 feet of drop per horizontal foot (a 2:1 or 3:1 slope). Steeper sides need to be stabilized with stone to slow the slumping. Slumping is exacerbated by muskrats, who love to dig into dams and pond shorelines. Muskrats can be controlled by a diligent hunter and trapper, but it is impractical to think they can be eliminated.

Maximize Edge

When designing a pond, it may be desirable to maximize the edge (shoreline) because that is the area that tends to be the most nutrient-rich and hospitable habitat. One way to increase the edge is by creating islands and peninsulas by digging holes and using the muck to create hummocks. This added edge can be planted with seeds or cuttings of existing species of value. In natural lakes and ponds, muskrats build such hummocks for their homes. These are then colonized by shrubs and other plants. Non-rampant species can be added to these constructed hummocks. Plants that thrive in this habitat include ostrich fern, elder, swamp rose, swamp blueberry, nannyberry, and bayberry.

The holes left after digging are habitat for emergent plants (especially arrowheads and water lilies) and frogs and other creatures. Emergent plants can be planted in submerged containers of soil for ease of tuber harvest and for control. This is an especially good strategy to use in new ponds of low fertility.

Deep, shaded pools and communities of submerged plants are good fish habitat. To encourage and maintain waterfowl populations, create shallow water communities of cattails, wild rice, reeds or bulrushes for cover and food; arrowhead, burr-reed, water plantain and associated insects and microorganisms are also good food sources. In deeper water, pondweeds, water lilies, lotus, duckweeds and others plants provide waterfowl forage. Islands provide safety for ducklings, as well as a good spot for growing potentially invasive species such as bamboo or Japanese knotweed. Shorelines can be planted with buttonbush, groundnut, nut sedge (chufa) and other waterfowl forage. Adjoining uplands can be planted with grains and legumes, mulberries and acorns to supplement waterfowl diet. Weeder geese can be integrated with agricultural systems.

Ponds and lakes, like any ecosystem, have a natural succession. A young pond is low in organic matter and nutrients. Pioneer lifeforms move in and proliferate, drawing in higher animals and plants. Gradually, nutrient levels and diversity increase. A mature pond accumulates organic matter on the bottom. Eventually, silt can fill in a pond or shallow lake, creating a swamp, a marsh or a bog. After that, trees move in, building up soil until what was once a pond becomes a wetland. By removing silt from ponds, one can gain valuable soil amendments and maintain an existing stage of succession.

Pond-like Areas

Drainage ditches, seasonal wet spots and stream banks can be planted with suitable pond bank, swamp or marsh plants. Swales and streams can be used to collect silt, decaying leaves and sand, which are excellent additions to the compost pile. Simple stone or log dams are good sediment traps.

The Pond at Three Sisters Farm and Nursery

The pond at Three Sisters Farm contains many species and serves a number of functions. It is one of several sources of irrigation water; its proximity to the west gardens and the amount of water it stores makes it important in the daily operation of the farm. It is a valuable addition to the farm ecosystem because it greatly enhances the diversity of life on the farm.

The pond was designed in 1990 with consultation from the local office of the Soil Conservation Service (SCS), now known as the Natural Resources Conservation Service. The SCS made a plan drawing for a teardrop-shaped pond. We adjusted it to accommodate a 6-foot wide and 50-foot long channel leading from the spring to the main pond. Our pond was initially mostly shallow, gradually sloping to 8 feet deep near the drainpipe. Later, we lowered the water level to 6 feet after our upstream neighbor determined the pond was making his yard too wet.

The shallow design limits the amount of water available for irrigation, not only because the pond simply holds too little water, but also because water for irrigation must be drawn from mid-level in a pond to get clear water. The top and the bottom layers of pond water may have debris and sediment that can clog intake filters and damage the pumps. Shallower water heats faster than deeper water; unfortunately, the warmer water encourages excess growth of undesirable plants, such as coontail and water milfoil.

A natural mix of plants that we *do* want has been allowed to grow around the pond, and we have added several species. A detailed list of pond vegetation is included below.

We planted the channel with watercress, where it grows in thick patches. Its dense root system absorbs excess nutrients from nearby agricultural fields. Because of this agricultural runoff, we do not eat the watercress, but we do occasionally harvest it to use in compost. Duckweed is a floating aquatic plant found growing abundantly in the stream channel and shallow areas of the pond. This plant has many functions: water purification, forage for ducks, chicken feed, and a supplement to compost. It can also serve as mosquito control; a thick growth in a pond will suppress mosquito larvae. Duckweed is fast growing (up to ten tons of it can be grown per acre per year), and it is protein rich (32-44%), which makes it an excellent livestock feed.

Ducks do, in fact, eat duckweed. We have not yet introduced ducks to the pond, but we plan to at some point so we can make better use of the pond's resources. I would like to try Cayuga ducks, a breed derived from the native black duck.

Solar Irrigation

Our pond water irrigation system consists of a 12-volt pump powered by a photovoltaic (PV) panel. System components include a 125-watt PV panel; a panel mounting system; a linear current booster to regulate electricity flow; a medium flow, 12-volt pump; two 1,200-gallon water storage tanks; and miscellaneous fittings and connectors for water lines. A hose with a foot valve and filter leads from the pond to the pump. A second hose leads from the pump to the sprinkler. This pump, when turned on at 9 a.m., can fill a 1,200-gallon tank by 6 p.m. We most often use the system to water newly planted crops with a rotating sprinkler. When running at full speed, in full sun, it will cover an area 25 feet in diameter.

Recreation

We and our neighbors use the pond regularly for fishing in the spring and summer. The neighbor's fishing rights were written into the purchase agreement when we purchased the property in 1998 (after ten years of leasing the farm), so this use must be preserved. We used to swim in the pond, but the excess growth of water milfoil and coontail make swimming less pleasant than it used to be.

Wildlife

Many species of wildlife use the pond at Three Sisters Farm. This satisfies our personal interest in living amidst natural diversity. More importantly, the diversity the pond helps support is a living connection that links our farm ecosystem to the surrounding landscape of forests, fields, lakes, ponds and streams.

Bullhead catfish, bluegill, smallmouth bass, largemouth bass and yellow perch have been introduced into the pond. The bluegill, catfish and smallmouth bass have become established as constant residents. The rest have not.

Great blue herons, green herons, ducks, Canada geese, kingfisher, swallows, and other fowl visit regularly. Red-winged blackbirds nest among the cattail stalks. We have observed osprey fishing in the pond on several occasions.

Muskrats have occupied the farm pond since 1994, when they moved in. They quickly reduced — and nearly eliminated — the cattails that had established themselves (though these have since returned). The muskrats dig their houses into the dam and other banks of the pond, destabilizing them and reducing pond depth. They also dig up the pond bottom, which seems to encourage the spread of water milfoil and coontail. With absolutely no regard for our careful shoreline plantings, they build their large houses out of piles of leaves and organic matter. (It is tempting to harvest their houses and use them for compost.) Sometimes, they stuff plant debris into the drainpipe, which raises the water level until I remove the clog.

We may need to control the muskrats by hunting them. But I have seen how shrubs such as alder, winterberry, ferns and other plants eventually become established on these muskrat-created hummocks. So perhaps we will wait and see what happens.

Two species of turtle have been observed in our pond: snapping turtle and painted turtle. The amphibian population includes gray tree frog, spring peeper, American bullfrog, northern leopard frog and American toad.

Many insects and other invertebrates live there as well, feeding on aquatic plants and in turn feeding higher organisms. Dragonflies and damselflies that breed in the pond forage for mosquitoes and other insects over the whole farm.

Honeybees

The farm's beehives sit the north edge of the pond, away from common use areas. The bees benefit from the reflected light off the pond because it helps warm the hive in the morning. They also have easy access to water and are somewhat sheltered by the rise of land to the northwest. The benefits of their location near the pond is an example of the permaculture principle of *relative location,* which teaches us to place elements in the right relationship to each other. The bee yard is in Zone 3 — far from the center of action; no one needs access to that side of the pond except when tending the bees.

Proposed Structural Adjustments to Pond

In order to maximize the usefulness of the pond and further develop it as an example of permaculture design, we plan to make a number of changes. Our plans were partially developed during a permaculture course held at the farm; we also consulted the NCRS for their input.

Deepening the pond should improve all of its existing functions. A depth of over 5 feet will reduce the habitat of the water milfoil and coontail, which will make the pond more useful for fishing and swimming. The increased depth will provide increased water retention for irrigation, and the resulting cooler water temperatures will provide improved fish habitat.

Simply draining and bulldozing the pond would devastate the pond ecosystem, which has developed for nearly 15 years. To mediate the effects of our renovation on the plants and wildlife, we will need to capture fish, turtles, and other organisms and hold them somewhere while the pond bottom dries enough for the equipment to enter the pond basin and dredge out the bottom. Therefore we have developed the following plan:

First we will lower the water depth 3 feet by cutting the pond's drainpipe. Then we will use an excavator to remove and compost the accumulated muck from the upper end of the pond and the inlet chan-nel. Next we will make a new dam for the upper portion of the pond, creating a small pond measuring roughly 40 feet by 30 feet. This will be the temporary holding pond for the aquatic life. When we drain the main portion of the pond, we will harvest some fish and transfer other fish and aquatic life to the upper pond. The pond will be allowed to dry for several weeks or months to allow equipment access. When we begin excavation, we will stockpile some of the muck for soil improvement to the West Garden.

As we excavate deeper, we will stockpile the clay subsoil at the eastern edge of the pond. Some of it will be used for a future projects — farm structures built with straw-bale insulation, clay plaster, and cob walls. Some of the excavated material will be added to the northwest bank, creating a gentle transition into the surrounding meadow. Care will be taken to preserve the alders, willows, witch hazel, cattails, and other plants now present on the pond edge.

After the pond is deepened and pipes replaced, we will remove the upper pond dam, allowing the water and pond life to re-enter the main pond. We will use the temporary dam material to make a small island in the middle of the pond. The creation of an island is being considered for the several reasons. We are considering the possibility of integrat-ing domesticated waterfowl into the farm system. Providing an isolated area may make them more likely to nest there, which would help ensure the viability of the flock. The island would also provide an area for the purposeful introduction of useful invasive plant species because any spreading by root shoots would be controlled on an island. The most obvious possibility is bamboo, which would benefit from the proximity to water and provide a dense stand of cover to protect ducklings. And there are many uses for bamboo on the farm.

Due to the natural slope ratio of our pond soil, a structural foundation may be needed to maintain the steeper slope we desire. This would require truckloads of stone to be brought in. The amount of stone needed and the cost of such a project is still being researched. We are considering recycling (free) broken concrete blocks from a nearby manufacturer for this purpose.

After the island is built, we will return the water level in the pond to its previous level and the shoreline will remain largely the same. We will then continue to add new plants to the system. Willow species to this area would perform multiple functions: bank stabilization, shade, bee forage, and basket material. Introduction of more sedges and bulrush species would provide duck forage and habitat. We will include propagation areas for our native plant nursery.

Wetland Species List

The "Temperate Zone Wetland Plants" list is included in an appendix. It can be used by the permaculture designer as a tool for assessment of existing wetlands and as a guide for selecting species to add to wetland areas.

Wetland Bee Forage

Many of the hundreds of wildflowers that grow well in wet areas are foraged by bees for pollen and nectar. While most of these flowers are only minor sources of pollen, they are nonetheless important to the hive for building strong colonies in the spring and sustaining them between major nectar flows of other flowers.

Plants near pond at Three Sisters Farm, 2000

Blue vervain, *Verbena hastata*	Mallow, *Althaea officinalis*
Boneset, *Eupatorium perfoliatum*	Marsh skullcap, *Scutellaria epilobifolia*
Bugleweed, *Lycopus* spp.	
Cattail, *Typha angustifolia*	Milkweed, *Asclepias syriaca*
Dogbane, *Apocynum* spp.	Queen Ann's lace, *Daucus carota*
Elder, *Sambucus canadensis*	Red clover, *Trifolium pratense*
Groundnut, *Apios americana*	St. John's wort, *Hypericum perforatum*
Heal-all, *Prunella vulgaris*	
Ironweed, *Veronia noveboracensis*	Umbrella sedge, *Cyperus difformis*
Jewelweed, *Impatiens pallida*	Yarrow, *Achillea millifolium*
Joe Pye weed, *Eupatorium purpureum*	Yellow iris, *Iris psuedoacorus*

Chapter 14

Education on the Farm

AFTER FOOD PRODUCTION AND PROMOTION of biodiversity, the third major service a farm can provide to the community is education. Many aspects of a market garden farm offer learning opportunities. The development of these opportunities into programs for the public can significantly contribute to farm income and balance physical workloads with teaching. This chapter looks closely at some considerations for developing on-farm educational programs. The programs at Three Sisters farm are discussed, including our farm tours, programs for school students and teachers, internships and permaculture design courses. A profile of the Hershey Montessori Farm School demonstrates how educational programs can be fully integrated with a small farm.

The theory of *Biophilia*, developed by biologist E.O. Wilson, describes the deep-seated relationship between humans and living systems in nature. But the increasing urbanization of culture and physical isolation of youth from nature does not allow a place or time for children to connect with her. Only when children have regular contact with nature can they develop a personal relationship to the earth.

It is easy for an adult to overlook the deep importance of simple experience with the soil, seeds and growing things. But the garden is a gateway to nature for many children. Whenever I work with children in the greenhouse or gardens, they are eager to participate in the planting of seeds. Young people, when not diverted by peer pressure, have an

DARRELL FREY

Young Chris Frey shows off his harvest.

inherent interest in the source of food and a natural curiosity to learn about the natural world. Not every child will become a gardener, but most find the intricacies of nature fascinating when witnessed firsthand. And it seems reasonable that a child exposed to the notion of stewardship will more likely grow up to be a responsible world citizen.

Benefits and Costs

Tours, workshops and classes on the farm provide a number of benefits to a farm — and they come with their own costs. Fees and donations bring in cash income. Promotion of programs through advertising or word of mouth not only brings in participants for the educational programs, but those participants also turn into customers. Visitors inspired by a farm's good work are the best advertisement one can get.

Hosting educational programs on the farm, though, takes up time that could otherwise be devoted to producing crops. Deciding to host programs requires some of the same "extra" tending that making the move to on-farm sales does. When the public is expected, the time required to maintain the landscape is increased. The Zone 1 areas, especially, should be inviting and attractive. In any public access area, everything should be neat and organized. A well-maintained driveway, well-defined parking, and well-mulched and weeded gardens in Zone 1 make an important good first impression. Developing and maintaining teaching gardens takes even more time away from crop production and requires additional labor to keep things tidy. Farm tours need to be supported by interpretive information, such as small posters placed around the farm to explain solar design, the role of chickens on the farm, and the bioshelter's design (upgrading this information is an ongoing and time-consuming task that should be acknowledged when making plans). The payback — in sales and fees — needs to compensate the extra labor if the endeavor is to work.

Public bathrooms, picnic areas, small group seating, and space to address groups of several dozen people are all needed to accommodate visitors. When visitors arrive, clear signage should lead them to parking and an information kiosk. A simple roofed structure with a safety-glass covered display can provide a quick orientation to the farm. Well-developed walkways and gates should be maintained to provide safe and easy access to the public areas of the farm and restrict access to private areas.

Many small farms either partner with a non-profit organization or establish their own non-profit to manage and facilitate educational programs. Although either option adds another layer of management, this approach often allows access to funding and tax breaks that make the effort worthwhile. Non-profit organizations usually expect their staff to develop programs and write grants to cover costs and their salary. An experienced educator and grant writer can match the farm's resources with the community's needs by developing pertinent programs and associated funding sources. It is important for the farm to have clear objectives and procedures that allow the for-profit farm and non-profit educational programs to co-exist and thrive.

At some point, the farm must decide whether it is an education facility with gardens or a farm with educational programs. Both are needed in the sustainable community. We have made the business of farming the priority. Our education programs provide supplemental income while allowing us to share the lessons we are learning.

The personality and skills of the farm staff as educators helps determine the ultimate direction. Visitors will be more attentive and learn much more when they sense an underlying passion for plants and biodiversity from the educator. Actual learning about nature proceeds from the natural curiosity of a healthy mind. Such curiosity can be sparked by the true interest in the subject on the part of the educator.

Three Sisters Farm Education Programs

At Three Sisters Farm, we have developed several approaches to providing education on the farm. We offer group tours (for schools, families, teacher training programs and garden clubs); host farm interns; conduct gardening and permaculture workshops; and administer a permaculture design course once a year. We also have a consultation service for clients who need help designing gardens and planning sustainable land use.

Our objective in offering the tours is to allow the opportunity for close up and detailed study of the farm ecology and the farm's production systems and processes. In order to address — in a conscious manner — the adjustments required to develop a sustainable community, people need to be exposed to the issues and to positive solutions to problems. And they need to learn some details about the intricate web of life in which we live. Tours and presentations about the land and its occupants provide the visitor with a sense of the complexity and beauty of nature and the importance of preserving native habitats in the midst of a working farm.

Teaching Gardens at Three Sisters Farm

All of our gardens — and the spaces in between — are used in courses and tours to provide lessons in permaculture design, intensive market gardening, and biodiversity. We have established several gardens specifically to provide learning opportunities for visitors. These include the Square-foot Gardens, the Spiral Garden and the Native Plant Garden.

The Square-foot Gardens: Useful Herbs

The Square-foot Garden is a series of well-defined framed beds that are planted (mostly) with herbs. The beds demonstrate intensive gardening techniques to visitors while also providing marketable materials. Plants in this garden include several types of thyme, purple sage, variegated sage, chives, golden oregano, Greek oregano, marjoram, lavender, rosemary, calendula, anise hyssop, nasturtium, and bush basil.

The Spiral Garden

The Spiral Garden is planted with trees, vines, bushes and herbs in wide beds arranged in a spiral pattern. It is primarily used to educate about habitat and diversity on the farm. Sitting areas allow for contemplation and relaxation. The Spiral Garden demonstrates a number of permaculture principles and practices: the zone system, forest gardening, microclimate modification, windbreak dynamics, beneficial insect habitat, and useful plants.

Native Plant Garden

The native plant garden, located at the upper end of the Pond Garden, is home to a variety of native shade- and sun-loving plants. Although native plants are incorporated throughout the farm, their concentration in one garden makes this a good spot for teaching native plant identification.

Shade Walk in the Tree Line

A short path through the tree line near the native plant garden is planted with a selection of spring "ephemerals" and shade-loving plants: trillium, Solomon's seal, false Solomon's seal, wild geranium, mayapple and ferns among them. This area provides propagation material for us and examples of these often unfamiliar but beautiful plants for us to show visitors.

Solar Energy

We want to make sure our visitors leave with a better understanding of how we use solar energy on the farm. First and foremost, we try to impress on them that the source of all we produce comes from the most ancient of processes — photosynthesis. We have a display providing a simple explanation of photosynthesis in the bioshelter.

Also posted in the bioshelter is information on how we use passive solar heating. A diagram shows how solar energy and photovoltaic cells pump irrigation water. Having this small display explaining how the system works aids in presentations to tour groups and other visitors.

Niche and Habitat on the Farm

A poster illustrates the various habitat niches on the farm and in the surrounding landscape. This display is used to teach the value of incorporating ecological agriculture into the natural landscape. We stress the fact that habitat enhancement of the farm is ongoing as we continue naturalizing native plants into the farm landscape, especially near the pond, the tree line and the edges of cultivated areas.

Tour for School Groups

We have developed a farm tour that teachers can use to address our state's Academic Standards for Environment and Ecology for school children (defined by the Pennsylvania Department of Education, see www.pa3e.ws/). Key terms in the standards (in italics, below) are used to connect the lessons of the farm to the state standards:

The Farm Tour

The tour begins with a general discussion about the farm's impact on the environment:

Because Three Sisters Farm is designed to demonstrate principles of *sustainable agriculture,* we can show students how farmland can work with nature to produce abundant crops.

By *composting* manure, nutrients are *recycled* and conserved. Composting and building the gardens on contour prevent nutrient run-off and subsequent pollution of *groundwater* and streams.

The farm is managed to allow and encourage *biological diversity.* Many *niches* have been created for wildlife to flourish among the gardens. Birdhouses and food and habitat plantings encourage a large number of songbirds. Wildflowers and insectary plants create habitat for many bees and beneficial insects. These in turn help keep pests in balance. Spiders, snakes, bats and other predators also help keep pests in check. Milkweed and other "weed" plants provide habitat for the migratory monarch and other butterflies species. The farm irrigation pond is habitat to migratory waterfowl and home to a healthy abundance of *lentic organisms.* In summary, the farm is managed as an agricultural *ecosystem* with a positive impact on the surrounding environment.

The greenhouse at Three Sisters Farm is managed as an indoor ecosystem. A form of *integrated pest management* is used to control insect pests. Habitat plantings, physical traps, release of beneficial organisms and soap sprays are used to keep pests in balance.

Farm Energy Systems

This section of the tour relates to a number of standards relating to societies' energy use:

Solar energy and biothermal energy from compost and fuel wood are all *renewable resources.* Use of these for greenhouse heating eliminates the need for *non-renewable* fossil fuels. A proper environment is maintained for year-round plant growth. Again, the model of an ecosystem is used to explain the energy interactions among farm components. The chickens provide CO_2 and fertilizer for the plants, and the plants provide food and oxygen for the chickens. This creates a *closed loop* of producers and consumers.

Nutrient Management

This part of the tour examines composting procedures at the farm, and other fertility considerations:

Manure, straw and leaves are removed from the *waste stream* of the community and recycled into fertilizer. The use of compost as farm fertilizer is a *renewable resource* that can reduce the need for nitrogen fertilizers produced with natural gas, a non-renewable resource. Large-scale adoption of composting can keep nutrients in cycle and reduce use of fossil fuels, helping provide for *societal needs* in a sustainable way with a positive impact on the environment.

Botany

This section can be used to augment our state's standards for science curriculum. The intention is to give the student hands-on experience in plant propagation. Students learn about crop planning and production, including soil mix ingredients, propagation, plant feeding and plant care. Students often get to make cuttings of scented geraniums, which root easily. They take these home or back to the classroom. They also get to taste edible flowers.

Final Discussion

We like to focus on how the farm provides the fresh fruit and vegetables needed in a healthy diet:

The USDA recommends three to five servings a day of vegetables. Pennsylvania farmers can best meet society's need for fresh produce by choosing crops adapted to our climate and growing them in healthy living soils. Healthy soils are achieved through the use of soil-building techniques. Crop succession and rotations help keep gardens productive and reduce pests.

Permaculture Internships

Internships continue to be used in many professions to provide entry-level training and obtain low-wage workers. This is especially true in market gardening. Permaculture internships offer design course graduates and those seeking a deeper understanding of ecological design an opportunity for hands-on experience.

Internships have become an important part of the small farm labor force in the US. The website of ATTRA (Appropriate Technology Transfer to Rural Areas), a project of the National Center for Appropriate Technology (NCAT), lists hundreds of farms hosting interns. (The term "intern" is widely used interchangeably with the term "apprentice" in these listings. But there is a difference. Traditionally, an apprentice is a worker under a long-term training program, leading to becoming a journeyman and then a master of a craft. An intern is someone who has been educated in a field and is undergoing practical training on the job.)

Many farms depend on the low-wage labor of interns as part of their business plan. The goal of the intern is presumed to be to gain practical work experience in permaculture and organic market gardening. Therefore, a farm offering an internship carries a responsibility to provide the intern with a full picture of the occupation. In the case of market gardening, the occupation requires long hours of physical labor in various weather conditions. Much of the learning is in the doing.

Hosting an intern brings a responsibility to provide a learning environment. Managing a market garden and permaculture consultancy is demanding work. Some years, the combination of workload and personal goals limits our ability to provide a quality internship, so we do not offer them every year. When our schedule and an applicant's match, we have found hosting interns to be a good arrangement to further our farm development and fill out our workforce.

As we expand both our market gardening and our permaculture teaching and consultation business, interns will continue to have a role. We anticipate constructing intern housing in the next year or two. Our experience with interns has been positive. We will continue to build on this to provide a quality experience for the intern.

Apprentices

Our three children went through an informal apprenticeship from age 12 to 18. They were offered as much farm work as they wanted during these years. In the process, they each gained seven years of organic landscaping and market gardening experience. They have a great deal of experience with all aspects of the farm, including operating the bioshelter, intensive organic gardening, raising poultry, managing compost, propagating plants, maintaining perennial plantings, pond management and so forth.

They are knowledgeable about a wide range of plants and the wildlife around us. They learned the delivery route and interacted with customers. They have also had the benefit of being associated with many of our interns and other workers. Almost all the interns have befriended at least one of our children. Since most interns bring an environmental ethic and interest in permaculture and other schools of sustainable design, our children's learning about our work in applying permaculture to our farm has been broadened and reinforced by our interns.

Annual Permaculture Course

I have been inspired by noted educator Dr. David Orr's ideas about how a classroom or other educational setting should embody the lessons being taught. I have heard him say this at lectures: "Classrooms should embody the lessons they would teach." At the least, the setting should not contradict them. This is easy at Three Sisters Farm — the bioshelter and gardens are the primary classrooms.

Once a year, we administer a course in permaculture for adults. The course includes presentations by several guest speakers and co-teachers, and field trips to local sites that demonstrate lessons in sustainable design. Participants camp at the farm for two weeks of immersion into intensive market gardening and permaculture design — and some great meals. Participants range in age from 18 to 80, though most students are in their 30s. Educational backgrounds range from high school graduates to folks with master's and doctoral degrees. All have had information and insights to share.

We strive to make the attitudes and atmosphere of the class embody permaculture principles. We foster this by encouraging mutual respect (teacher is well prepared; students are on time; everyone is rested, well fed and ready to actively participate); promoting earth care (minimize paper use, arrange car pooling, use materials efficiently, and use recycled materials when possible); and building community (shared meals, class discussion, group design projects).

We study nutrient cycling and gardening, water use, energy management, sustainable forestry, beekeeping, green

Permaculture course.

building and other aspects of sustainable design. Lessons begin with permaculture concepts and both global and local ecology. Because it is a design course, hands-on activities include research, mapping and drawing.

Each course has a specific design project. By the fourth or fifth day the project is presented and participants develop designs working in small groups. On the last day of the course, participants present their designs. After several days of extra hours, shared computer time, networking and barter among participants, surprisingly good design work emerges.

These courses give me hope for a better world. Meeting and sharing our work with dedicated and thoughtful individuals who want to put permaculture design into practice is gratifying. The process of teaching, learning and working with course participants is part of our own ongoing learning about sustainability and ecology. We look forward to continuing these courses in the years ahead.

PROFILE

Hershey Montessori School

In the soul of the adolescent, great values are hidden, and in the minds of these boys and girls there lies all our hopes of future progress.

— Maria Montessori

No place demonstrates the value of the farm as an education facility better than Hershey Montessori School. Located near rural Huntsburg, Ohio, this 97-acre facility houses the adolescent program for Hershey Montessori School. The school's 50 students engage in a variety of farm and facility management activities; their curriculum is integrated into the seasonal cycle of farm activities. Together, the students and staff manage a micro-economy of school activities and farm enterprises.

Hershey Montessori School's design and development is intended to be a realization of the vision of Maria Montessori for the education of 12- to 18-year-old students. In her writing on education, Montessori stressed the adolescent's need to begin to integrate themselves into the world around them. She envisioned learning communities, where real life occupations would provide the context for multilevel learning to occur. Montessori named this community *Erdkinder,* from the German for *earth children.* She only provided an outline of the education model because she believed the details would be developed uniquely for each farm. Regardless of future career paths, the time spent at an Erdkinder school, Montessori believed, "would provide an opportunity to study civilization through its origin in agriculture. She suggested students should live away from home in the country, where they would learn to manage a hostel and

The school building at Hershey Montessori Farm School.

open a shop where sale of produce would bring in the fundamental mechanics of society, production and exchange on which economic life is based." (from www.montessori-ami.org/montessori/erdkinder.htm).

Building on Maria Montessori's vision of Erdkinder, the founders of Hershey Montessori School envisioned a farm school that would grow and develop as students and teachers explored ways to sustainably interact with the land and the local community. Debbie Guren, president of the Hershey Foundation, agreed to fund the adolescent program in 1996, after attending an International Montessori Adolescent Colloquium.

Today the school's facilities include the Farmhouse and the Community Center, two livestock barns, a poultry house and yard, a program barn with wood shop, pottery shop and craft work areas, bioshelter/science class room, gardens, pastures, apiary and a maple sugar house. A bed and breakfast cottage is attached to the house that serves as office and residence of the farm manager. Under the guidance of teachers and staff, students manage most of the school's enterprises. Students prepare and serve 22,000 meals each year.

In many ways, this school exemplifies environmental educator and author David Orr's premise that a school's buildings and management should embody the lessons the school would teach, becoming "a pedagogy of place." Environmental awareness and action inform all aspects of the school. Hershey Montessori School presents thoughtful design using natural materials, abundant natural light, energy conservation measures, native landscaping and an ethic of land stewardship through animal husbandry, intensive gardening and sustainable resource management. Linen napkins are preferred over disposable paper napkins to reduce paper waste. Ongoing efforts to reduce fossil fuel use and study alternatives include a 2,000-watt grid-tied photovoltaic system, geothermal heating and cooling, and plans for production of biodiesel fuel for farm use. A managed nutrient cycle provides for composting kitchen wastes and animal bedding to create organic fertilizer for the gardens and landscape.

The Farmhouse is a 22,000-square foot structure serving as both home for 24 residential boarding students and classroom space. The building's design is based on the traditional New England farm house, which has a "Big House" built on the east end, a barn on the west end and work and living spaces in the middle. This beautiful structure is built of stone and wood and many other natural materials, such as hardwood and tile flooring. The "Big House" on the east end includes a common room, library, dining room and commercial kitchen. The middle section includes a mud room/coat room, classroom space and the main entrance. The three-story west end includes laundry and food service areas, utility rooms, and more classrooms. The building also includes boys and girls dormitories and additional classrooms and day use areas.

In 2009, the school opened a sustainably designed Community Center that houses classrooms, a performance space, computer research center, art and music rooms, and offices. This building is heated and cooled by a geothermal energy system, is built into a north-facing berm, and has a south-facing passive solar exposure. Carpeting is made of recycled materials and classroom floors are bamboo. Lighting systems are automatically controlled and the landscaping, which was researched and designed by students, contains primarily indigenous species of trees, shrubs, and flowers. The view from this new building completes a triangle of the Farmhouse, the older Community Center and the barns.

The school is landscaped with herb and flower gardens and with plants native to northeast Ohio. Wastewater from the building is treated with a system that includes an aerated settling tank, sand filters, sterilizer, cattail filtration beds and an evaporative pond. As the school has developed, student teams have participated in planning the school's various systems and projects.

A courtyard on the northwest side of the schoolhouse overlooks a four-acre farm, with green pastures, intensive gardens, red livestock barns, and a bioshelter. The pastoral scene speaks of good husbandry and planned stewardship. A winding walkway leads across a seasonal stream, past wood-fenced pastures, through the livestock barn and on to the gardens and distant farm buildings. Livestock includes a dairy cow, often a beef cow, goats, two or more pigs, half a dozen sheep, a horse and flock of chickens. The animals have ready access to the good-sized pastures. Fencing keeps the animals away from the stream, maintaining the integrity of the stream banks and preserving water quality. In the distance, evidence of other enterprises are glimpsed: the wood-framed sugar house just inside the forest, the bed and breakfast behind the barns, assorted season-extension tunnels and frames, and beehives tucked into a sunny forest edge.

Bedding and manure from the farm animals is gathered and composted in a designated space near the gardens. This composting facility has an equipment storage shed and concrete block wall that shelters and contains the compost, and controls nutrient runoff. Composting materials are easily accessible for turning by tractor or by hand. The storage shed and compost yard also provides a critical wind deflector for the bioshelter.

The farm's 91-acre forest is sustainably managed as part of the farm's micro-economy. Some timber used to build school structures came from trees removed for site development; some from trees removed from the forest as part of planned forest management. The forest also provides wood for use in the school's woodshop to produce cutting boards, turned bowls and other products. When necessary, lumber is milled onsite with a portable sawmill.

All these activities and more are integrated into the school's curriculum. In addition, the school's summer programs provide year-round opportunities for learning. Student-managed enterprises include custodial duties, food service work, bed and breakfast management, maple syrup production, forest management, animal husbandry, woodworking, pottery, apiculture, bicycle repair, bioshelter management and market gardening. Farm products are combined in value-added gift baskets each year. Baskets may contain maple syrup, honey, pancake mix, a cutting board and beeswax candles. The school's micro-economy generates about $30,000 annually; a similar amount is spent on materials and supplies for the various enterprises.

Many agricultural enterprises are regulated by state and federal laws, limiting some farm production abilities. For example, before they could be sold to the public, farm manager Jim Ewert-Krocker needed to be licensed to process high-acid foods, such as pickled vegetables and tomato sauces.

An important feature of the farm school is its students' engagement with the local community. Each Friday is a community service day, when students participate in various service activities, including involvement at a local nursing home and elementary school. Some students run a farm stand at a local farmers' market one day a week, spring through fall. In addition, a group of students host an annual colloquium on an issue such as local food systems or our environmental impact. This event provides the students with an ideal opportunity to fulfill the Montessori vision: the application of critical analysis and thinking skills and engagement with the surrounding community in the ongoing discovery of sustainable community development.

A Bioshelter/Science Classroom

The development and design of the bioshelter for the farm school is an example of the school's philosophy in action. The project began when educator David Kahn led a group of students through a design process to plan a greenhouse for the school. After studying season extension, high tunnel and greenhouse design, David and the students visited Three Sisters Farm to consult on various design possibilities. They left wanting to build a bioshelter. Because a bioshelter is a long-term investment requiring site-specific planning, the school took several years to study options, consult with designers and review plans. Farm manager and teacher Jim Ewert-Krocker, drawing on his prior horticulture experience and the school's goals for the integration of lessons in biology, botany, physics and other sciences into the building, provided the design criteria.

Completed in 2004, the farm school bioshelter functions as a working science classroom and plant production facility. It is a two-bay, gutter-connected DeCloet greenhouse that has 1,920 square feet of interior space. The steel-framed support structure is glazed with 8-mil polycarbonate. Thermostatically controlled roof vents allow passive ventilation, eliminating the need for fans. The building is oriented with the long sides facing north and south. Because of substantial winter cloud cover, a fully glazed roof was preferred over an insulated roof on the north side. This allows more incidental light to enter the bioshelter. The basic greenhouse kit was modified by the addition of an insulated concrete wall on the north side and on the north half of the east and west end walls. The north half of the building has a heated concrete floor. The south half has a gravel

The bioshelter at Hershey Montessori Farm School.

DARRELL FREY

floor. A coatroom provides an airlock entranceway on the east end. The space includes a science lab, a composting toilet, sinks and a compost bin connected through peat moss biofilters to a deep-soil growing bed. There are a number of large planters and tables for growing plants. The exterior base of the north walls is protected by earthen berms. The exterior of the south wall is bermed with wood-framed herb beds sitting on top of the berms. Three doors on the north side provide access for loading and unloading the compost chambers.

The bioshelter's mechanical systems include heating and ventilation systems, compost chambers and data collection systems. Because the school is located in an intensive snowbelt, (averaging 120 inches of snow annually), and winter lake effect clouds are common, a heating system is required to supplement the thermal mass of the north concrete wall. The heating system was sized to keep the workplace comfortable for students in the winter months. A 350,000 Btu Central Boiler Classic dual fuel boiler located outside the bioshelter heats the cement slab floor and the planter benches. Glycol is circulated from the boiler stove through plastic tubing to transfer to the concrete floor, plant tables and other sections of the building. The heating system has a backup fuel tank, but as of the winter of 2009–2010, only wood has been used to heat the bioshelter. Students and staff have developed the capability to produce biodiesel for filling the backup fuel tank.

Crops and projects in the bioshelter change through the year. At various times, students may be assigned the role of manager, mechanical systems manager, or garden manager. Plant production in the fall is planned to provide salad greens and herbs to the school's kitchen. Spring production provides plants for an annual plant sale and for the school's market gardens. Alyssum, nasturtium and other flowers provide nectar for aphid predators. Science experiments involve physic projects, biological water filtration for goldfish and various botany projects.

Intensive outdoor gardens complete the farm's circle of life. The gardens are planned each year by teams of students and teachers and worked by the students and farm manager. The 1.5 acres of gardens are planted each year in a variety of crops, including vegetables for the school's use and crops for market. Farm-produced compost provides the basis for maintaining rich, productive soil. Organic gardening practices ensure a healthy environment for plants and people.

The Hershey Montessori School sells produce through a local farmers' market and a Community Supported Agriculture (CSA) program. Student garden managers assist in the organization of the program. Members usually come to the farm to pick up their share of the weekly harvest from July through the middle of October; they are also are encouraged to help in the gardens at scheduled special events. Crops have included salad greens, green beans, yellow beans, peas, carrots, beets, Swiss chard, garlic, cucumbers, zucchini, summer squash, winter squash, decorative gourds, tomatoes, sweet peppers, hot peppers, cut flowers, honey, maple syrup, and eggs.

According to Montessori teacher David Kahn, the students are as likely to have dirt under their fingernails as geometry problems in their heads. As with the farm school's other enterprises, gardening offers the students endless opportunities for learning. Beyond the curricula embedded in the garden enterprise, a deeper learning occurs, wherein the student feels the personal satisfaction of learning practical skills and completing tasks — and is rewarded with the fruits of his or her labor.

CHAPTER 15

Home Sweet Home

We are stardust. We are golden. We are billion-year-old carbon.
And we've got to get ourselves back to the garden.
—Joni Mitchell, *Woodstock*

A FTER 10 YEARS OF LIVING AT THE FARM, we left our son (and later, our daughter) to be the on-farm resident manager, and we moved back to our woodland home five miles away. While we were developing Three Sisters Farm, we focused our energy on expanding the market gardens and eventually moved to the farm. So the gardens at our woodland home had to wait.

We had begun to develop gardens at our woodland home property in the mid 1980s. The landscape plan was designed to preserve the forest setting, minimize grass, and to create garden space near the house. A garden clearing was reclaimed from the brush filling the old cornfield southeast of the house. We sought to maximize the solar dynamics of the property, laying out a cultivated clearing in the forest on the most level part of the property.

A key element of permaculture home design is the placement of the home in the landscape. (Although here I am discussing our home landscape, on the farm, it is a very important element in a successful farm design.) On the southwest, west and north sides of the house, we removed only a few trees that were too close to the house and have preserved the natural forest landscape. This provides protection from strong winds and cooling shade in the summer. It also eliminated the need to manage the landscape. To the east of the house we have a driveway and parking, beyond which we retained more forest cover. A

417

DARRELL FREY

View of Linda Frey's Zone one garden and house.

30-foot-wide area with a fire circle was cleared beyond the parking area. This shady space is the only grassy area.

The best homestead gardens mature over time. The interweaving of the gardener's spirit, the soil's fertility, and the garden's unique characteristics manifests as a new tapestry each year. A garden landscape often tells the tale of a gardener's journey. Favorite flowers hide under berry bushes. Garden fences are edged with flowering groundcovers and herbs collected over many years. Favorite varieties of lettuce and kales fill some beds. Tomatoes are planted and tended in the gardener's personal style. Masses of daylilies and iris fill the odd corners or frame the garden shed. One garden may burst forth in spring poppies, another foxglove. Self-seeded plants move about among managed beds.

Our home gardens are actually only a couple of years old but, because of the collection of plants at Three Sisters Farm and our experience, the gardens already give the visitor an impression of long and careful tending.

Placing the main kitchen garden directly to the south of the house provides easy access to herbs and other plants used daily. The house provides wind protection to this garden and reduces the amount of fencing required. This space receives about eight hours of sunlight a day in summer and six hours in March and September.

When we began, the soil was only a few inches thick over a compacted hardpan. Basically, it was marginal cropland subject to erosion and heavily damaged by excess cropping of corn prior to 1970. The soil's natural pH of 6.5 had become acidic. Most nutrients were locked up or leached away. When we purchased the site, the crop field and pastures had been abandoned and were overgrown with 20-year-old crabapples and hawthorn trees. These provided protection for a number of young hardwood trees. They, along with deer berries, blueberries and various brambles also provided food for deer, rabbits, birds and other wildlife. The wildlife was in turn contributing their nutrient-rich droppings. Beginning gardening with a soil in these early stages of recovery required the importation of compost, leaf and straw mulch and other organic matter. The garden soil was amended with essential minerals, rock lime, rock phosphate, greensand and seaweed. A simple outdoor compost bin cycles kitchen and garden waste back into the garden soil.

Taking the Lessons Learned Back Home

Since 2003, Linda has been rehabilitating the long-neglected gardens. With much manual labor, she has wrested the clutching roots of hawthorns from the ground and forked and loosened the soil. Rocks (gathered from the stream and carried 50 feet uphill) line and stabilize terraced beds along a west-facing slope. Compost, made from horse manure and leaves, builds on the fertility created between 1983 and 1993.

Some perennials survived the decade of neglect. Good King Henry (a perennial relative of lambsquarters), Egyptian onions, Swensen's red grape, and red currants have all been rejuvenated and given many new neighbors.

Linda's leaning is toward a cottage garden, an exquisite blend of color and texture, food and flower. The garden begins at the front door. An alfresco dining table catches the morning sun reflecting off the southeast wall of the house, but is shaded from the heat of midday by a young wild black cherry tree.

A 5-foot-tall potted bay tree puts out its new green leaves beside the table. Other potted herbs and houseplants are set about the garden. Stones lining the beds seem to emerge from the ground, a natural frame for soil and plants. Each bed is planted with a combination of edible flowers, kitchen herbs, vegetables and flowers. Columbine and forget-me-nots fade as light purple foxgloves raise their spires under a young lilac bush. One bed features lavender, delicate red coralbells, purple flowering sage and Johnny-jump-ups. Another bed combines yellow pansies with pink dianthus, chives, golden oregano, and mint. A perennial bed holds a colorful mix of foxglove, lupine, Echinacea, sweet woodruff, bleeding hearts and lily. Cabbages are interplanted with more Echinacea. Rosemary planted in pots and buried in the beds for the summer shares a bed with chard, beets and lettuce. One bed is edged with onions and planted with 3 square-foot blocks of lettuce, carrots, beets and onions. In another, spring chervil gives way to summer's bergamot and parsley.

Tomatoes, surrounded by carrots, a favorite companion, climb a trellis along the long edge of a bed of basil with calendula planted at the rounded corners. A three-sided pyramidal bamboo trellis in a triangular bed is edged with scarlet runner beans interplanted with early summer cabbage. Yet another bed is planted with native columbine, Johnny-jump-ups, iris, and violets. Wild cranesbill geranium and violets grow between the beds and among the edging stones. The western edge of this garden is a sloping woodland of maple, black cherry, oak and aspen trees. These keep the house very cool all summer and provide the perfect habitat for trailing arbutus, and rattlesnake plantain, a native orchid.

Across the driveway, a children's garden provides our grandchildren their own special place. The centerpiece is an igloo-shaped trellis of saplings. This, when covered with pole beans, cherry tomatoes and nasturtium vines, provides an edible playhouse. Stepping-stones around the trellis lead to a sand pit and small gardens planted with flowers and plants with interesting shapes and textures.

Further up the driveway, longer beds form the Zone 2 gardens. These stone-lined beds are planted with main crops of potatoes, cabbages, strawberry and flowers. Compost piles and heaps of gathered leaves are located here. Red currants edge the west side of this triangular garden, and a red grape climbs a trellis. The eastern edge is bordered by native low bush blueberry and planted rose bushes. Rhubarb, asparagus and strawberry all have a place in this garden.

Stinging nettles have a home on the edge of the woods to the east. They are harvested for spring tonic soup or collected and dried for use as an ingredient in nourishing teas.

Further down the driveway are wild blueberry, red, black and yellow raspberry, blackberry and a flavorful and productive wild grape.

This garden will require a continued input of organic matter for a number of years. The south edge of the garden, with the easiest access to the driveway, is used to stockpile mulches and composting manure from other farms. Our dog Forest, an active English setter, happily patrols the unfenced garden day and night to deter animal invaders. Before Forest joined our household, our crop gardening was limited to a few small beds and container-grown vegetables on the second floor deck.

Linda expands her garden each year, digging and preparing half a dozen new beds. Perennial flowers and herbs are divided each year and used to add color and accent to the landscape. The lines between forest and garden interweave. Wildflowers and native trees and shrubs surround the beds. Some larger trees need to be thinned as the gardens expand. These provide posts and firewood.

Linda still provides backup work at the farm and helps train interns and new workers each spring. Her expertise in plants and intensive gardening is invaluable to the farm. The lessons she learned managing the gardens at Three Sisters Farm and from years of visiting the gardens of friends have been brought home. The resulting mix of perennials and annuals, crops and flowers, is probably much closer to a permaculture landscape than the intensively managed production gardens at Three Sisters Farm.

A morning breakfast on the edge of the cottage garden is a totally relaxing experience. You are surrounded by bird song and many colors of flowers and vegetables as you sit and watch the hummingbirds sip nectar from the foxglove and coralbells. After years of managing two acres of market garden beds, Linda's labor in the home garden, though no less demanding physically, is a welcome change of scale.

View of Linda Frey's Zone one garden from door step.

DARRELL FREY

Epilogue: The Farm Ecosystem Evolves

*Slowly, but with no doubt or hesitation whatever, and in
something of a solemn expectancy, the two animals passed
through the broken tumultuous water and moored their boat
at the flowery margin of the island. In silence they landed, and
pushed through the blossom and scented herbage and under-
growth that led up to the level ground, till they stood on a little
lawn of a marvelous green, set round with Nature's own
orchard-trees — crab-apple, wild cherry, and sloe.*

— KENNETH GRAHAME, "THE PIPER AT THE GATES OF DAWN,"
THE WIND IN THE WILLOWS

Feeling Part of the Cycle

I RECENTLY TOOK AN AFTERNOON OFF from the double-duty effort of
overseeing the planting of the spring gardens and finishing this manuscript. I walked along a woodland trail leading to the banks of French
Creek. The early summer heat was moderated in the shade of the forest.
The arching vases of the ostrich ferns were fully open along the river
bank. The scent of Dame's rocket filled the air. Wild onions blossomed
among fading violets under the shade of basswood and maple.

A great blue heron took flight, disturbed by my arrival. As I sat under
the shade of a truly old maple tree, a kingfisher flew by, returning to a
nest dug into the bank upstream. Bass jumped and splashed from the
stream to catch insects. Dragonflies darted about to capture any mosquito that strayed from the moist forest beyond the bank. The timeless
calm and ancient vital energy of the place touched a sympathetic reso-
nance within me.

I slipped off my shoes and waded into the shallows along shore. The water was cool and refreshing. A breeze blew against the current as the clear water rolled over stones and mussel shells. I sat a while in the foot-deep water along the river's edge under the overarching maple tree. Eventually I lay back into the cool water, letting the quietly rippling waters wash away the sweat and dust of a morning in the garden.

French Creek, which drains 1,235 square miles of northwest Pennsylvania and southwest New York, is home to 28 species of fresh-water mussels, dozens of species of fish, amphibians and birds, and hundreds of species of insects, plants and trees. The continued health and vitality of this river is dependent on the care taken by inhabitants of the watershed to keep it clean.

One day, many years ago, I was driving home from a produce delivery to Pittsburgh. Ahead, I could see a thunderstorm racing to cross my path. Dark clouds of the thunderhead, dimly backlit by a hidden setting sun, rose tens of thousands of feet. Jagged fingers of lightning clawed at the roiling clouds. As the rain poured down, I had a clear vision of the cycle of water between western Pennsylvania and the Midwest. Air masses rolling over the Ohio Valley gathered moisture and the dust of coal power plants — and the emissions of millions of cars. This tainted rain falls on the Upper Ohio Watershed. In turn, our wastewater, industrial and sewage treatment plant effluent and the run-off from parking lots, highways and agricultural fields taints the water as it travels downstream through the Ohio River to the Mississippi and on to the oxygen-depleted dead zone in the Gulf of Mexico.

What happens in our little field connects to so many things. The water leaving our pond or running down the driveway drain enters Mill Creek directly. After passing through one or two beaver dams, it meanders down a quiet agricultural and forested valley to French Creek. Tonight, my 5-year-old grandson has shown me a mussel shell he picked up on a canoe trip from the shallows of French Creek, near where Mill Creek enters this river. "Something used to live in here," Adrian explains as he opens and closes the shell. The mussel shell's dark and rough exterior, tinged with mossy green algae, looks like a river stone. The smooth, pearly interior is tinted pale green and stained with river mud. We both are fascinated with the shell for a few moments. I trust he too will be able to share such a moment with his grandson in 45 years.

Biodiversity and environmental quality are not just nice things to learn about on cable TV or in magazines. The natural world is the foundation upon which we all depend. Stewardship begins in our yards and gardens, and it extends to the choices we make in what we consume. The decade or two ahead are critical. Humans have managed and impacted bioregions for tens of thousands of years. But the choices we are making now have a stronger and longer-lasting impact on the planet than at any time in human existence.

Biodiversity: Farm Community

The speed with which biological diversity has returned to our five-acre field is amazing. Create an ecological niche and something will occupy it. As the complexity of the landscape increases, microclimates develop and new species enter the system. The non-native invasives, including multiflora rose, Canada thistle, garlic mustard and house finches will be controlled and — we hope — removed. Others, such as wild elderberry, black raspberry, serviceberry, common yellow throat and indigo bunting we are happy to see.

Catbirds and other New Neighbors

All my prior experience with the sleek gray catbird led me to believe they were shy and secretive. Their preference for living and foraging among shrubs on the edge of the forest and stream banks apparently limited my experience with them. As the windbreaks and tree crops grew, they provided cover, allowing the catbird to forage more in the gardens, first from the hazelnuts along our south border, then from the mulberry trees behind the bale-shelter, and finally from the shrubs to the west and north of the Spiral Garden.

This year (2009) we have been surprised at the appearance of a particularly friendly catbird nesting in the Spiral Garden. The amicable fellow with a stylish black cap flits about the garden as we weed, plant and mulch. No doubt he is keeping an eye on us as he hops about the branches of the Dolgo crabapple, his mate hidden on a nest in the tangle of rugosa rose and honeysuckle. Pruning and re-trellising the honeysuckle will have to wait until they are done nesting.

The catbird seems to go out of his way to cross our path as we tend the garden or water plants in the seedling frames near the Spiral Garden.

Perhaps he made a human friend in his winter travels between the Gulf Coast and Central America. He walks around quite close to us, usually with a small caterpillar in his black beak. He flicks his tail and cocks his head from side to side, as if curious about our work.

Brown thrashers have taken up residence in the expanding tree line, as have the Baltimore oriole, which formerly stayed in the small, forested floodplain just south of the farm. Bluebirds have virtually disappeared from the farm, and we have finally gotten around to removing the house finch nests from the nest boxes. We hope that by reducing the number of house finches, we will encourage the bluebird to return.

Another creature startled me a bit. As I set up the solar irrigation for the first time this year, I disturbed a common water snake. He seemed to stand his ground (water, actually) among the cattails as I cleaned the screened bucket that holds the intake pipe. I am not normally concerned by snakes, but these are known for their aggression and non-poisonous bites.

When we thin the lambsquarters growing thickly among the crops in the South Garden, we allow some to remain and set seed. The oak leaf variety is being allowed to spread a little because of the unique shape of its leaves. The bright pink magenta lambsquarters has spread throughout the farm and has created some interesting shades as it crosses with the common white-tipped varieties. Somewhere along the years, we have lost most of the yellow- and copper-tinged varieties. These were never numerous, but they added an interesting contrast to the salad of the season's summer mix. The few plants we do find among the crops are left to self-sow.

As the yard has reverted to a meadow, and the mulberries have spread their branches wide, common blue violets have multiplied both among the asters and under the mulberry trees. Their leaves are a welcome addition to the spring salad. Gathering them is much easier now that they have claimed their niche.

This description of the farm diversity could continue for pages to come. Each year brings a different mix of pests and their predators. Currently, our gardens are home to a lot of toads. Wasps are in decline after a few very wet years. Snake, slug and snail populations fluctuate. We are on the lookout for these animals because it is a diversity of predators that brings stability to the farm's pest management.

The uncertain climate can bring a June frost, but it may also provide a growing season six weeks longer than usual. These changes certainly effect the emergence times of both plants and insect and blooming times of fruit and flower. As ecological farmers, we need to be sensitive to all these complexities and help our landscape adjust to the changes.

A Hopeful Enterprise

Design and planning for small-scale intensive agricultural enterprises is a hopeful and creative process. My childhood memories of walking the nearly dry bed of Glade Run, picking mulberries and raspberries while the hummingbirds and honeybees foraged the orange blossoms of jewelweed, is also a vision I have for urban children today. Parks, garden lots and other green spaces in a permanent culture will provide wild places for children drawn to foraging — places where they can endure thorn-scratched arms and legs and purple-stained fingers and lips in exchange for eating their fill of abundant blackberries.

The myth of Eden is a primal memory of our evolutionary past, at a time when we foraged the seashore, rivers and forest edge for Nature's abundant food. We must begin to forge a new myth that allows humans to be redeemed in the eyes of the creation (or manifest universe, as you will) and regain our place in the garden of our own design.

Appendix A
Applicable Regulations for Agricultural Composting in Pennsylvania

AGRICULTURAL COMPOSTING IS DEFINED as the composting of agricultural wastes and certain other organic materials on a farm to produce compost for beneficial use. Agricultural composting is regulated at the state level by the Department of Environmental Protection (DEP) and may require a permit in certain situations. (There are exemptions for water quality permits for manure management systems if they comply with state guidelines.) The Pennsylvania Department of Environmental Protection allows non-livestock (crop) farms to "bring in manures and/or other organic materials to compost and use as a source of organic matter for their soils. They may sell surplus compost off the farm."

Relevant Pennsylvania DEP rules excerpted from "Agricultural Composting of Manures: Supplement to Manure Management for Environmental Protection," 1997:

- Compost Sources
  - Bedding materials and livestock manures (which are assumed to be from farm operations) must comply with the sizing, location and sales criteria listed below.
  - Off-farm "co-products" such as bark, shredded paper, and clean, unpainted and untreated sawdust are considered part of normal

farming operations and do not require permitting or approval.

- Food processing residuals of fruits and vegetables (this includes leaves, pomace, and other unused portions) are considered part of normal farming operations and do not require permitting or approval to compost on-farm. However, if this material is destined for off-site, non-farm usage (e.g., sales) then a General Permit for Residual Waste Composting (WMGR025) is required from the DEP.

- Spent mushroom substrate (mushroom soil or mushroom compost) is considered part of normal farming operations and does not require permitting or approval.

- Off-farm materials such as clean newspaper, cardboard, and high-grade office paper are considered part of normal farming operations and do not require permitting or approval.

- Leaves, grass clippings or other yard and garden residues added to compost do not require permitting as long as the "Guidelines for Yard Waste Composting Facilities" are followed (including registration with DEP). (Contact your regional DEP office for a copy).

- Biosolids, residual and municipal wastes require permits.

- Sizing

- If the composting activity disturbs five or more acres, a National Pollution Discharge Elimination System (NPDES) permit is required for construction activities. (Contact your local DEP regional office.)

- Location

- Compost sites should not be established or operated:

- Within 100' of surface waters and sinkholes.

- Within 300' of a water supply.

- Within Zone 1 of a public water supply (may be up to 400').

- In or within 300' of an exceptional value wetland.

- In or within 100' of a wetland other than an exceptional value wetland.

- In a 100-year floodway or below the 100-year flood elevation.

- On sites with impermeable surface soils (e.g., clay).

- On sites with very rapid permeability (sand or gravel).
- On sites where the seasonal high water table or bedrock is within four feet from the surface (must be avoided or modified with an impervious surface or liner).
- Within 300' of an occupied dwelling, unless the owner has provided written consent.
- Within 100' of any property line, unless the landowners otherwise agree and execute a waiver.
- Sales
  - If a claim is made that the finished material is a specialty fertilizer, plant growth substance, or soil conditioner, a license must be obtained and the product be registered with the PA Dept of Agriculture.

Appendix B
Compost Calculations

FOLLOWING ARE SOME OF THE CALCULATIONS I used in the examples cited in the chapter. Because they are typical of the type of calculations needed on a market garden farm, I am including them in detail. This should allow you to apply them to your own situation.

Compost
How much finished compost do you need annually?:

1. First, figure out the square footage requiring compost:
    Total square footage of garden space:

10,000	sq. ft. South Garden
12,000	sq. ft. Pond Garden
12,000	sq. ft. West Garden in transition to Forest Garden
10,000	sq. ft. East Garden
44,000	sq. ft. total

    Square footage of bed space available (after subtracting pathways between beds):

    30,000 sq. ft.

    Square footage of bed space needing annual application of compost (non-root crops):

    30,000 sq. ft. x 50% = 15,000 sq. ft.

    So, we have 15,000 sq. ft. of beds that need compost.

2. Figure how many cubic yards are needed for your square footage (final answer may be a large range because the depth of compost has to be incorporated into the calculation. Depth will vary depending on the crop, conditions, etc.)

   Amount of finished compost needed:

   One cart holds 9 cu. ft.

   2 to 3 carts are needed annually for every 200 square feet of garden space.

   2 to 3 carts x 9 cu. ft./cart = 18 to 27 cu. ft.

   So, we need 18 to 27 cu. ft. of compost for every 200 square feet of garden per year.

3. Calculate how much finished compost you will need for your beds:

   [Set up the equation, and solve for "x" (in this case, 18 and 27 x 15,000 divided by 200).]

   18 to 27 cu. ft.: 200 sq. ft. as "x" cu. ft.:15,000 sq. ft.

   "x" = 1,350 to 2,025 cu. ft. of finished compost

   [Then, divide cubic feet by 27 to get cubic yards]:

   1,350 cu. ft divided by 27 = 50 cu. yd.

   2,025 cu. ft. divided by 27 = 75 cu. yd.

   So, Three Sisters farm needs between 50 and 75 cubic yards of finished compost for our 15,000 square feet of bed space.

How Much Raw Material Do You Need to Produce the Finished Compost?

If you are making your own compost, you will need to know how much raw material will produce the amount of finished compost you need. You should assume that the volume will shrink by 2/3 between raw and finished compost (a 3:1 finished compost ratio),

Raw compost delivered requirements:

   3 x (50 to 75 cu. yd.) = 150 to 225 cu. yd.

In our case, a delivery of raw compost is 25 cubic yards, so:

   150 cu. yd./25 cu. yd./delivery = 6 deliveries

   225.0 cu. yd./25 cu. yd./delivery = 9 deliveries

So, we require at least 6 and a maximum of 9 deliveries of raw compost per year.

Storage Space for Finishing Compost

Compost piled in windrows takes up takes up significant space. You need to know how much space you need to accommodate the amount of compost you'd like to produce.

Space Occupied by Windrows:

> Raw Compost: 8 rows, each 10 ft. x 30 ft. = 2,400 sq. ft.
> Finished Compost: 22 rows, each 10 ft. x 10 ft. = 2,200 sq. ft

So, we need 4,600 square feet of open, sunny space for our windrows. (The area also has to be large enough and situated in a place that allows trucks enough room to come in, dump their load and turn around.)

Heat Gain and Loss

Heat Contribution from Indoor Compost Chambers

1. Figure out cubic yards of raw compost you will have:

   2.5 cu. yd. raw compost/chamber section x 8 sections = 20 cu. yd. maximum capacity

2. Figure the maximum heat contribution of that amount of raw compost (using the figure mentioned in the text: 10^6 Btus/cu. yd./month):

   20 cu. yd. x 10^6 Btus/cu. yd./month = 20 x 106 Btus/month.
   20 x 10^6 Btus/month x 6 months of heating season = 120 x 10^6 Btus/6-mo. heating season.
   120 x 10^6 Btus is maximum heat contribution for 6 months. Dividing by 180 days (in 6 months) gives: 666,700 Btus/day maximum heat contribution.

Heat Contribution from Indoor Compost Chamber Relative to Bioshelter Solar Gain

[To calculate, subtract Btus of solar gain from Btus of compost heat and multiply by 100 to get percentage.]

Month	Avg. daily solar gain (Btus/day)	Compost heat as % solar gain
Dec.	1,920,775	35%
Mar.	3,026,000	24%
Sept.	3,831,300	17%

Heat Contribution from Indoor Compost Chamber Relative to Bioshelter Heat Loss

In our bioshelter, 675,000 Btus are required to maintain 50° inside temperature over 14 hrs on an average winter night with outside temperature of 25° (23 percent of the building's thermal storage capacity).

1,620,360 Btus are required to maintain 50° inside temperature over 14 hrs on a cold winter night with outside temperature of -10°. (could go for 1.9 days before needing backup heat).

Cold winter conditions stated above are expected to occur
  10–14 times/yr.
10 days x 24/14 x 1,620,360 Btus = 27,777,600 Btus.
14 days x 24/14 x 1,620,360 Btus = 38,888,640 Btus.
Therefore, annual heat loss is 28 to 39×10^6 Btus.

What all these calculations show is that our maximum annual heat generation from compost is 120×106 Btus – which is more than 3 times the maximum expected annual heat loss. So, for our situation, compost chambers are very worthwhile.

Heat Contribution from Indoor Compost Chamber Relative to Backup Heat

The maximum annual heat generation from compost is also more than 3 times the heat provided by our backup heating systems:

3 cords of hardwood are burned each year in the bioshelter as
  backup heat.
3 cords x 24 x 10^6 Btus/cord x 0.50 stove efficiency = 36×10^6 Btus.

CO_2 Generation from Indoor Compost Chambers

Our compost chambers can hold 20 cu. yd. This amount of raw compost will produce an amount of carbon dioxide that can be calculated as follows:

1 cu. yd. of wet compost weighs 1,200 lbs.
20 cu. yd. wet compost x 1,200 lb/cu. yd. = 24,000 lbs wet compost.
1 kg wet compost generates 10 to 50 grams of CO_2/day
  (= .01 to 0.05 lbs CO_2/day).
24,000 lbs wet compost x 0.01 to 0.05 lb/day of generated CO_2
  = 240 to 1,200 lbs CO_2/day.

240 to 1,200 lbs CO_2/day x 30 days/mo
   = 7,200 to 36,000 lbs CO_2/mo.

One pound of CO_2 is 454,000 mg of CO_2. So the amount of CO_2 generated by the bioshelter's compost chambers is:

240 to 1,200 lbs CO_2/day x 454,000 mg/lb = 1.1 to 5.4 x 10^8 mg
   CO_2/day.

Our bioshelter volume growing area is (roughly) 47,565 cu. ft. (1,344,872 liters).

If we take mg CO_2 generated per day and divide it by bioshelter volume in liters, we can estimate CO_2 concentration = 80 to 400 mg/L (assume same as ppm):

1.1 x 10^8 mg CO_2/day divided by 1,344,872 liters
   = .00000008 mg/L.
5.4 x 10^8 mg CO_2/day divided by 1,344,872 = .000004 mg/L.

Because we typically operate the compost chambers at half capacity, we can expect half that amount, or 40 to 200 mg/L CO_2 generation per day.

(Note: This calculation does not account for air infiltration/losses, nor does it account for the CO_2 generated by the chickens that live in the bioshelter.)

Appendix C
Greenhouse Heat Dynamics:
Figuring Solar Gain, Solar Storage,
and Heat Loss

B Y PROVIDING FORMULAS ALONG WITH THE ACTUAL NUMBERS I used
for making calculations for our bioshelter, I intend to show how you
can apply the information in Chapter 5 to your own situation.

Solar Gain

To determine net solar heat gain, subtract heat loss from expected solar
gain. If solar gain is less than heat loss, you will need a backup heating system to make up the difference.
The larger the difference, the larger the
backup system you will need.

Net solar gain is determined by these
factors:

- the position of glazing relative to south
- the average daily gain figured in Btus
  per square foot
- the type of glazing
- the angle of glazing relative to solar
  angle above horizon
- the square footage of glazing

   **The position of glazing relative to
south:** Solar south is determined by first

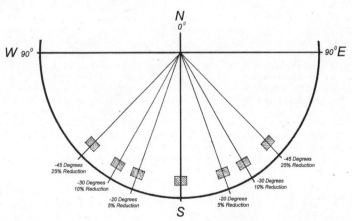

Percent decrease of solar gain as a building is faced away from south.

locating *magnetic* south with a compass and than adjusting for magnetic declination for the specific location and date. See data at www.ndgc.noaa.gov.

The average daily gain figured in Btus per square foot: To determine daily gain, first determine expected solar heat gain for a given month. To do this, you must know two things: the transmissivity of the glazing and the average daily Btus available per square foot (this is dependent on the angle of the glazing relative to the position of the sun during that month).

The type of glazing: Glazing materials vary in light transmission. All glazing materials reflect or block some percent of sunlight and some wavelengths of light. The amount of light that passes through a glazing material is called *transmissivity*. The following are typical percentages of light transmitted by common glazing material. (Note that transmissivity does not necessarily stay the same over time. All plastic-based glazing will degrade, so transmissivity is highest when glazing is first installed. Dirt and scratches will also reduce transmissivity):

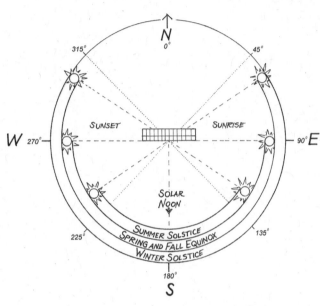

Solar angle (at solar noon) above the horizon at Solstices and Equinoxes for 45 degrees north latitude as example.

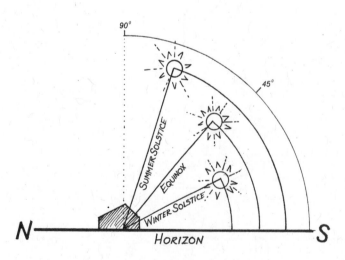

- glass, double pane, with ½ inch air space: 90%

- acrylic, double pane, with ½ inch air space: 86–91%

- polycarbonate, with ½ inch airspace: 82–89%

- fiber-reinforced plastic (FRP): 85–90% (when new — it degrades over time)

- 6-ml UV resistant polyethylene: 85% (when new — and it only lasts 4–5 years)

The angle of glazing relative to solar angle above horizon: The angle of the glazing affects solar gain. Light passing perpendicularly though glass (at a 90-degree angle) provides the most gain. As the glazing tilts away from perpendicular, gain is reduced because some light is deflected. Glazing should be angled for the season during which the most gain is desired.

Figuring Average Daily Gain

Assuming a building faces south, solar heat gain can be calculated using the following formula:

square feet area of glazing x daily Btus per square foot
(at the glazing angle for a time of year) x percent transmissivity

Example: Three Sisters bioshelter has 2,600 square feet of glazing x 1,000 Btus per square foot (in January) average x 90% transmissivity = 2,340,000 Btus average gain daily in January.

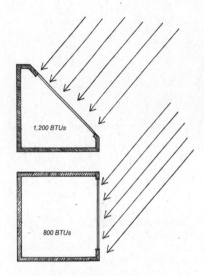

Expected average Btu per square foot at 90 degrees and 45 degrees.

Heat Storage

The amount of thermal mass in a solar structure should balance the expected heat gain each day. The figures are used to plan for adequate heat storage.

Thermal Mass: Storage Capacity of Various Materials

	Specific heat capacity*	Btu/cu. ft. per degree F. temperature change
Water	1.0	62.5
Stone	.21	34.6
Concrete	.23	34.5
Soil	.21	20
Wood	.33	10.6
Steel	.12	59

(*specific heat capacity is a material's ability to absorb heat in Btus per pound)

Heat Loss: Conduction, Radiation, Convection and Infiltration

Heat *conducts* through building walls, roof, floor and glazing, and it radiates away from the building. *Convection* is the movement of heat in air and water. Heat *radiates* from thermal mass into the air. *Infiltration* is the entry of cold air from outside and the escape of hot air from inside.

> Total heat loss per hour = infiltration loss plus conduction loss x hours of heating period

Conduction

The conduction rate of a material is given as *U value*. (U value is the inverse of the insulation, or *R-value*.

To calculate heat loss from a building per hour, use the formula:

$$Q = U \times A \times \triangle T$$

Q = heat in Btus; U = inverse of R-value; A = area (of wall, glazing, floor or roof); T = change $\triangle$ of temperature inside to outside

So, heat loss (Q) through a wall each hour is calculated by multiplying the U-value of the wall by the area of the wall by the temperature difference between inside and outside.

Insulation values (R-values) of various materials

- R-values:
  - straw bale: R-40
  - fiberglass: 5 inches, R-19
  - polystyrene (expanded and extruded), per inch: R-5
  - wood, per inch: R-1
  - double-pane glass with 1-inch air space: R-1
- U-values (inverse of R-values):
  - straw bale: U-1/40
  - fiberglass: U-1/19
  - expanded polystyrene, per inch: U-1/5
  - wood, per inch: U-1/1
  - double-pane glass with 1-inch air space: U-1/1

Convection

Air holds heat. When warm air leaves a building it takes heat with it. Precise estimates are difficult because moist air holds more heat than dry air. For the purposes of planning, this is disregarded.

The formula for the heat contained in a cubic foot of air is:

$Q = \triangle T \times V \times ACH \times .018$

Q = heat in Btus; $\triangle T$ = temperature difference between inside and outside; V = volume of air in the building; ACH= air changes per hour; .018 represents the amount of heat in a cubic foot of air. (This number is based on the weight of air multiplied by the specific heat capacity of air.)

Resources: www.doe.gov, www.noaa.gov

Worksheet: Sizing a Heating System

To decide on the size of a heating system, you first have to know how much heat will be lost. The rate at which heat leaves a building is determined by:

• the rate of heat movement through walls, glazing, ceiling and floors
• the rate of infiltration, or air exchange with outside
• temperature difference between inside and outside.

When calculating, you want to plan for the *worst case scenario* of the coldest and longest night.

Below are the steps to follow. As an example, numbers are given for the bioshelter for a 14-hour heat loss period (night) with outside temperatures of –10°F and desired inside temperatures of 50°F.

1. *Determine temperature difference between inside and outside.*
   Temperature inside building: 50° F.
   Lowest temperature in the area: -10° F.
   Temperature difference is 60 degrees.

2. *Determine heat loss from entire structure. Because the materials used for walls, roof, and glazing each have a different insulation value, determine heat loss though each material based on area and U-value, then add the numbers together to get total heat loss.*

   Glazing:
   In the bioshelter, we have 2,600 square feet of glazing with a U-value of 1/1.

   $Q = U \times A \times \triangle T$
   Q = 2,600 square feet glazing x 1/1 x 60 degree difference
   Q = 156,000 Btus per hour lost through glazing

Walls:

In the bioshelter, we have 2,100 square feet of R-19 wall

$Q = U \times A \times \triangle T$

$Q = 1/19 \times 2{,}100$ square feet walls $\times$ 60 degrees difference

$Q = 6631.6$ Btus per hour lost through walls

Roof:

In the bioshelter, we have 1,696 square feet of R-40 insulated roof.

$Q = U \times A \times \triangle T$

$Q = 1{,}696$ square feet of insulated roof $\times$ 1/40 $\times$ 60 degrees

$Q = 2{,}540$ Btus per hour lost through roof

Floor:

You must include the amount of heat lost through the (insulated) floor by finding the difference between desired temperature and ground temperature. Because the ground in our area maintains an average temperature of 50 degrees F year-round, there would be *no* heat loss in this instance.

3. *To calculate the rate of infiltration, first calculate the volume of air in a building.* (The bioshelter has 43,700 cubic feet of air space.) The air exchange rate in a well-sealed building can be assumed to be one air exchange per hour.

$Q = \triangle T \times V \times ACH \times .018$

$Q = 60$ degrees $\times$ 43,700 cubic feet $\times$ 1 air exchange per hour $\times$ .018

$Q = 47{,}196$ Btus of heat lost via air exchange per hour

4. *Now, add total Btus of heat loss:*

Glazing: 156,000 Btus per hour

Walls: 6,631.6 btu per hour

Roof: 2,540 Btus per hour

Floor: 0

Infiltration: 47,196 Btus per hour

Total loss per hour: 212,367.6 Btus per hour

5. *Multiply hourly heat loss by the amount of time of no solar gain (generally 14 hours in mid winter) to determine heating needs.*

212,367.6 Btus lost per hour $\times$ 14 hours = 2,973,146 Btus required to maintain 50°F inside our bioshelter on the coldest night expected.

Temperate Zone Wetland Plants

Form: ah: annual herbaceous. d: deciduous. em: emergent. ev.: evergreen fl: floating-leaved. ph: perennial herbaceous. sh: shrub. smt: small tree. sub: submerged. t: tree. Ht: height.

Habitat: b: bog. bl: bottomland. btl: burnt land. d: ditch. dl: dry land. l: lake. m: marsh. ms: moist soil. p: pond. sh: shore. shw: shallow water. sst: slow stream. sp: spring. stb: stream bank. sw: swamp. wd: well drained. wm: wet meadow. ww: wet woods. qw: quiet water. Z: Zone.

Propagation: c: cutting. d: division. l: layering rh: rhizome. s: seed. t: tuber

Notes: ^ after species name indicates that some species in this genus are on the Federal Noxious Weed List (as of May 1, 2010). Take care to plant only the non-invasive species. See usda.gov.

\# after species name indicates that plant has been listed as weedy, invasive, or noxious in N. America. See www.invasive.org.

Genus /species	Family	Common Name	Form	Habitat	PROP.	Human Uses	Animal Uses	Functions/ Properties	Problems
Acer, spp.	Aceraceae	maple, silver, black, red, box elder	d, t	sw, bl, stb, ls	S	lumber, fuel, syrup, pulpwood, woodenware	edible seed, browse	nesting site for birds, squirrels; shade, mulch,	---
Acnidia	Amaranthaceae	water hemp	ph	ms	S	---	edible seed (waterfowl)	---	---
Acorus calamus	Acoraceae	calumnus, sweet flag	ph	sw, ms, sp, sst, dl, Z: 1-6	S	medicinal; food: salad (leaf), candy (root), flavoring; perfume	---	---	invasive, can spread to dry land
Alisma gramineum* A. lanceolatum A. plantago A. triviale	Alismataceae	water plantain	em, ph Ht: 1-3' (*A. gram.: sub)	shw, d, Z: 3-6 *A. .gram northern: alkaline)	S	edible fleshy base, rich in starch, must be dried	browse, edible seed for waterfowl	bee forage	invasive
Alnus, spp.	Betulaceae	alder	d, t, sh	wm, sw, stb	C, S	medicinal, edible	browse, beaver food, habitat	nitrogen fixation, bee forage, orna- mental, mulch (chipped),	invasive
Althea, spp.	Malvaceae	marshmallow	ph	sw, wm, Z: 3-7	S	medicinal, edible leaves,	---	bee forage, ornamental	---
Amelanchier, spp.	Rosaceae	Juneberry	sh, smt	sw, wd Z: 3-7	S	fruit, fuel, specialty wood	forage and browse; fruit (birds)	bee forage, ornamental	---
Anacharis canadensis	Hydro- charitaceae	elodia, frog bit, waterweed	sub, fl	slow, fresh water, 1-10' deep (mod. or sandy loam); Z: 3-6	S	aquarium plant	forage: leafy stem and rare seed (waterfowl); snail forage	ornamental aquarium plant; fish habitat; oxygenates water; compost plant	very invasive (do not plant without means to control or release into waterways)

Genus /species	Family	Common Name	Form	Habitat	PROP.	Human Uses	Animal Uses	Functions/ Properties	Problems
Angelica, spp. *A. atropurpurea*	Apiaceae	angelica	ph	wm, d Z: 2 or 3-7	S	medicinal root, leaves and seed; food: leaf and stem	---	---	---
Apios americana	Fabaceae	groundnut	vining ph	sw, ws, p, sh	T, S	food: tuber rare seed; latex; medicinal	forage, vines, tubers (hogs, rabbits, squirrel, mice); see (turkey,quail)	erosion control; nitrogen fixation; biomass production; bee forage	---
Aponogeton distachyos	Aponogeton-aceae	cape pond weed, water hawkweed	ph, em	---	---	food: roots, flower spikes	---	---	---
Aralia racemosa	Araliaceae	spikenard	ph	shady stb, s Z: 3-7	---	medicinal root, edible root, berries	---	---	---
Asarum canadense	Aristolochiaceae	wild ginger	ph	stb, b	---	medicinal, spice, scent	---	---	endangered
Asclepias incarnata	Asclepiadaceae	swamp milkweed	ph	wm, stb, p, b	S	medicinal; food; latex; fiber "down"	---	bee forage	---
Azolla, spp.^	Azollaceae	water fern, water velvet, mosquito fern	ph, free-fl	qw	C	---	---	host of nitrogen-fixing *Anabaena* algae; mulch	invasive (though may die in cold weather);: *Note associated algae can be toxic to fish,live-stock and humans*
Bidens connata	Asteraceae	swamp beggar's tick	ah	sw, wm, d	S	medicinal; dye; dry flowers	edible seed (waterfowl, game birds) and leaves (small mammals)	---	---
Brasenia schreberi	Cabombaceae	watershield	fl, ph	acid, water, sand or silt bottom Z: 4-7	S	food: stem, leaves, root (starchy)	edible seed (water fowl)	---	---
Butomus umbellatus	Butomaceae	flowering rush	ph	sw, psh Z: 3-7	S	food: rhizome and seed	---	ornamental	---
Calla palustris	Araceae	water arum	ph	sw, sp, cold psh, brooks, wm Z: 3-5	RH	food: rhizome and seed	---	pollen	poisonous berries

Genus /species	Family	Common Name	Form	Habitat	PROP.	Human Uses	Animal Uses	Functions/ Properties	Problems
Callitriche stagnalis #	Callitrichaceae	common water starwort	sub, fl	p, l, d Z: 3-7	C	---	---	oxygenates water; fresh-water shrimp and fish habitat	may be invasive
Caltha palustris	Ranunculaceae	cowslip, marsh marigold	ph	sp, sw, psh, stb Z: 2-5	---	food: leaves and buds	---	abundant nectar and pollen	poisonous raw
Carex, spp.	Cyperaceae	pond sedge	ph	m, psh Z: 2-5	---	---	edible seed (water fowl)	water fowl cover	invasive
Chamaedaphne calyculata	Ericaceae	cassandra, leatherleaf	sh Ht: 4'	b	S, C	---	browse (birds and small game)	stabilizes bog	---
Cephalanthus occidentalis	Rubiaceae	button willow, buttonbush	sh Ht: 6-7'	stb, sw, b Z: 3-7	S, C	medicinal, tannin	browse (water fowl)	bee forage (nectar)	---
Ceratophyllum demersum	Cerato phyllaceae	hornwort, coontail	sub, fl, ph	l, p 1-6' deep Z: 3-8 tolerates change of water level	C	aquarium	edible seed and leaves (water fowl)	oxygenates water; habitat: fish, insects, crustaceans	---
Chara fragilis	Chlorophyceae	muskgrass, stonewort	sub, (algae)	limey, p, l, sst; alkaline Ht: 30" deep in clear water	c	used in the manufacture of polishes (has a high calcium content)	all parts edible (water fowl); food: fish and livestock	habitat; water purification; fertilizer (when composted)	Invasive; strong, foul odor (can flavor water fowl)
Chelone glabra	Scrophulari-aceae	white turtlehead	ph	ms, stb, sw, d	S	medicinal	---	ornamental	---
Cyperus, spp. *C. esculentus*	Cyperaceae	nutgrass, ground almond, chufa, nutsedge	ph	ms-wd, m, mudflats Z: 3-8	D	medicinal; food: tuber (raw, cooked, roasted for coffee)	edible tuber and seed (water fowl and other wildlife	seeds rich in oil	invasive; allelopathic (reduces growth and germination of surround-ing plants)
Cornus florida	Cornaceae	flowering dogwood	smt	stb-wd, sw, wm, Z: 3-7	S	medicinal, tool handles' woodenware	edible fruit (birds)	ornamental; nutrient pump; minor bee forage	---
Dentaria, spp. *D. laciniata D. maxima*	Brassicaceae	toothwort	ph	shady stb Z: 2-7	D	food: spicy relish from root	---	---	---
Dionaea muscipula	Droseraceae	Venus flytrap	---	b Z: 6-7	---	bioshelter plant	---	carnivorous (insects)	endangered
Drosera D. rotundifolia	Droseraceae	sundew	ph	b	---	medicinal, bioshelter plant	---	carnivorous (insects)	endangered
Echinochloa	Poaceae	wild millet, barnyard grass	ph	mudflats, psh, lsh, wd	s	---	edible seed: (water fowl and other birds)	habitat: water fowl and wildlife	invasive in wet soil

Genus /species	Family	Common Name	Form	Habitat	PROP.	Human Uses	Animal Uses	Functions/ Properties	Problems
*Eichhornia crassipes^*	Pontederiaceae	water hyacinth	ph	qw, p, l, sst, d Z: 5-9	C, S?	food: cooked flower, leaf stock young leaves	edible leaf and flower (livestock); rich in nutrients	ornamental, methane feed stock, fertilizer, water purification	invasive in warm climates; low temps will kill it
Eleocharis	Cyperaceae	spikerush	em, ph	psh, lsh	S, T	*E. tuberose* has edible root (called water chestnut)	edible seed and tuber: (water fowl)	---	---
Epilobium angustifolium E. hirsutum	Onagraceae	fireweed willow weed willow herb	sh	wm, sw, btl, b, lsh, psh, stb Z: 3-7	S, T	food: leaves, shoots, rhizomes: tea	browse: seeds (sm. mammals)	ornamental, pioneer, bee forage	---
Equisetum, spp. E. palustre E. fluviatile E. arvense	Equisetaceae	horsetails, scouring rush	ph	wm, sw, wm, d Z: 3-8	D	medicinal; scours wood and metal	food: (water fowl, bear, moose, muskrat)	---	invasive, poisonous to livestock
Eupatorium perfoliatum	Asteraceae	boneset, Joe-Pye weed	ph	wm, sw, stb Z: 3-7	S, D	medicinal	---	bee forage; ornamental (Joe-Pye weed)	---
Fraxinus, spp. F. lanceolata F. nigra F. pennsylvanica	Oleaceae	ash: green, black, red	t	stb, lsh, sw Z: 3-7	S	medicinal; lumber; handles; fuel; furniture; basketry (*N. nigra*)	---	bee forage (propolis)	---
Geum rivale	Rosaceae	water avens	ph	p, l, stb, sw, shw	---	edible roots	---	---	---
Glyceria maxima	Poaceae	reed sweet grass	em, ph Ht: 6'	psh, lsh, sw Z: 1-7	RH, D	thatch	winter fodder for livestock	decomposes rapidly	invasive (overshadows other em)
Hydrastis canadensis	Ranunculaceae	goldenseal	ph	mw, wm Z: 3-6	S	medicinal	---	---	rare in the wild
Hydrophyllum, spp.	Hydro-phyllaceae	waterleaf, Indian salad	ph	---	---	edible roots and shoots	---	---	---
Ilex, spp.	Aquifoliaceae	holly, *winterberry, **yaupon	sh st: 6-16'	sw, wm, stb *Z: 3-5 **Z: 6-8	S, C	tea (yaupon has caffeine)	browse; berries (birds)	---	---
Impatiens capensis I. pallida	Balsaminaceae	jewelweed	ah	shady stb, wm–wd	C	medicinal; food: cooked shoots	food (wildlife, humming-birds)	bee forage	poisonous raw
Iris pseudacorus I. versicolor	Iridaceae	yellow iris, wild iris	ph	lsh, psh, wm, m Z: 3-7	D	medicinal; perfume	---	ornamental; bee forage (nectar)	---
Juncus, spp.	Juncaceae	rushes	ph	b, wm, m, sw, lsh, psh-wd Z: 2-9	S	weaving	edible seed (waterfowl)	erosion control; wildlife cover; aq. habitat; does not clog waterways	weed in pasture

Genus /species	Family	Common Name	Form	Habitat	PROP.	Human Uses	Animal Uses	Functions/ Properties	Problems
Lepidium	Brassicaceae	peppercross, pepperweed	ah	stb, sp Z: 3-6	S	edible leaves, seeds, stem and roots	---	---	---
Ledum groenlandicum L. latifolium	Ericaceae (Heaths)	Labrador tea	sh Ht: 1-6'	b, sw, wm, tundra Z: 1-5	S, C	medicinal leaves (tea)	---	---	---
Leersia oryzoides	Poaceae	rice cutgrass	---	b, sw, wm, tundra Z: 1-5	S	---	edible seed and roots (waterfowl, birds and muskrat)	---	invasive
Lemna, spp. #	Lemnaceae	duckweed	ph, fl	p, l, d, sst Z: 4-9	D	food: dried plant	food (waterfowl, livestock, gamebirds)	aquatic habitat; water purification; some types thrive in low-O_2 water	can limit light penetration and use up nutrients; *L. minor* has been listed as invasive in N. America
Lindera benzoin	Lauraceae	spicebush, wild allspice	sh Ht: 6-15'	wooded, sw, stb Z: 4-9	S, C	tea; medicinal (bark, leaf, twig); spice (berry)	edible fruit (birds)	male and female plants; bee forage	---
Lobelia, spp.	Campanulaceae	lobelia, cardinal flower	ph	stb, sw, lsh, psh	S	medicinal; edible root?	---	ornamental	rare
Lonicera, spp.	Caprifoliaceae	honeysuckle	d vine, sh	sw, wm-wd	S, C	some varieties have edible fruit	browse; edible fruit (wildlife)	ornamental, bee forage	---
*Lythrum salicaria^*	Lythraceae or Primulaceae	purple loosestrife	d, sh	m, sw, wm, d Z : 3-7	S, C	medicinal	edible seed (birds)	major bee forage; bird nesting habitat	very invasive; listed as a noxious weed in many states and sale may be regulated
Mentha, spp. *M. piperita M. spicata M. aquatica*	Lamiaceae	mints peppermint spearmint watermint	ph	wm, stb, sp-wd Z: 3-7 tolerates low pH	S, C	medicinal, tea oil, flavoring	Bee forage; attracts beneficial insects, repels others; peppermint contains salicylic acid; various mints contain citronella, menthol, camphor	---	invasive; *M. perperita* will dry off livestock

Genus /species	Family	Common Name	Form	Habitat	PROP.	Human Uses	Animal Uses	Functions/ Properties	Problems
Monarda, spp. M. didyma	Lamiaceae	bee balm	ph	stb, d, wm, ww Z: 4-7	S, D	medicinal, tea	nectar (hummingbirds)	bee forage	---
Myrica cerifera	Myricaceae	bayberry, wax myrtle	sh (low thicket) Ht: 40'	sw, m, ww, cool areas Z : 6-8	S, D,L	medicinal (root and bark); food: berry, seed, leaf; wax	edible berry (game birds, waterfowl, other birds)	tolerate salt spray and wide soil variation	doesn't like heavy soils
Najas, spp. N. flexilus N. gracillima N. minor#	Najadaceae	naiads, water nymphs	sub ph Ht: 6-10"	deep fresh water: 1-8'	D	---	edible tubers, nutlets, seed, foliage (waterfowl): fish forage	fish cover; associated with pond- weed and wild celery	may be invasive
Nasturtium aquaticum N. officinale	Brassicaceae	Watercress	ph	sp, st, shallow, cold, water, ws Z: 3– tropical	D, S	medicinal (root and leaf); edible: shoots and leaves	---	used for wastewater treatment; rich in vit. A, B, C, iron and iodine	---
Nelumbo lutea#	Nymphaceae	American lotus	fl, p	water 2-12' deep Z: 5-7 tolerates changing water level	D	---	edible seed, root and leaf (wildlife)	fish habitat, ornamental	plant in tubs to control and harvest
Nitella gracilis	Chlorophyceae	stonewort	sub	fresh water 2-10' deep pH 5.5-6.5	D	---	forage (fish and water- fowl)	---	---
Nuphar advena N. lutea N. polysepalum	Nymphaeaceae	yellow pond-lily, spatterdock	fl, ph	l, p, sunlit, fresh water, 3-10' deep; tolerates changing water levels	S, D	food: root and seed (nutritious flour or popped)	---	aquatic habitat	---
Nymphaea tuberosa N. odorata#	Nymphaeaceae	white water lily	fl, ph	qw, p, sst 1-8' deep Z : 3-8	D, D	medicinal; food: young leaf, bud, seed, roots (rich in starch, oil, protein)	edible plant (moose, muskrat, porcupine); edible seed (waterfowl)	aquatic habitat; ornamental	shades water
Nymphoides peltata#	Menyanthaceae	fringed water lily	fl, p	p, l	S, D	---	---	aquatic habitat; ornamental; bee forage (nectar)	shades water
Nyssa sylvatica	Cornaceae	pepperidge, blackgum, sourgum	d, t	sw, bl–wd Z: 4-7	S	food: fruit; woodenware; rollers; crates; wheel hubs	browse, wildlife forage	bee forage, ornamental	hard to split and work
Oryza sativa^	Poaceae	rice	a grass	paddy to dry land Z: 7-8 (some varieties) Z: 6	S	edible seed (staple), straw	edible seed	---	---

Genus /species	Family	Common Name	Form	Habitat	PROP.	Human Uses	Animal Uses	Functions/ Properties	Problems
Onoclea sensibili	Dryopteridaceae	sensitive fern	ph	m, wm, sw Z: 3-8	D	food: young shoots (cooked)	fronds: turkey food	---	---
Osmunda regalis	Osmundaceae	royal fern	ph	m, wm, sw Z: 3-9	D	food: young fiddleheads; cloth (fiber); medicinal	---	---	---
Oxydendrum arboreum	Ericaceae	sourwood, lily-of-the-valley tree	smt, t	sw, wd	S	medicinal, handles, woodenware	---	ornamental, bee forage	---
Peltandra virginica	Araceae	arrow arum	ph	sw, b, stb, p, l (to 1' deep water) Z: 4-8	S, D	food: fruit, flower spike, dried root (up to 4 lbs)	edible seed (wood ducks and water fowl)	tolerates shade	poisonous before drying
Phalaris canariensis P. arundinacea	Poaceae	canarygrass, reed canarygrass	ph	d, m, psh, lsh	S, D	edible seed, mats, weaving, thatch	edible seed (game birds and waterfowl)	---	clogs channels
Phragmites communis P. phragmites	Poaceae	reed grass, common reed	ph grass Ht: 6'	d, sw, m, psh, lsh, Z: 3-8	S, D	food: stems (sugar, flour) and young shoots, leaves, seed (cereal); thatch; paper; cellulose; insulation; weaving	grazing	erosion control, wildlife cover and habitat	low value to system; slow to decompose; invasive (replaces cattails); seeds rare
Platanus occidentalis	Platanaceae	American sycamore	t	stb, bl Z: 3-7	S	syrup; sap (emergency water); lumber, crates; butcher block; furniture	---	bee forage	---
Polygonum, spp. P. lapathifolium P. persicaria# P. muhlenbergii	Polygonaceae	smartweeds, knotweeds	ph or ah	d, psh, lsh, m, seasonal ws Z: 3-7	S	food: shoots, leaf, roots, medicinal dye	edible seed (waterfowl, songbirds, sm. mammals)	aquatic habitat; bee forage; rich in vit. P, C and rutin	some species weedy or invasive
Polystichum acrostichoides	Dryopteridaceae	Christmas fern	ev. ph	sw, wm, stb	D	Christmas decoration	browse (deer)	---	---
Pontederia cordata	Pontederiaceae	pickerel plant	em, ph Ht: 3'	p, sst, m, rich soil Z: 2-8	S, D	food: young leaf, fruit	edible seed (waterfowl)	fruit and seed highly nutritious, fish habitat, ornamental	invasive
Populus, spp.	Salicaceae	cottonwood, poplar, aspen	t	lsh, bl, mw Z: 2-7	S, C	medicinal; lumber, fuel; pulpwood alcohol; food: inner bark, catkins	browse and forage	pioneer tree	---

Genus /species	Family	Common Name	Form	Habitat	PROP.	Human Uses	Animal Uses	Functions/ Properties	Problems
Potamogeton, spp. # *P. pectinatus* *P. americanus* *P. richardsonii*	Zosteraecae or Potamogeton-aceae	pondweed, Sago pond weed	ph	sub, fl, p, sst, up to 12' Z: 3-tropics	S, D	food: tubers, stems, leaves	edible seeds and plant (waterfowl) esp. *P. pectinatus*	aquatic habitat	may be weedy or invasive
Pteris pensylvanica *P. nodulosa*	Pteridaceae	ostrich fern	ph Ht: to 6'	stb, m, wm	C	food: fiddle heads under 6" (excellent raw)	---	---	---
Quercus bicolor *Q. michauxii*	Fagaceae	swamp white oak swamp chestnut oak	d, d	sw, stb	S	food: sweet acorns; lumber	edible acorns	---	---
Ranunculus peltatus	Ranunculaceae	water crowfoot, buttercup	fl, ph	p, l, sst	D	---	---	bee forage (pollen and nectar); aq. habitat	---
Rosa palustris	Rosaceae	swamp rose	sh	sw, b, wm	S, D, C	food: fruit, flowers, shoots seed, leaf	browse	bee forage; fruit rich in vit. C, calcium, iron and phosphorus; seed rich in vit. E	---
Ruppia maritima	Ruppiaceae	widgeon-grass	em	l, p; alkaline water	S, D	---	edible seed (waterfowl)	---	---
Sagittaria ssp. ^ *S. cuneata* *S. latiofolia*	Alismataceae	arrowhead, duck potato, wapato	em	p, l, sst. m, d Z: 2-9 tolerates changing water levels	S, D	food: tubers (rich in starch)	edible tubers (water fowl, muskrat, fish, good pig forage porcupine); edible seed	---	*S. sagittifolia* L. has been listed as invasive in N. America
Salicornia, spp.	Chenopodi-aceae	glasswort pickleweed,	ph	p, l; tidal flats, alkaline, Z : 1-9	---	food: tender tops	edible seed, leaf, stem	thrive in salt	---
Salix, spp. *S. discolor* *S. viminalis* *S. babylonica*	Salicaceae	willow pussy willow osier willow weeping willow	sh-t	stb, lsh' psh, m, d, wm-wd	S, C	food: leaves medicinal, charcoal, basketry	browse and forage	bee forage; erosion control; fire resistant hedge; transpiration pump (weeping)	can invade waterlines
Sanbucus canadensus	Caprifoliaceae	elder	ph	stb, lsh, psh, d, wm to wd; full sun to partial	S, C, D	food: fruit, flower; syrup taps; flutes; medicinal	edible fruit (birds and wildlife); browse, forage; insect	indicates a high water table	---

Genus /species	Family	Common Name	Form	Habitat	PROP.	Human Uses	Animal Uses	Functions/ Properties	Problems
Sanbucus canadensus				shade Z: 2-7			eat buds, branches, blossoms		
Sarracenia, spp.	Sarraceniaceae	pitcherplant	ph	b, m, wm Z: 2-9	---	bioshelter plant	---	carnivorous (insects)	endangered
Saxifraga	Saxifragaceae	lettuce saxifrage, deer tongue	ph	sp, sw, m, damp slopes Z: 2-8	---	food: leaves (raw or cooked) rich in vitamins	browse and forage	---	---
Scirpus, spp. *S. fluviatilis* *S. paludosus*	Cyperaceae	bullrush	em, ph, aw Ht: 1-7'	lsh, psh, sw, m, d	S, D	food: tubers, rhizomes, seeds, pollen; thatch	edible tubers and plant (waterfowl and muskrat); edible seed	waterfowl cover, erosion control, water purification	causes silting of ponds and lakes
Sparganium, spp.^ *S. americanum* *S. eurycarpum* *S. antipodium*	Sparganiaceae	burr reed	ph	sw, m, psh, lsh, d Z: 1-7 tolerates changing water levels	S, D	food: seeds and tubers	edible seed (waterfowl, game birds, mammals); edible plant (sm. mammals)	waterfowl cover, erosion control	handling plant can cause health problems
Sphagnum, spp. *S. obtusifolium*	Sphagnaceae	sphagnum moss bog moss	ph	b, sw, m	D	medicinal; food: entire plant; peat; soil mix	---	---	handling plant can cause health problems
Spirodela polyrrhiza	Lemnaceae	duckmeat, duckweed	ph, (minute) fl	sw, p, l, sst Z: 5-8	D	food: dried	edible plant (waterfowl and livestock)	water purification, aq. habitat, methane digester, fertilizer	dense growth can block light and kill sub plants
Spirulina, ssp.	Cyanophyceae	spirulina, blue-green algae	algae	p. l. sw. sst	D	food: dried and powdered	edible (fish and fowl)	rich in amino acids, vitamins and minerals	---
Symplocarpus foetidus	Araceae	skunk cabbage	ph	sw, ww, stb Z: 2-7	S, D	food: leaf and dried root; medicinal	---	bee forage (earliest pollen)	foul odor
Trapa, spp. *T. natans* #	Trapaceae or Hydracaryaecae	water chestnut	fl, ph	l, p, d Z: 6-tropical	S, D	food: seed (starchy)	---	aquatic habitat	needs high nutrient level
Typha latifolia *T. angustifolia*	Typhaceae	cattail, bullrush, reed mace	em, ph	sw, m, lsh' psh, d	S, D	food: shoots, roots, pollen spikes; weaving fiber; insulation; tinder; paper; fish bait	edible tubers (waterfowl, sm. mammals, pigs)	waterfowl and wildlife cover, water purification	invasive; silting crowds out more valuable em; large quantities of leaves are poisonous to livestock

Genus /species	Family	Common Name	Form	Habitat	PROP.	Human Uses	Animal Uses	Functions/ Properties	Problems
Utricularia, spp. *U. vulgaris* *U. minor*	Lentibulariaceae	bladderwort	ph, fl and sub	alkaline, p, l, d, sst *U. minor* likes acid	S, D	---	browse (birds and deer)	carnivorous (insects)	---
Vaccinium corymbosum	Ericaceae	swamp blueberry	sh Ht: 6'	sw, wm	S	medicinal; food: berry	browse and forage	hedge	---
Vaccinium macrocarpon	Ericaceae	cranberry	ev. sh Ht: 6'	b, m-dry peat, acidic soil, conifer woods	D	medicinal; food: berry	forage	---	needs symbiotic fungus on roots
Vallisneria spiralis	Hydrocharitacea	wild celery, eel grass	sub, ph	p, l, sst, alkaline Z: 4-7	---	---	edible plant and seed (water fowl)	oxygenates water, aq. habitat	---
Veratrum album *V. viride*	Liliaceae	American hellebore	ph	sw, wm, stb Z: 2-7	Rh, D	medicinal, insecticide, rat poison	---	---	---
Viburnum *V. opulus* *V. prunifolium*	Caprifoliaceae	highbush cranberry, nannyberry	sh ht. 6-10'	sw, m, b, wm, stb	S, C	medicinal food: fruit	---	rich in vit. C, K; erosion control; hedge	---
Wolffia, spp.	Lemnaceae	duckweed, watermeal	ph, (minute) fl	p, l, d, sst Z: 3-9	D	food: dried plant	edible seed: (water fowl, livestock, game birds)	succeeds azolla in late summer; aq. habitat, water purification	---
Zannichellia palustris	Zannichelliaceae	horned pondweed	sub, ph	p, l, sst	S, D	---	food: waterfowl	aq. habitat	---
Zizania aquatica	Poaceae	wild rice	ah	sw, m, p, l, sst, needs currents Z: 3-9	S	food: grain, stem base	edible seeds: (waterfowl and birds)	---	---
Zizaniopsis miliacea	Poaceae	giant cutgrass	ph	psh, lsh	S	food: rhizome tips, young shoots	edible seeds: (waterfowl and birds); forage	---	---

Index

About the Author

Darrell Frey is the latest in a long line of gardeners and farmers, from both sides of his family tree. Raised to have a deep appreciation for nature's beauty and the fruits of good labor, he found in Permaculture design a way to be close to the earth while obtaining a yield from sun and soil and rain.

Along with Linda Frey and their three children, and inspired by other gardeners and many sustainable living writers, researchers and teachers, he developed Three Sisters Farm and Bioshelter. The five-acre plot of season extenders, gardens, uncultivated areas and pond, is an ongoing research project in right livelihood and Permaculture design.

As a sustainable community development consultant, Darrell offers a wide range of services to numerous clients. When not working on a new project or tending the landscape, he enjoys time with friends and family, including five grandchildren.

For more information about Darrell Frey and Three Sisters Farm, visit Bioshelter.com.

If you have enjoyed *Bioshelter Market Garden*, you might also enjoy other

BOOKS TO BUILD A NEW SOCIETY

Our books provide positive solutions for people who want to
make a difference. We specialize in:

**Sustainable Living • Green Building • Peak Oil • Renewable Energy
Environment & Economy • Natural Building & Appropriate Technology
Progressive Leadership • Resistance and Community
Educational & Parenting Resources**

For a full list of NSP's titles, please call 1-800-567-6772 *or check out our website* at:

www.newsociety.com

NEW SOCIETY PUBLISHERS
Deep Green for over 30 years